Chemistry of the Mediterranean Diet

Amélia Martins Delgado •
Maria Daniel Vaz Almeida •
Salvatore Parisi

Chemistry of the Mediterranean Diet

Amélia Martins Delgado
Consultant for Food Safety
and Nutrition
Lisbon, Portugal

Maria Daniel Vaz Almeida
Faculty of Nutrition and Food Sciences
University of Porto
Oporto, Portugal

Salvatore Parisi
Associazione 'Componiamo il Futuro' (CO.I.F.)
Palermo, Italy

ISBN 978-3-319-29368-4 ISBN 978-3-319-29370-7 (eBook)
DOI 10.1007/978-3-319-29370-7

Library of Congress Control Number: 2016945570

Photographs and cover illustration by Tobias N. Wassermann.

Printed on acid-free paper

This Springer imprint is published by Springer Nature
The registered company is Springer International Publishing AG Switzerland

Acknowledgments

AMD wishes to express her gratitude to Prof. Pedro Louro, Head of the Research group on Dairy Technology, INIAV—IP, for his review of 2.3 and his valuable suggestions; to Eng. Vitor Barros, Principal Researcher of INIAV, IP, and coordinator of the Portuguese committee for the application to UNESCO's MD's representative list, for the supply of useful documentation and other resources; to Eng. Fernando Severino, Regional Director of Agriculture, and member of the Portuguese committee for the application to UNESCO's MD's representative list, for his friendly support;

"Let food be your medicine and medicine be your food." (Hippocrates)

Contents

Abbreviations

ABA	Abscisic acid
ALA	α-Linolenic acid
a_w	Water activity
bw	Body weight
Ca	Calcium
CHD	Coronary heart disease
CIHEAM	International Centre for Advanced Mediterranean Agronomic Studies
CLA	Conjugated linoleic acid
Con A	Concanavalin A
Cu	Copper
DGAC	Dietary Guidelines Advisory Committee
DHA	Docosahexaenoic acid
DNA	Deoxyribonucleic acid
DPA	Docosapentaenoic acid
EC	European Council
EFSA	European Food Safety Authority
EGCG	(−)-epigallocatechin-3-gallate
EOL	Encyclopedia of Life
EPA	Eicosapentaenoic acid
EU	European Union
EUFIC	European Food Information Council
EVOO	Extra virgin olive oil
F	Fluorine
FAO	Food and Agriculture Organization of the United Nations
FBS	Food balance sheet
FDA	US Food and Drug Administration
Fe	Iron
FIL/IDF	Fédération International de Laiterie/International Dairy Federation
G6PD	Glucose-6-phosphate dehydrogenase
GI	Gastrointestinal
GLA	Cis-γ-linolenic acid
GRAS	Generally recognised as safe

HDL-c	High-density lipoprotein cholesterol
ICO	International Coffee Organization
INSA	Instituto Nacional de Saúde Doutor Ricardo Jorge
IOOC	International Olive Oil Council
IPP	Isoleucine-proline-proline, tripeptide
IUPAC	International Union of Pure and Applied Chemistry
KIDMED	Mediterranean diet quality index for children and adolescents
LA	Linoleic acid
LAB	Lactic acid bacteria
LD_{50}	Lethal dose
LDL	Low-density lipoprotein
LDL-c	Low-density lipoprotein cholesterol
LPS	Lipopolysaccharide
MAI	Mediterranean adequacy index
MD	Mediterranean diet
MDS	Mediterranean diet score
MEDAS	Mediterranean diet adherence screener
Med-DQI	Mediterranean dietary quality index
MEFA	(n-3) fatty acid
Mg	Magnesium
Mn	Manganese
MUFA	Monounsaturated fatty acid
Na	Sodium
NaCl	Sodium chloride
NCBI	National Center for Biotechnology Information
NCD	Non-communicable disease
NHS	National Health Service
NLM	National Library of Medicine
OA	Oleic acid
OVOO	Ordinary virgin olive oil
P	Phosphorus
PAF	Platelet activating factor
PDO	Protected designation of origin
PEITC	2-phenethyl isothiocyanate
PhIP	2-amino-1-methyl-6-phenylimidazo(4,5-b)pyridine
PortFIR	Portuguese Food Information Resource
PUFA	Polyunsaturated fatty acid
RAE	Retinol activity equivalent
ROS	Reactive oxygen species
Se	Selenium
SFA	Saturated fatty acid
t11	Δ11 trans
t9	Δ9 trans
TCA	Tabela da Composição de Alimentos
TFA	Trans-fatty acid(s), trans-fats

UN	United Nations
UNESCO	United Nations Education, Scientific and Cultural Organization
USA	United States of America
USDA	United States Department of Agriculture
USDHHS	United States Department of Health and Human Services
VOO	Virgin olive oil
VPP	Valine-proline-proline, tripeptide
WHO	World Health Organization
WHO-ROEM	World Health Organization-Regional Office for the Eastern Mediterranean
Zn	Zinc

Part I

Introduction to the Mediterranean Diet

The Mediterranean Diet: What Is It? 1

Abstract

The Mediterranean basin and the Iberian Peninsula constitute a vast geographical area where three continents intercept. Ancient civilizations characterised by cultural and religious diversity flourished in the region. The Mediterranean diet (MD) represents unity in diversity, integrating food habits with cultural habits (such as the convivial aspects of meals), landscapes (such as the presence of olive orchards and vineyards), and food preservation methods. The concept of the MD was first coined by Ancel Keys, an American physician who highlighted the health benefits of the food pattern of southern Europeans after World War II. The MD is now recognized as one of the most healthy food patterns in the world. This book takes as reference the evolution of the original concept by Ancel Keys, as well as the countries that are currently included in the United Nations Educational, Scientific and Cultural Organization Representative List (Portugal, Spain, Morocco, Italy, Greece, Croatia and Cyprus), which classifies the MD as an 'intangible heritage of humanity'. This chapter discusses the origins and primary features of the MD, mainly from a dietary perspective.

1.1 The Mediterranean Diet: An Introduction

The Mediterranean diet (MD) constitutes a paradigm that inspires healthy dietary recommendations worldwide. The concept of the MD, 'diet' being from the Greek *diaita* ('way of life') or the Latin *diaeta* ('prescribed way of life'), is wider than just a food pattern and includes lifestyle and traditions. Ancel Keys and co-workers in the 1950s were the first to establish the link between the MD and health by showing an inverse correlation between adherence to the MD and the incidence of coronary heart disease. Keys described the MD as a dietary pattern and lifestyle observed in southern Europe just after World War II, consisting of frugal meals with wheat, wine and olive oil as key elements. He described meals as communal events that included many vegetables and herbs and very small amounts of meat and fish, with

A.M. Delgado et al., *Chemistry of the Mediterranean Diet*,
DOI 10.1007/978-3-319-29370-7_1

pulses and cheeses as the preferred sources of protein. Cooking methods were simple, despite the resulting variety of flavours and colours. Seasonal fruits were the preferred desserts, and nuts and olives were eaten as snacks. Coffee and tea played an important role in these communal meals, and sweet desserts were reserved for festivities, when the intake of meat and fish was also increased. The United Nations Education, Scientific and Cultural Organization (UNESCO) classified the MD as an 'intangible heritage of humanity', aiming to call attention to and preserve this pattern. Countries that make up the Representative List in 2015 are Portugal, Spain, Morocco, Italy, Greece, Cyprus and Croatia. This chapter discusses time trends in dietary habits, based on data from the United Nations Food and Agriculture Organization's Food Balance Sheets and literature reviews of diet indexes and epidemiological and cohort studies. A Westernisation of food habits has been recognized in the area, characterised by a high-energy diet, with increasing consumption of industrially processed foods. These foods usually contain large amounts of salt, simple sugars, saturated and trans fats, which industries offer in response to consumers' demands. Consequently, the intake of complex carbohydrates, fibres, fruits and vegetables has decreased. The energy and animal proteins consumed largely exceed World Health Organization recommendations, while, generally, a smaller variety of foods is being consumed. Adherence to the MD dietary pattern has been rapidly decreasing in the area since 2000, particularly in Greece, Portugal and Spain. These observations point to a nutrition transition period that encompasses considerable changes in diet and physical activity patterns, which may be leading to an increase in the incidence of chronic and degenerative diseases. Recent epidemiological and metabolic studies support that the adoption of MD-like dietary patterns results in better overall health status and self-perception of well-being. A reversal of the decreasing adherence to an MD will require an approach at various levels and in a wide range of settings. The acquisition of healthy food habits during childhood and the development of cooking skills may contribute to ensuring the long-term implementation of MD.

1.2 The Concept of the Mediterranean Diet

The Mediterranean basin is the region surrounding the Mediterranean Sea, where Europe, Asia and Africa intercept. There are 23 internationally recognised countries in the Mediterranean area: Portugal, Spain, France, Monaco, Italy, Malta, Slovenia, Croatia, Bosnia-Herzegovina, Montenegro, Albania, Greece, Cyprus, Macedonia, Syria, Turkey, Lebanon, Israel, Egypt, Libya, Tunisia, Algeria, and Morocco. Figure 1.1 shows the Mediterranean region, highlighting the countries that currently represent the 'Mediterranean Diet' of UNESCO: Portugal, Morocco, Spain, Italy, Greece, Croatia and Cyprus.

Prominent ancient civilizations ascended in the region. The mild climate is ideal for the cultivation of olive trees and vineyards, which shaped the landscape, culture and traditions, including food habits. Braudel, a recognised French historian (1912–1985), approached history from the perspective of the common man. His

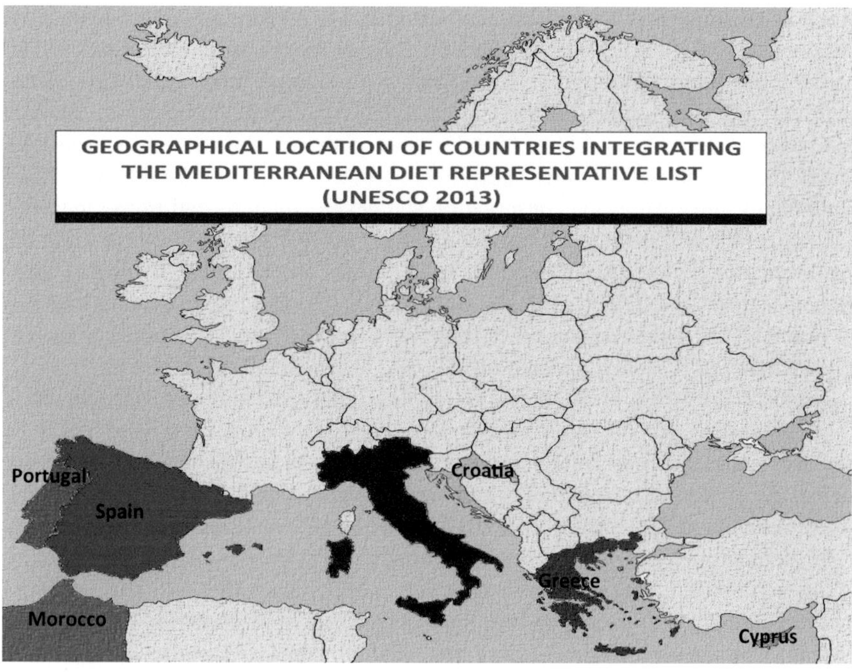

Fig. 1.1 Mediterranean Region and countries integrating the representative list of 'Mediterranean Diet' of UNESCO (UNESCO 2013)

notable work about the geohistory of the Mediterranean region remains a reference. In Braudel's approach, the region is treated, in an interdisciplinary manner, as a whole, irrespective of religious and national divides (Piterberg et al. 2010). Braudel and his followers consider that the Mediterranean region spans from the first olive tree in the north to the first compact palms in the desert. The area surrounding the Mediterranean Sea exhibits large geographical, economic, political, cultural, ethnic and religious diversity which, in turn, influences the food practices and habits of the region's inhabitants. The dietary patterns of Mediterranean peoples and their association with health, wellbeing and longevity have stimulated much research from different scientific disciplines, such as biochemistry, nutrition, genetics, general medical sciences, sociology, anthropology and history.

As Trichopoulou and Lagiou wrote, 'The Mediterranean diet and lifestyle were shaped by climatic condition, poverty and hardship rather than by intellectual insight or wisdom. Nevertheless, results from methodological superior nutritional investigations have provided strong support for the dramatic ecologic evidence represented by the Mediterranean natural experiment' (Trichopoulou and Lagiou 1997).

The broader definition of the MD is (UNESCO 2013): 'a set of skills, knowledge, practices and traditions ranging from the landscape to the table, including the crops, harvesting, fishing, conservation, processing, preparation and, particularly, consumption of food'. Substantially, the MD is based on a nutritional model without important temporal or geographical variations: three pilasters—wheat, olive oil and wine—must be considered (CIHEAM 2012). In addition, the MD encompasses more than simply food because of the indubitable correlation with social life and cultural heritage. The system is rooted in respect for the territory and biodiversity and ensures the conservation and development of traditional activities and crafts linked to fishing and farming. The key role of women in transmitting the expertise, rituals, traditional gestures, celebrations, and the safeguarding of techniques, is to be highlighted. The practical demonstration of these assumptions can be observed in Mediterranean cities such as Tavira in Portugal, Koroni in Greece, Chefchaouen in Morocco, Cilento in Italy and other sites (UNESCO 2013).

When, in the 1950s, Keys started his studies in Italy and later published the book *How to Eat Well and Stay Well, the Mediterranean Way*' (Keys and Keys 1975), he probably could not have anticipated that the concept he and his co-workers coined as the 'good Mediterranean diet' would be considered, half a century later, patrimony of mankind. This occurred in 2010, when the MD was classified by UNESCO as an 'Intangible Cultural Heritage of Humanity' to help demonstrate the diversity of this heritage and to raise awareness about its importance, thereby contributing to its safeguarding. The corresponding Representative List of countries (Spain, Italy, Greece, Cyprus and Morocco) was amended in 2013 to include Portugal and Croatia.

An intangible cultural heritage is 'traditional, contemporary and living at the same time'; it is *inclusive* because of the preservation of identity, generation after generation, sometimes influenced by migratory flows and the effect of different (non-native) environments; *representative* because of the historical knowledge of community rituals and behaviours and their adoption by other communities; and *community based* because of the conscious awareness of community members (UNESCO 2013).

The concept of the MD is thus multidisciplinary, encompassing culture, climate, history, and sociology, as well as food habits. Approaches to the MD beyond nutritional aspects, dietary patterns, and food composition, and their relation to health and wellness, fall outside the scope of this book.

References

CIHEAM (2012) The Mediterranean diet for sustainable regional development. Presses de Sciences Po, Paris

Keys AB, Keys M (1975) How to eat well and stay well the Mediterranean way. Doubleday, London

Piterberg G, Ruiz TF, Symcox G (eds) (2010) Braudel revisited—the Mediterranean world 1600-1800, vol 13, UCLA Clark Memorial Library series. University of Toronto Press, Toronto

Trichopoulou A, Lagiou P (1997) Healthy traditional Mediterranean diet: an expression of culture, history, and lifestyle. Nutr Rev 55(11):383–389. doi:10.1111/j.1753-4887.1997.tb01578.x

UNESCO (2013) Intangible Cultural Heritage. Representative List. Intergovernmental Committee for the safeguarding of the Intangible Cultural Heritage. Convention for the safeguarding of the Intangible Cultural Heritage. EN Title: Mediterranean diet. Nomination file no. 00884 for inscription in 2013 on the representative list of the Intangible Cultural Heritage of Humanity, Baku. RL 2013:1–30. Available at http://www.unesco.org/culture/ich/doc/download.php?versionID=20926. Accessed 30 Oct 2015

WHO (2015) Programmes and projects. Nutrition. Nutrition health topics. 3. Global and regional food consumption patterns and trends. Available at http://www.who.int/nutrition/topics/3_foodconsumption/en/. Accessed 30 Oct 2015

WHO-ROEM (2012) Promoting a healthy diet for the WHO Eastern Mediterranean Region: user-friendly guide. WHO Regional Office for the Eastern Mediterranean, Cairo. Available at http://applications.emro.who.int/dsaf/emropub_2011_1274.pdf?ua=1. Accessed 30 Oct 2015

2 Food and Nutrient Features of the Mediterranean Diet

Abstract

This chapter describes the Mediterranean food pattern as rich in vegetables and characterised by a high consumption of olive oil and a reduced intake of meat and dairy products, particularly liquid milk. The so-called good Mediterranean diet corresponds to the dietary pattern found in the olive-growing areas of the Mediterranean region. The concept is linked to rural communities experiencing a period of economic depression after World War II and before the wide dissemination of the fast-food culture. Despite regional variations, common components and cultural aspects can be identified, namely olive oil as the main source of lipids, the consumption of large amounts of seasonal vegetables, fruits and aromatic herbs (some of them gathered from the wild), as well as small intakes of meat and fish, often replaced or complemented with pulses, as sources of protein. Several global and governmental organizations acknowledge the Mediterranean diet as nutritionally adequate, health-promoting and sustainable because of its emphasis on biodiversity and the intake of small meat portions. In short, Mediterranean-style dietary patterns score highly for health, as well as for estimated sustainability scores, and can be followed in Mediterranean as well as in non-Mediterranean countries.

2.1 The Mediterranean Diet: Food and Nutrient Features

The Mediterranean diet (MD) as a dietary pattern, and its relation to public health, was first noticed and extensively studied by Ancel Keys, an American medical doctor who travelled to Naples in the early 1950s, establishing the concept of what he later called the 'good Mediterranean diet' (Grande et al. 1965, 1972; Keys 1995; Keys et al. 1980). Keys and co-workers conducted an extensive epidemiological study known as the 'Seven Countries Study' from the middle 1950s to the late 1970s in seven countries: the USA, Finland, Netherlands, Italy, Greece, Japan and former Yugoslavia—now Croatia and Serbia. The study established a correlation

A.M. Delgado et al., *Chemistry of the Mediterranean Diet*,
DOI 10.1007/978-3-319-29370-7_2

between blood cholesterol levels and the risk of coronary heart disease (Keys and Fidanza 1960; Keys et al. 1980). In the 1960s, coronary deaths in the USA and northern Europe greatly exceeded those in southern Europe, even after controlling for age, cholesterol and blood pressure levels, smoking, physical activity and weight. The Seven Countries Study also showed that cardiovascular risk factors in midlife are significantly associated with increased risk of dementia later in life (Keys et al. 1980). The importance of eating patterns became clear, and Keys described 'the good Mediterranean diet' as mainly vegetarian, characterised by a high consumption of olive oil and reduced intake of meat and dairy products, particularly liquid milk, when compared with the dietary habits of northern Europe and the USA. The diet characterised by Keys (Grande et al. 1972; Keys and Keys 1959; Keys et al. 1980; Keys 1995) and other authors (Bach-Faig et al. 2011; Georgoulis et al. 2014; Trichopoulou and Lagiou 1997; Trichopoulou et al. 1995) mainly corresponds to the dietary patterns found in the olive-growing areas of the Mediterranean basin, mainly of rural communities experiencing a period of economic depression after World War II and before wide dissemination of the fast-food culture. There are several variants in the region, but some common components and cultural aspects can be identified, namely olive oil as the main source of lipids; the consumption of large amounts of seasonal vegetables, fruits and aromatic herbs (some of them gathered from the wild); as well as commensality since meals are a communal event.

According to several authors (Keys 1995; Trichopoulou and Lagiou 1997), this dietary pattern included the daily consumption of olive oil, which accounted for most of the energy intake. Tree nuts and table olives were also commonly consumed. Large quantities and varieties of vegetables, legumes and fruits supplied vitamins, fibres and antioxidants. Beans, peas, and cheese were important sources of protein. Meat and fish were consumed in very small amounts. Wheat, potatoes and rice (mostly minimally processed) constituted the carbohydrate sources. Liquid milk was not commonly consumed by adults. It is noteworthy that Trichopoulou and Lagiou (1997) stressed the role of moderate wine consumption during meals, as have other authors more recently (Covas et al. 2010; Jordão et al. 2010; Nishizuka et al. 2011; Opie et al. 2011).

As observed by Keys and co-workers (Grande et al. 1972; Keys 1995), a classical meal always included a large amount of cooked and/or raw vegetables. Typical examples are salads that include a large variety of leaves and herbs, seasoned with olive oil. Meat was absent or consumed only in very small amounts. Red wine was most often present in adult's meals, except in Muslim countries. Cakes and other sweet desserts were reserved for special occasions, and seasonal fruit was the typical dessert. Besides olive oil, bread, cheese and wine are described as playing central roles in this diet (Keys 1995; Keys et al. 1980). For cultural and religious reasons, green tea with mint is most consumed in Muslim countries, and may, in some aspects, act as wine's counterpart due to its composition, as we show in Part II.

To lay people, the term 'diet' generally means the food and drink consumed by individuals or population groups, but it is even more commonly associated with voluntary food restriction. However, the original Greek word *diaita* meant 'way of life' and the Latin word *diaeta* 'prescribed way of life', therefore encompassing food habits, daily activities, culture and lifestyle. When the pioneering works of Keys found an association between several health aspects (longevity, low morbidity and mortality from coronary heart disease and cancer) and what they later coined as the good Mediterranean diet, such characteristics were also registered. Therefore, occupational and leisure activities, adaptation to geographical and weather conditions as well as dependence on local resources and balance between people and the ecological system were as important to the broad concept of the MD as the food and drink included in the daily choices of individuals. It is worth mentioning that the communities investigated by Keys lived simple lives with hard occupational activities leading to high energy expenditure within a framework of food scarcity shaped by seasonal variances. Scarcity was the rule; abundance was the exception that led to festivities (cultural, religious) when people indulged in eating and drinking. Therefore, engaging in demanding occupational activities, under the direct influence of weather conditions and adapting to seasonal variations, constitute a common ground for the food and nutrient features of the MD.

As an expression of culture, history and lifestyle, several elements characterise the MD:

- Daily food intake distributed as four or five meals according to season and in proportion to labour intensity
- First and second daily meals (breakfast and lunch) were more important than the evening meal (dinner)
- Meal sharing, in a calm and peaceful environment
- A large diversity of foods, in small quantities, constituting a variety of textures and tastes
- Seasonal, locally produced and minimally processed foods
- Simple cooking methods
- Marked distinction between common days and festivities

The food features of the MD include the following:

- High fruit and vegetable consumption (unprocessed)
- High intake of wholegrain cereals, pulses and nuts
- Garlic, onions and olives all year round
- Olive oil as the 'central' fat
- High fish intake depending on proximity to the sea
- Low intake of red and processed meats
- Preference for white meat, especially poultry
- Moderate intake of dairy foods, with a preference for cheese and yoghurt
- Regular but moderate intake of alcoholic drinks, particularly wine at meal times

The analysis of such food patterns reveals the nutritional characteristics described in Table 2.1.

Table 2.1 Main nutritional features of the Mediterranean Diet

Nutrients	% Total energy intake	Particularities
Carbohydrates	60–70	Of which 50 % starch
Protein	Around 10	Of high biological value; pulses and other vegetables as relevant sources
Lipids	20–32	Monounsaturated fatty acid: oleic acid from olive oil and nuts
		Polyunsaturated fatty acid ratio n-6:n-3 = 1–2:1 from fatty fish, nuts versus vegetable seed oils, margarine
		Modest saturated fatty acid intake
Alcohol	Null	Alcoholic drinks are forbidden in the Muslim religion
	4–7	Mainly from wine, during meals
Fibre	Not applicable	Rich in soluble and insoluble fibre, from fresh fruit, vegetables, wholegrain cereals and nuts

As mentioned above, seven countries are included in the United Nations Education, Scientific and Cultural Organization (UNESCO) MD Representative List in 2015: Portugal, Spain, Morocco, Italy, Greece, Cyprus and Croatia. Data from the corresponding Food Balance Sheets (FBS), obtained from the UN Food and Agriculture Organization (FAO), were compared to illustrate the above observations and to obtain information on time trends in food consumption, merging information with studies that infer deviations by applying diet quality indexes (FAO 2015a). The evolution of dietary patterns, and tools available to assess such changes, are the object of the next chapter. FAO FBS from 1961 until 2011 are publicly available. No information about Croatia exists before 1992, thus reducing the time span under analysis in that country, as shown in Figs. 2.1, 2.2, 2.3, 2.4, 2.5, 2.6 and 2.7.

Food availability compiled by the FAO in FBS provides an estimate of the food available for human consumption in a country for a certain period of time, usually 1 year. Total food availability is computed from statistical data on supply (internal production, imports and stock changes), utilisation (exports, feed, seed, industrial use and non-food uses), and changes in stocks during the same period. The per capita value is obtained by dividing the annual quantity of each food group by the total population of the country in the same period. Therefore, the daily energy availability (kjoules or kcal/person/day) is an indirect estimation of food available for human consumption (FAO 2015b).

The FAO and the World Health Organization (WHO) define energy requirement as "the amount of food energy needed to balance energy expenditure in order to maintain body size, body composition and a level of necessary and desirable physical activity consistent with long-term good health. This includes the energy needed for the optimal growth and development of children, for the deposition of tissues during pregnancy, and for the secretion of milk during lactation consistent with the good health of mother and child" (FAO 2001).

Energy for metabolic and physiological functions is derived from the chemical energy bound in food and its macronutrient constituents. As human energy and

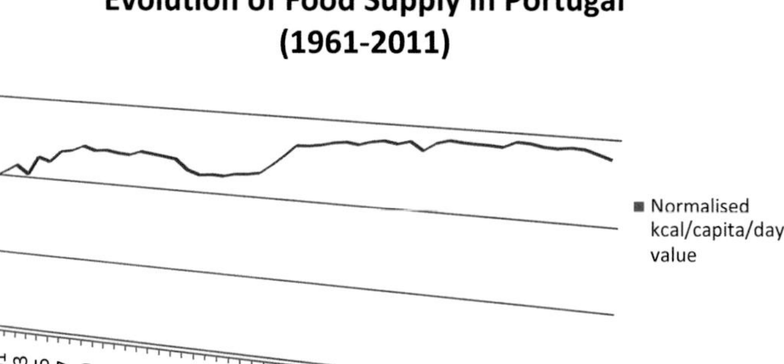

Fig. 2.1 Evolution of Food Supply in Portugal from 1961 to 2011. The graph shows the observed and normalised trend on the basis of FAO data (FAO 2015a) as kcal/capita/day values. The 1961 value (2476.0 kcal/capita/day) is assumed to be 100. In accordance with FAO criteria, 'food supply' corresponds to 'average food available for consumption', which differs from actual average food intake, due to losses and waste at various levels of the food chain before reaching individual consumers

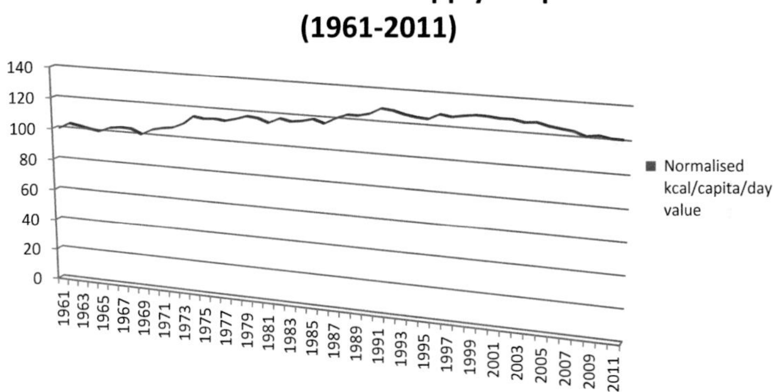

Fig. 2.2 Evolution of Food Supply in Spain from 1961 to 2011.The graph shows the observed and normalised trend on the basis of FAO data (FAO 2015a) as kcal/capita/day values. The 1961 value (2632.0 kcal/capita/day) is assumed to be 100

nutritional requirements vary widely according to age, sex, physical activity, body size and composition and health/disease status, we have considered the theoretical recommendations for an 'average person' (that is, a healthy adult with moderate physical activity, irrespective of sex) of 1750–2750 kcal/day, in which the WHO reference value of 2000 kcal/day falls, to illustrate the extent to which national energy availability meets the population's requirements.

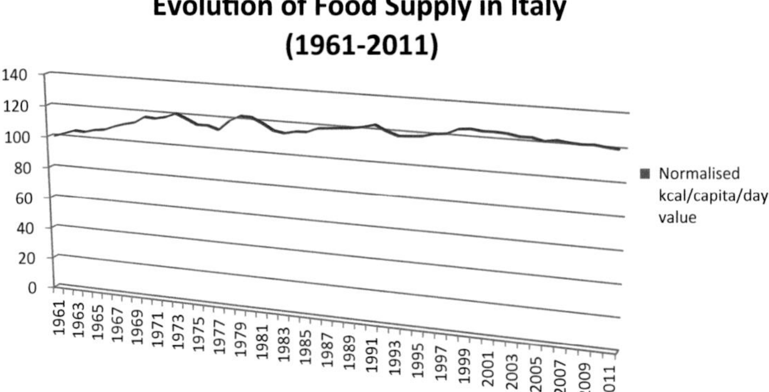

Fig. 2.3 Evolution of Food Supply in Italy from 1961 to 2011. The graph shows the observed and normalised trend on the basis of FAO data (FAO 2015a) as kcal/capita/day values. The 1961 value (2955.0 kcal/capita/day) is assumed to be 100

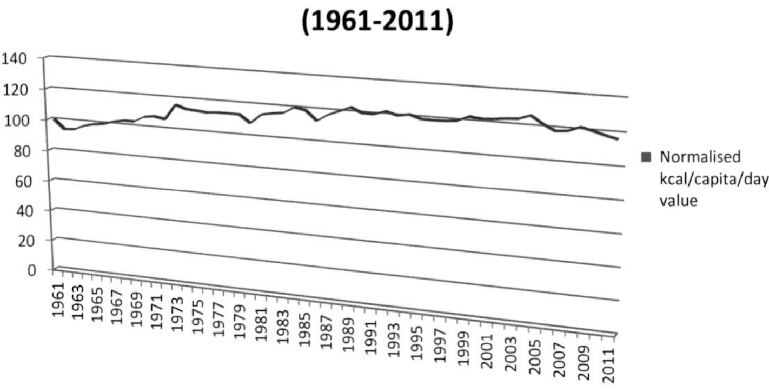

Fig. 2.4 Evolution of Food Supply in Greece from 1961 to 2011.The graph shows the observed and normalised trend on the basis of FAO data (FAO 2015a) as kcal/capita/day values. The 1961 value (2824.0 kcal/capita/day) is assumed to be 100

Figures 2.1, 2.2, 2.3, 2.4, 2.5, 2.6 and 2.7 indicate that, in 1961 and subsequent years, each average apparent food consumption or food availability at the national level was about 2000–2500 kcal/person/day, falling within the range of the referred ideal energy intake. The lowest value was registered for Morocco in 1961 (Fig. 2.6), followed by Cyprus (Fig. 2.7) and Portugal (Fig. 2.1). However, a marked increase in total energy available for consumption of approximately 450 kcal/capita/day was observed globally in subsequent years (WHO-ROEM 2012), and Mediterranean countries also followed this trend, reaching levels of 3500 kcal/capita/day and higher. More recently, a downward trend in the average total energy available has been registered in the region, except for Morocco and Cyprus (Figs. 2.1, 2.2, 2.3, 2.4, 2.5, 2.6 and 2.7).

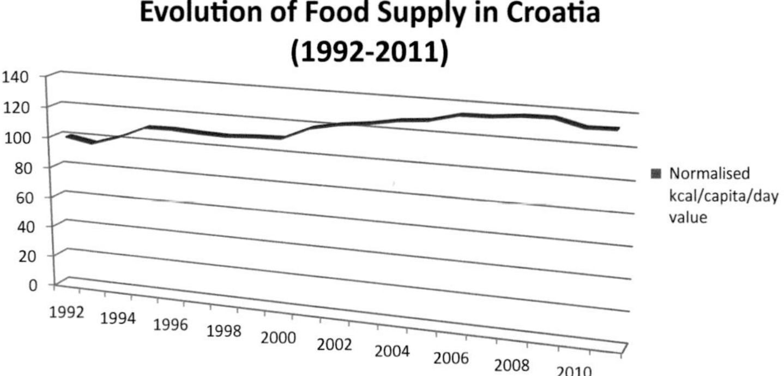

Fig. 2.5 Evolution of Food Supply in Croatia from 1992 to 2011. The graph shows the observed and normalised trend on the basis of FAO data (FAO 2015a) as kcal/capita/day values. The 1992 value (2312.0 kcal/capita/day) is assumed to be 100

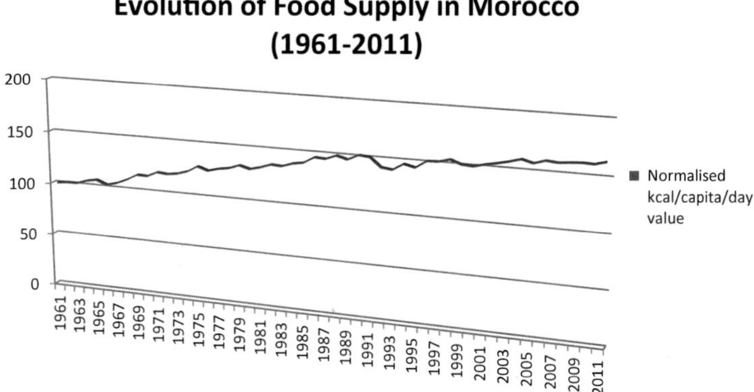

Fig. 2.6 Evolution of Food Supply in Morocco from 1961 to 2011. The graph shows the observed and normalised trend on the basis of FAO data (FAO 2015a) as kcal/capita/day values. The 1961 value (2047.0 kcal/capita/day) is assumed as 100

By the end of the Seven Countries Study, Keys and colleagues (1980) observed a westernization of food habits in the region, which has recently been confirmed by other authors (da Silva et al. 2009) in Mediterranean European countries. This involves an increased consumption of meat, milk, animal fats, vegetable oils (excluding olive oil) and sugars and a decreased consumption of cereals, legumes and wine and other alcoholic beverages (Vareiro et al. 2009). These aspects are discussed in more detail in Part II.

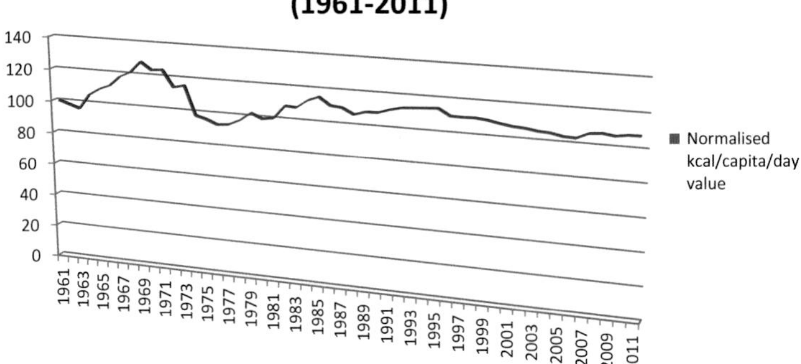

Fig. 2.7 Evolution of Food Supply in Cyprus from 1961 to 2011. The graph shows the observed and normalised trend on the basis of FAO data (FAO 2015a) as kcal/capita/day values. The 1961 value (2478.0 kcal/capita/day) is assumed as 100

References

Bach-Faig A, Berry EM, Lairon D, Reguant J, Trichopoulou A, Dernini S, Medina FX, Battino M, Belahsen R, Miranda G, Serra-Majem L, Mediterranean Diet Foundation Expert Group (2011) Mediterranean diet pyramid today. Science and cultural updates. Public Health Nutr 14(12A): £2274–2284. doi:10.1017/S1368980011002515

Covas MI, Gambert P, Fitó M, de la Torre R (2010) Wine and oxidative stress: up-to-date evidence of the effects of moderate wine consumption on oxidative damage in humans. Atherosclerosis 208(2):297–304. doi:10.1016/j.atherosclerosis.2009.06.031

da Silva R, Bach-Faig A, Quintana BR, Buckland G, Almeida MDV, Serra-Majem L (2009) Worldwide variation of adherence to the Mediterranean diet, in 1961–1965 and 2000–2003. Public Health Nutr 12(9A):1676–1684. doi:10.1017/S1368980009990541

FAO (2001) Human energy requirements. Report of a Joint FAO/WHO/UNU Expert Consultation. Food and nutrition technical report series 1, Rome, 17–24 Oct 2001. FAO, Rome. Available at http://www.fao.org/3/a-y5686e.pdf. Accessed 30 Oct 2015

FAO (2015a) Statistics Division. Food Balance/Food Balance Sheets. Available at http://faostat3.fao.org/browse/FB/FBS/E. Accessed 30 Oct 2015

FAO (2015b) Statistics Division. Glossary. Available at http://faostat3.fao.org/mes/glossary/E. Accessed 30 Oct 2015

Georgoulis M, Kontogianni MD, Yiannakouris N (2014) Mediterranean diet and diabetes: prevention and treatment. Nutrients 6(4):1406–1423. doi:10.3390/nu6041406

Grande F, Anderson JT, Chlouverakis C, Proja M, Keys A (1965) Effect of dietary cholesterol on man's serum lipids. J Nutr 87(1):52–62

Grande F, Anderson JT, Keys A (1972) Diets of different fatty acid composition producing identical serum cholesterol levels in man's. Am J Clin Nutr 25(1):53–60

Jordão AM, Gonçalves FJ, Cporreia AC, Cantão J, Rivero-Pérez MD, Gonzalez-SanJosé ML (2010) Proanthocyanidin content, antioxidant capacity and scavenger activity of Portuguese sparkling wines (Bairrada Appellation of Origin). J Sci Food Agric 90(12):2144–2152. doi:10.1002/jsfa.4064

Keys A (1995) Mediterranean diet and public health: personal reflections. Am J Clin Nutr 61(6 suppl):1321s–1333s

Keys A, Keys M (1959) Eat well and stay well. Doubleday & Company, New York

Keys A, Fidanza F (1960) Serum cholesterol and relative body weight of coronary patients in different populations. Circulation 22:1091–1106

Keys A, Aravanis C, Blackburn I, Buzina R, Djordjevic BS, Dontas AS, Fidanza F, Karvonen MJ, Kimura N, Menotti A, Muhacek I, Nedeljkovic S, Puddu V, Punsar S, Taylor HL, van Buchem FSP (1980) Seven countries—a multivariate analysis of death and coronary heart disease. A Commonwealth Fund Book. Harvard University Press, Cambridge, MA

Nishizuka T, Fujita Y, Sato Y, Nakano A, Kakino A, Ohshima S, Kanda T, Yoshimoto R, Sawamura T (2011) Procyanidins are potent inhibitors of LOX-1: a new player in the French Paradox. Proc Jpn Acad Ser B Phys Biol Sci 87(3):104–113. doi:10.2183/pjab.87.104

Opie LH, Lamont K, Lecour S (2011) Wine and heart health: learning from the French paradox. SA Heart J 8(3):172–177

Trichopoulou A, Lagiou P (1997) Healthy traditional Mediterranean diet: an expression of culture, history, and lifestyle. Nutr Rev 55(11):383–389. doi:10.1111/j.1753-4887.1997.tb01578.x

Trichopoulou A, Kouris-Blazos A, Wahlqvist ML, Gnardellis C, Lagiou P, Polychronopoulos E, Vassilakou T, Lipworth L, Trichopoulos D (1995) Diet and overall survival in elderly people. BMJ 311(7018):1457–1460. doi:10.1136/bmj.311.7018.1457

Vareiro D, Faig AB, Quintana B, Bertomeu I, Buckland G, Almeida MDV, Majem LS (2009) Availability of Mediterranean and non-Mediterranean foods during the last four decades: comparison of several geographical areas. Public Health Nutr 12(9A):936–941. doi:10.1017/S136898000999053X

WHO-ROEM (2012) Promoting a healthy diet for the WHO Eastern Mediterranean Region: user-friendly guide. WHO Regional Office for the Eastern Mediterranean, Cairo. Available at http://applications.emro.who.int/dsaf/emropub_2011_1274.pdf?ua=1. Accessed 30 Oct 2015

3 Adherence to the Mediterranean Diet

Abstract

This chapter discusses the desirable features of the Mediterranean diet (MD) and the current evolution of food habits in the countries forming the United Nations Education, Scientific and Cultural Organization MD Representative List. Several dietary indexes that measure adherence to the MD are presented and discussed for the countries of the area, as well as from a global perspective. The concept of 'dietary pattern' is used here as an integrated approach enabling the identification and quantification of associations between the 'overall diet' and specific health/disease outcomes. Some negative effects of adopting a so-called Western or globalized diet are highlighted, namely the increased proportion of processed energy-dense foods in the daily diet, along with a decreasing trend in the consumption of green vegetables and fresh fruits. These behavioural changes can be the result of a dietary transition, in this case with deleterious consequences. On the other hand, the analysis of nutritional epidemiology studies, complemented with information provided by studies at cellular and/or molecular levels, enables the discussion of the multiple associations between the MD, health, well-being and longevity.

3.1 Measuring Adherence to the Mediterranean Diet

Data from the United Nations (UN) Food and Agriculture Organization (FAO) Food Balance Sheets (FBS) have been used to illustrate time trends in total food supply (measured as total energy availability) in the seven countries of the UN Education, Scientific and Cultural Organization (UNESCO) Representative List (Chap. 2). Data from FBS are also applied to compare food, nutrient and energy availability against theoretical models of nutritional requirements, thereby highlighting the risks of excesses in and/or shortages of foods, nutrients and energy. In conjunction with health data (morbidity and/or mortality rates), they also help to establish possible associations between food and health/disease patterns. FBS

A.M. Delgado et al., *Chemistry of the Mediterranean Diet*,
DOI 10.1007/978-3-319-29370-7_3

provide very useful information, albeit with limitations, at different levels—national (e.g. country), regional (e.g. Europe, Asia) and global (e.g. world, developed vs. underdeveloped regions)—and therefore enable comparisons between countries/regions in a specific year and/or across time. The data compiled by the FAO for FBS are a valuable resource for research and for policy purposes. The longitudinal and joint analysis of food availability and morbidity and mortality data reveals the positive and/or negative consequences of dietary changes, thereby constituting a base for the development of nutrition policies. The aim is to improve the health and well-being of populations.

When comparing the availability of energy provided by 'Mediterranean and non-Mediterranean' foods in five geographic areas, within a time interval of 43 years, Vareiro and co-workers (2001) showed that an increased energy availability of the so-called non-Mediterranean foods (animal fats, vegetable oils, sugars and meat) occurred in European Mediterranean countries (Albania, Cyprus, France, Greece, Italy, Malta, Portugal, Spain, Turkey and Yugoslavia). On the other hand, a corresponding decrease in 'Mediterranean' foods (cereals, alcoholic beverages—including wine—and legumes) was observed (Vareiro et al. 2009). Along the southern shores of the Mediterranean Sea (Algeria, Egypt, Israel, Lebanon, Libyan Arab Jamahiriya, Morocco, Syrian Arab Republic and Tunisia), changes were dissimilar to those of their European counterparts. In detail, a certain increase in most food groups was observed, particularly in terms of energy provided by nuts, vegetable oils, fish, meat, vegetables and sugar. However, a food or dietary pattern such as the Mediterranean diet (MD) is a complex association of foods and drinks distributed across time (daily, weekly, monthly and yearly). The evaluation of single foods or nutrients as unrelated entities has obvious limitations, as foods and drinks are consumed in different combinations and are prepared and cooked in various ways with numerous ingredients. Therefore, a more complete and accurate picture requires the use of thorough approaches. Accordingly, several researchers have formulated indexes that combine foods and/or nutrient features of the MD. Indexes built to measure the quality of a certain food pattern were initially based on recommendations for food/nutrient intake according to the available scientific evidence. This theoretical a priori approach aims to classify the quality of a diet on a single score based on different components: foods, food groups or combination of foods and nutrients (Bach et al. 2006). Later, the application of statistical methods (e.g. factorial analysis, principal component analysis or cluster analysis) to actual food/nutrient intake enabled the generation of food patterns, thus constituting an a posteriori approach. Scores developed by several researchers are presented in Table 3.1.

The widely used original MD Score, either the original version or its variants (Trichopoulou et al. 1995, 2003), have contributed to the establishment of several relationships between adherence to the MD and health/disease in population groups and settings as different as Greece, Denmark, Italy, Israel, Spain, China (and Chinese individuals overseas, in Australia and the USA), Australia (Anglo-Celts and Greco-Australians), Finland and the Netherlands. Other indexes have also

Table 3.1 Indexes developed to measure adherence to the Mediterranean diet, according to different authors

Index	Main characteristics	Authors
Mediterranean Diet Score—MDS—version 1	8 components: food items (vegetables/potatoes, legumes/nuts/seeds, fruits, cereals, dairy products, meat and poultry) + moderate alcohol + ratio MUFA/SFA	Trichopoulou et al. (1995)
Mediterranean Diet Score—MDS—version 2	10 components: food items (vegetables, legumes, fruits and nuts, cereals/potatoes, poultry, fish, dairy products, meat products) + moderate alcohol + ratio MUFA/SFA	Trichopoulou et al. (2003)
Mediterranean Adequacy Index—MAI	Quotient of % of energy from Mediterranean food groups (bread, cereals, legumes, potatoes, vegetables, fruit, fish, red wine, vegetable oils) by % of energy from non-Mediterranean food groups (milk, cheese, meat, eggs, animal fats and margarines, sugar, cakes, pies and cookies)	Alberti-Fidanza et al. (1999)
Mediterranean Dietary Quality Index—Med-DQI	7 components: food items (cereals, meat, fish, olive oil, vegetables and fruit) + % SFA + cholesterol intake	Gerber (2006)
Mediterranean Pattern Score—MDP score	8 components: food items (vegetables, fruits, olive oil, fish, meat, foods of high glycaemic index[a]) + fibre + alcohol intake	Martínez-González et al. (2012)
Mediterranean Score—MS	11 components: food items (cereals, fruits, vegetables, legumes, nuts, seeds, olive oil, fish, poultry, dairy products, eggs, sweets, meat)	Goulet et al. (2003)
Alternate Mediterranean Diet Score—aMED	9 components: food items (vegetables, except potatoes, legumes, fruits, nuts, whole grains, red and processed meat, fish) + ratio MUFA/SFA + ethanol	Fung et al. (2005)
Mediterranean Diet Quality Index in Children and Adolescents—KIDMED	16 components: food items and practices (fruit/fruit juices, vegetables, fish, legumes, pasta/rice, nuts, olive oil, yogurts or cheese, cereals or grains for breakfast, milk products for breakfast, frequency of fast-food consumption + omitting breakfast + pastries for breakfast + sweets/day)	Serra-Majem et al. (2004)
Mediterranean Diet Adherence Screener—MEDAS	14 components: food items (olive oil for seasoning and frying, white meat, vegetables, fruit, red meat or sausages, animal fat, sugar sweetened beverages, pulses, commercial pastries, nuts, red wine) + olive oil as main cooking fat + consumption of dishes with traditional sauce of tomatoes, garlic, onion or leeks sautéed in olive oil	Schröder et al. (2011)

[a]White bread, pasta and rice

strengthened the evidence of the benefits of the MD in preventing cardiovascular diseases, obesity, asthma and some types of cancer, as reported in Sect. 3.3.

In spite of several limitations, there is a wealth of evidence indicating the positive effects of the MD in protecting health, in preventing disease and also in reversing some disease statuses. Observed limitations concern the accuracy and validity of the data on food and/or nutrient consumption used to build the MD indexes, and the application of different indexes to investigate the relationships between the MD and health/disease indicators.

3.2 Global Adherence to the Mediterranean Diet

In 2009, da Silva and colleagues compared the worldwide adherence to the MD in two time periods, highlighting the variations that have occurred within approximately 4 decades. Based on FAO FBS data, the authors computed the Mediterranean Adequacy Index (MAI) for 41 countries representing five continents: Europe, Africa, Asia, America (North and South) and Oceania in the periods 1961–1965 and 2000–2003 (da Silva et al. 2009). Worldwide adherence was also computed using data from 169 countries for both periods. In theory, the MAI may range from zero (no adherence to MD at all) to infinite (positive) values. Three levels of adherence to MAI are considered: low score $\leq$2, high score $\geq$4 and intermediate score $2 \leq \text{MAI} \leq 4$ (Alberti-Fidanza et al. 1999).

da Silva and co-workers observed a clear and general decrease in the MAI scores over time, at either the global (from 2.86 to 2.03) or the country level, ranging from 5.54 to 0.63 in the first period (1961–1965) and from 4.09 to 0.64 in the second time interval (2000–2003). In the 1960s, the five countries with the highest scores were located in the Mediterranean basin (Greece, Albania, Turkey, Egypt and Tunisia). It is interesting to note that Japan ranked sixth in the same period of time. In fact, the Japanese diet (and, indeed, other food patterns) includes several features of the MD and has also been considered a ‘healthy dietary pattern’ (da Silva et al. 2009).

The findings clearly demonstrate a general trend of lower adherence to the MD, which is more noticeable at the turn of the twentieth to the twenty-first century. It should be noted that the largest disparities were found in Greece (5.54–2.04), Japan (4.11–1.51), Albania (5.07–2.51), Turkey (5.03–2.80), Spain (3.35–1.19) and Portugal (3.39–1.27).

The observed shifts and the rapid pace at which they took place, reveal a state of general ‘nutrition transition’. This concept refers to large changes in the structure and composition of diet and physical activity patterns that occur/have occurred to humankind at different times (Popkin 2006). Popkin identified five patterns of nutrition transition:

1. Collecting food
2. Famine
3. Receding famine
4. Degenerative disease

5. Behavioural change

As well as shifts in diets and physical activity, changes also occur at the demographic, socio-economic and health levels. Thus, it seems that most of the previously analysed countries would have experienced a shift from pattern three to four.

Rodrigues and colleagues (2008) analysed trends in Portuguese household diet quality by applying simultaneously the MAI and a revised Healthy Diet Indicator [originally developed by Huijbregts and co-workers (1997)] that measures compliance with World Health Organization (WHO) dietary goals. Median MAI was found to decrease between 1990 and 2000 (from 2.23 to 2.09); consequently, the proportion of those results with a high MD adherence (MAI $\geq$ 4) decreased from 21.6 to 16.2 % of households in 10 years. According to Rodrigues and colleagues, compliance with WHO dietary recommendations was also low in Portuguese households between 1990 and 2000. Lower adherence to MD was more likely to occur in households in which the responsible adults were young, more educated, lived in an urban area and had a non-manual occupation and a higher income. These findings may indicate this population is living in a situation of nutrition transition as proposed by Popkin (2006), that is, they are in pattern 4: increase in degenerative diseases and diet-related non-communicable diseases (NCD).

Another study in elderly Portuguese households in 1990 and 2000 revealed a similar or higher adherence to the MD in this age group compared with the national median and with adults (Santos et al. 2009). According to these authors, adherence was considered to be intermediate, as they scored 2.2 and 2.4 in the MDI median in 1990 and 2000, respectively. It is worth noticing that the proportion of elderly with high scores remained relatively stable (24.6 vs. 24.0 %) in the same time period. Moreover, this proportion increased in the intermediate score and decreased in the low score (Santos et al. 2009).

Satalic and colleagues (2004) evaluated the MD adherence level of Croatian university students via the 'Mediterranean-Dietary Quality Index' (Med-DQI) and registered poor adherence scores. These authors observed a higher correlation for Med-DQI with macronutrient intakes than with micronutrient intakes.

Mariscal-Arcas and co-workers (2008) used the Med-DQI for children and adolescents (KIDMED) instrument to evaluate the level of adherence of young populations to this model. In a sample of 3190 schoolchildren aged 8–16 years from Granada, Southern Spain, the authors registered an average to good level of adherence to the dietary model. Major deviations resulted from the regular consumption of fast food in restaurants (weekly intake by >60 %) and the consumption of sweets and candies several times a day (>20 %). It is noteworthy that no regular consumption of nuts was registered. Moreover, the youngest children in particular tended to eat pastry at breakfast. As a consequence of this food pattern, their protein intake was more than double the recommended levels, and mean energy intake exceeded the mean theoretical energy requirements.

According to Satalic and colleagues (2004) and Mariscal-Arcas and co-workers (2008), and following Keys's (1995) recommendations, children and adolescents are the most receptive age groups and should be targeted in nutrition education programs. The acquisition of healthy food habits in childhood will certainly have beneficial effects in later life. The period of education can be critical in the development of desirable lifestyle habits, even in young adults.

Nutrition interventions that solely targeted knowledge acquisition proved to be of limited impact, as lifestyle changes require competencies and skills, primarily in food and eating. This is the case for cooking skills, which are usually passed on through the socialisation process (mainly directed at girls and young women). The ability to control the quality, composition, diversity and quantity of what is eaten is hugely important in terms of food preparation, combination and cooking. Cooking skills have been reported as decreasing in the past century, partly due to demographic, family structure and economic changes. The lack of cooking skills and the wide availability of pre-prepared and ready-to-eat foods and meals is a two-way process; the food industry responds to a societal need, which in turn leads to decreased competencies in this area.

In a study of adolescents, Leal and co-workers (2011) reported adherence to MD (measured by the KIDMED index) to be positively associated with cooking habits and skills. Such findings draw attention to the importance of acquiring such skills to affect food choice and consumption.

It is worth remembering that the MD refers to dietary patterns found in olive-growing areas of the Mediterranean region and described in the 1960s and beyond. Several variants of the MD exist, but some common components can be identified:

- High monounsaturated/saturated fat ratio
- Moderate levels of wine consumption and mainly during meals
- High consumption of vegetables, fruits, pulses and grains
- Moderate consumption of milk and dairy products, mostly in the form of cheese
- Low consumption of meat and meat products

3.3 Evidence of the Health Benefits of the Mediterranean Diet

The association of the MD with better health has been found through different epidemiological studies, from ecological to case-control and cohort studies. Clinical trials have also been conducted testing the positive effects of several MD components. Further evidence of the health benefits of the MD has been obtained via systematic reviews and meta-analyses of prospective cohort studies performed by Sofi and colleagues. A reduction in overall mortality, in incidence or mortality from cardiovascular diseases, in cancer incidence or mortality and in incidence of neurodegenerative diseases such as Parkinson's and Alzheimer's were shown to be associated with adherence to the MD (Sofi et al. 2008, 2010). Another systematic review of observational and intervention studies revealed a possible role for the MD in overweight/obesity prevention, despite some inconsistent results attributable to methodological differences in the studies analysed (Buckland et al. 2008).

Both epidemiological and metabolic studies suggest that individuals can greatly benefit from adopting elements of the MD. Several authors have applied the above-mentioned instruments in epidemiological studies (Chrysohoou et al. 2004; Fung et al. 2005, 2009; Georgoulis et al. 2014; Martínez-González et al. 2012; Rodrigues et al. 2008), mainly seeking associations between diet and health or to substantiate evidence on previously reported health benefits of the MD.

The reduced risk of coronary disease when the MD is adopted, extensively studied by Keys, has continued to be confirmed by several authors in different populations (Fung et al. 2009).

A modified MD Adherence Screener (MEDAS) index was used to show an inverse relationship between adherence to MD and obesity indexes (Martínez-González et al. 2012). Other researchers applied a modified Mediterranean Diet Score (MDS) to young and active US adults; as a result, they registered significant inverse associations with metabolic syndrome, low-density lipoprotein cholesterol and reported weight gain (Yang et al. 2014).

Georgoulis and colleagues reviewed data from the literature, exploring MD as a whole dietary pattern rather than focusing on the effect of its individual components. These authors presented evidence of reduced risk of developing type II diabetes mellitus as well as evidence from interventional studies assessing the effect of the MD on diabetes control and the management of diabetes-related complications (Georgoulis et al. 2014).

Trichopoulou and co-workers assessed the effect of MD pattern on the overall survival of elderly inhabitants from rural Greek villages using the previously mentioned MDS. These authors observed that a 1-unit increase in diet score was significantly associated with a 17 % decrease in overall mortality, showing the benefits of the MD on longevity and quality of life at older ages (Trichopoulou et al. 1995).

The self-perception of well-being and quality of life is rapidly becoming an important issue, especially in the case of aged populations. One of the most accredited hypotheses is that the MD is positively associated with better overall health status and reduced risk of major chronic diseases, such as some types of cancer and neurodegenerative diseases. These effects are attributed to the high intake of different beneficial compounds, such as antioxidants, and positively correlated to the level of adherence to the MD (Bonaccio et al. 2013, 2014).

On the other hand, Nordmann and colleagues observed more favourable changes for the group under an MD regimen than for those following low-fat diets in a comparative study. In particular, modifications were observed in body weight, body mass index, systolic blood pressure, diastolic blood pressure, fasting plasma glucose, total cholesterol and high-sensitivity C-reactive protein. These authors confirmed that the MD appears to be more effective than low-fat diets in inducing clinically relevant long-term changes in cardiovascular risk factors and inflammatory markers (Nordmann et al. 2011).

The MD has also been found to improve asthma control (Arvaniti et al. 2011; Barros et al. 2008; Garcia-Marcos et al. 2007), to have anti-inflammatory effects

(Mena et al. 2009), to reduce plasma oxidative stress (Dai et al. 2008) and to reduce the risk of metabolic syndrome (Kastorini et al. 2011).

According to the WHO, economic changes are usually associated with changes along the food chain, from the production and processing sectors to the distribution and marketing of foods. The worldwide increase in urbanisation that has occurred in past decades has been shown to influence food habits and lifestyle, and therefore contribute to shifts in health and disease patterns (WHO 2015). Changes in diets, patterns of work and leisure time—often referred to as the 'nutrition transition'—contribute to the factors underlying NCD at the global level, including the less developed and poorest countries. Moreover, the pace of these changes is fast in the Mediterranean countries.

According to WHO (2015), dietary changes that characterise the 'nutrition transition' include both quantitative and qualitative changes to a higher energy-dense diet, with a greater role for the following:

- Addition of fat and sugars in foods
- Greater saturated fat intake (mostly from animal sources)
- Reduced intakes of complex carbohydrates and dietary fibre
- Reduced intakes of fruits and vegetables

The previous points aimed to introduce and discuss the concept of the MD, to document time trends in its evolution, and to summarise the research results regarding the MD. Nutritional epidemiology research has been crucial in establishing associations between the 'prodigious Mediterranean diet' (Peres 1997), well-being and health, and longevity of the populations following this food pattern.

Part II brings a complementary view of the MD by focusing on its chemical aspects. Organised by food groups, several components—identified and quantified through chemical analysis—are described in conjunction with the mechanisms involved in biological reactions. The basic aim is to provide evidence that the MD constitutes a paradigm and a valuable resource for the formulation of nutritional theoretical models and applied healthy eating patterns.

References

Alberti-Fidanza A, Fidanza F, Chiuchiù MP, Verducci G, Fruttini D (1999) Dietary studies on two rural Italian population groups of the Seven Countries Study. 3. Trend of food and nutrient intake from 1960 to 1991. Eur J Clin Nutr 53(11):854–860

Arvaniti F, Priftis KN, Papadimitriou A, Papadopoulos M, Roma E, Kapsokefalou M, Anthracopoulos MB, Panagiotakos DB (2011) Adherence to the Mediterranean type of diet is associated with lower prevalence of asthma symptoms, among 10-12 years old children: the PANACEA study. Pediatr Allergy Immunol 22(3):283–289. doi:10.1111/j.1399-3038.2010.01113.x

Bach A, Serra-Majem L, Carrasco JL, Roman B, Ngo J, Bertomeu I, Obrador B (2006) The use of indexes evaluating the adherence to the Mediterranean diet in epidemiological studies: a review. Public Health Nutr 9(1A):132–146. doi:10.1079/PHN2005936

Barros R, Moreira A, Fonseca J, Ferraz de Oliveira J, Delgado L, Castel-Branco MG, Haahtela T, Lopes C, Moreira P (2008) Adherence to the Mediterranean diet and fresh fruit intake are associated with improved asthma control. Allergy 63(7):917–923. doi:10.1111/j.1398-9995.2008.01665.x

Bonaccio M, Di Castelnuovo A, Bonanni A, Costanzo S, De Lucia F, Pounis G, Zito F, Donati MB, de Gaetano G, Iacoviello L (2013) Adherence to a Mediterranean diet is associated with a better health-related quality of life: a possible role of high dietary antioxidant content. BMJ 3(8): e003003. doi:10.1136/bmjopen-2013-003003

Bonaccio M, Di Castelnuovo A, De Curtis A, Costanzo S, Persichillo M, Donati MB, Cerletti C, Iacoviello L, de Gaetano G (2014) Adherence to the Mediterranean diet is associated with lower platelet and leukocyte counts: results from the Moli-sani study. Blood 123(19): 3037–3044. doi:10.1182/blood-2013-12-541672

Buckland G, Bach A, Serra-Majem L (2008) Obesity and the Mediterranean diet: a systematic review of observational and intervention studies. Obes Rev 9(6):582–593. doi:10.1111/j.1467-789X.2008.00503.x

Chrysohoou C, Panagiotakos DB, Pitsavos C, Das UN, Stefanadis C (2004) Adherence to the Mediterranean diet attenuates inflammation and coagulation process in healthy adults: the ATTICA study. J Am Coll Cardiol 44(1):152–158. doi:10.1016/j.jacc.2004.03.039

Dai J, Jones DP, Goldberg J, Ziegler TR, Bostick RM, Wilson PW, Manatunga AK, Shallenberger L, Jones L, Vaccarino V (2008) Association between adherence to the Mediterranean diet and oxidative stress. Am J Clin Nutr 88(5):1364–1370

da Silva R, Bach-Faig A, Quintana BR, Buckland G, Almeida MDV, Serra-Majem L (2009) Worldwide variation of adherence to the Mediterranean diet, in 1961–1965 and 2000–2003. Public Health Nutr 12(9A):1676–1684. doi:10.1017/S1368980009990541

Fung TT, McCullough ML, Newby PK, Manson JE, Meigs JB, Rifai N, Willett WC, Hu FB (2005) Diet-quality scores and plasma concentrations of markers of inflammation and endothelial dysfunction. Am J Clin Nutr 82(1):163–173

Fung TT, Rexrode KM, Mantzoros CS, Manson JE, Willett WC, Hu FB (2009) Mediterranean diet and incidence of and mortality from coronary heart disease and stroke in women. Circulation 119(8):1093–1100. doi:10.1161/CIRCULATIONAHA.108.816736

Garcia-Marcos L, Canflanca IM, Garrido JB, Varela AL-S, Garcia-Hernandez G, Grima FG, Gonzalez-Diaz C, Carvajal-Urueña I, Arnedo-Pena A, Busquets-Monge RM, Suarez Varela MM, Blanco-Quiros A (2007) Relationship of asthma and rhinoconjunctivitis with obesity, exercise and Mediterranean diet in Spanish schoolchildren. Thorax 62(6):503–508. doi:10.1136/thx.2006.060020

Georgoulis M, Kontogianni MD, Yiannakouris N (2014) Mediterranean diet and diabetes: prevention and treatment. Nutrients 6(4):1406–1423. doi:10.3390/nu6041406

Gerber M (2006) Qualitative methods to evaluate Mediterranean diet in adults. Public Health Nutr 9(1A):147–151. doi:10.1079/PHN2005937

Goulet J, Lamarche B, Nadeau G, Lemieux S (2003) Effect of a nutritional intervention promoting the Mediterranean food pattern on plasma lipids, lipoproteins and body weight in healthy French-Canadian women. Atherosclerosis 170(1):115–124. doi:10.1016/S0021-9150(03)00243-0

Huijbregts P, Feskens E, Räsänen L, Fidanza F, Nissinen A, Menotti A, Kromhout D (1997) Dietary pattern and 20 year mortality in elderly men in Finland, Italy, and The Netherlands: longitudinal cohort study. BMJ 315(7099):13–17. doi:10.1136/bmj.315.7099.13

Kastorini CM, Milionis HJ, Esposito K, Giugliano D, Goudevenos JA, Panagiotakos DB (2011) The effect of Mediterranean diet on metabolic syndrome and its components: a meta-analysis of 50 studies and 534,906 individuals. J Am Coll Cardiol 57(11):1299–1313. doi:10.1016/j.jacc.2010.09.073

Keys A (1995) Mediterranean diet and public health: personal reflections. Am J Clin Nutr 61(6 suppl):1321s–1333s

Leal F, Oliveira B, Rodrigues S (2011) Relation between cooking habits and skills and Mediterranean diet in a sample of Portuguese adolescents. Perspect Public Health 131(6):283–287. doi:10.1177/1757913911419909

Mariscal-Arcas M, Rivas A, Velasco J, Ortega M, Caballero AM, Olea-Serrano F (2008) Evaluation of the Mediterranean Diet Quality Index (KIDMED) in children and adolescents in Southern Spain. Public Health Nutr 12(9):1408–1412. doi:10.1017/S1368980008004126

Martínez-González MA, García-Arellano A, Toledo E, Salas-Salvadó J, Buil-Cosiales P, Corella D, Covas MI, Schröder H, Arós F, Gómez-Gracia E, Fiol M, Ruiz-Gutiérrez V, Lapetra J, Lamuela-Raventos RM, Serra-Majem L, Pintó X, Muñoz MA, Wärnberg J, Ros E, Estruch R (2012) A 14-item Mediterranean diet assessment tool and obesity indexes among high-risk subjects: the PREDIMED trial. PLoS One 7(8):e43134. doi:10.1371/journal.pone.0043134

Mena MP, Sacanella E, Vazquez-Agell M, Morales M, Fitó M, Escoda R, Serrano-Martínez M, Salas-Salvadó J, Benages N, Casas R, Lamuela-Raventós RM, Masanes F, Ros E, Estruch R (2009) Inhibition of circulating immune cell activation: a molecular antiinflammatory effect of the Mediterranean diet. Am J Clin Nutr 89(1):248–256. doi:10.3945/ajcn.2008.26094

Nordmann AJ, Suter-Zimmermann K, Bucher HC, Shai I, Tuttle KR, Estruch R, Briel M (2011) Meta-analysis comparing Mediterranean to low-fat diets for modification of cardiovascular risk factors. Am J Med 124(9):841–851. doi:10.1016/j.amjmed.2011.04.024

Peres E (1997) Bem comidos e Bem Bebidos. Editorial Caminho, Lisboa

Popkin BM (2006) Global nutrition dynamics: the world is shifting rapidly toward a diet linked with noncommunicable diseases. Am J Clin Nutr 84(2):289–298

Rodrigues SSP, Caraher M, Trichopoulou A, Almeida MDV (2008) Portuguese households' diet quality (adherence to Mediterranean food pattern and compliance with WHO population dietary goals): trends, regional disparities and socioeconomic determinants. Eur J Clin Nutr 62(11):1263–1272. doi:10.1038/sj.ejcn.1602852

Santos D, Rodrigues S, Trichopoulou A, Almeida MDV (2009) Adherence to the Mediterranean diet by Portuguese elderly. J Nutr Health Aging 13:S21–S22

Satalic Z, Baric IC, Keser I, Maric B (2004) Evaluation of diet quality with the Mediterranean dietary quality index in university students. Int J Food Sci Nutr 55(8):589–595. doi:10.1080/09637480500086141

Schröder H, Fitó M, Estruch R, Martínez-González MA, Corella D, Salas-Salvadó J, Lamuela-Raventós R, Ros E, Salaverría I, Fiol M, Lapetra J, Vinyoles E, Gómez-Gracia E, Lahoz C, Serra-Majem L, Pintó X, Ruiz-Gutierrez V, Covas MI (2011) A short screener is valid for assessing Mediterranean diet adherence among older Spanish men and women. J Nutr 141 (6):1140–1145. doi:10.3945/jn.110.135566

Serra-Majem L, Ribas L, Ngo J, Ortega RM, García A, Pérez-Rodrigo C, Aranceta J (2004) Food, youth and the Mediterranean diet in Spain. Development of KIDMED, Mediterranean Diet Quality Index in children and adolescents. Public Health Nutr 7(7):931–935. doi:10.1079/PHN2004556

Sofi F, Cesari F, Abbate R, Gensini GF, Casini A (2008) Adherence to Mediterranean diet and health status: meta-analysis. BMJ 337:a1344. doi:10.1136/bmj.a1344

Sofi F, Abbate R, Gensini GF, Casini A (2010) Accruing evidence on benefits of adherence to the Mediterranean diet on health: an updated systematic review and meta-analysis. Am J Clin Nutr 92(5):1189–1196. doi:10.3945/ajcn.2010.29673

Trichopoulou A, Kouris-Blazos A, Wahlqvist ML, Gnardellis C, Lagiou P, Polychronopoulos E, Vassilakou T, Lipworth L, Trichopoulos D (1995) Diet and overall survival in elderly people. BMJ 311(7018):1457–1460. doi:10.1136/bmj.311.7018.1457

Trichopoulou A, Costacou T, Bamia C, Trichopoulos D (2003) Adherence to a Mediterranean diet and survival in a Greek population. N Engl J Med 348(26):2599–2608. doi:10.1056/NEJMoa025039

Vareiro D, Faig AB, Quintana B, Bertomeu I, Buckland G, Almeida MDV, Majem LS (2009) Availability of Mediterranean and non-Mediterranean foods during the last four decades: comparison of several geographical areas. Public Health Nutr 12(9A):936–941. doi:10.1017/S136898000999053X

WHO (2015) Programmes and projects. Nutrition. Nutrition health topics. 3. Global and regional food consumption patterns and trends. Available at http://www.who.int/nutrition/topics/3_food consumption/en/. Accessed 30 Oct 2015

Yang J, Farioli A, Korre M, Kales SN (2014) Modified Mediterranean diet score and cardiovascular risk in a North American working population. PLoS One 9(2):e87539. doi:10.1371/journal.pone.0087539

Part II

Facts on the Composition of 'Mediterranean Foods'

4 Olive Oil and Table Olives

Abstract

Olive orchards shape the Mediterranean landscape. Cultivars with larger fruits and lower oil content are selected to produce table olives, whereas cultivars with smaller fruit and higher oil content are selected for olive oil production. Olive oil is thus obtained by gently squeezing a fruit at the environmental temperature, as opposed to most edible oils, which are extracted from seeds through more elaborate processes that generally involve refining. The final composition of a seed oil is mainly a mixture of fatty acids (despite a few reports of seed oils containing reduced amounts of hydrophobic beneficial compounds). On the other hand, virgin olive oil is always a complex mixture of lipids and different proportions of a wide variety of compounds, including vitamins, flavonoids, simple phenols, chlorophylls, squalene and others. Olive oil is the main dietary fat in the Mediterranean diet, and some of its well-proven effects are substantiated in a health claim from the European Food Safety Authority regarding cardiovascular health. As olives are not edible when they are just harvested, many preparation methods exist to obtain and preserve table olives. The most widely distributed and best studied is the Sevillian style, which uses fruits harvested at their maximum size, albeit still green. Olives are debittered and fermented, mostly by lactic acid bacteria, among which some probiotic strains can be found. Olives are a good source of vitamins, a wide variety of phenolic compounds, some oil and dietary fibres. To ensure preservation, the levels of salt are quite significant; consequently, olives should be eaten in moderate portions as appetizers, or the level of salt should be reduced accordingly if included in cooked dishes.

A.M. Delgado et al., *Chemistry of the Mediterranean Diet*,
DOI 10.1007/978-3-319-29370-7_4

4.1 Olive Oil and Table Olives: An Introduction

The origin of the olive tree is lost in time. According to the International Olive Oil Council (IOOC), the wild olive tree originated in Asia Minor, where it is extremely abundant and grows in thick forests (IOOC 2015a). It may have spread from Syria to Greece. Olive trees (*Olea europaea*) are mentioned in Greek mythology and are still associated with symbols of peace and wisdom. From the sixth century BC onwards, the olive spread throughout the Mediterranean countries, reaching Tripoli, Tunis and the island of Sicily. From there, they moved to southern Italy. The Roman Empire continued the expansion of the olive tree to the countries bordering the Mediterranean Sea, providing resources for their legions and promoting settlement. Later, during Arab invasions into the Iberian Peninsula, their olive varieties were introduced to the south of Spain and Portugal. The Arab influence is so strong that the Portuguese and Spanish words for olive are azeitona/aceituna and for olive oil are azeite/aceite (IOOC 2015a). The Arab roots of these words stress the clear distinction that exists locally between olive oil and other oils (mainly seed oils, known under the generic designation of 'oils').

Some specimens of *O. europaea* more than 2000 years old have been reported. Olive trees grow slowly, under natural xerophytic conditions, taking about 20 years to attain maturity; however, they may start to produce fruits after 2–5 years. The productive period extends to about 150 years, after which the yield slows with ageing.

The olive is the emblematic tree of the Mediterranean region, shaping its landscape: 'the Mediterranean ends where the olive tree no longer grows' (IOOC 2015a). The harsh soil and difficult agricultural conditions were overcome with the use of terraces for the cultivation of olive groves, vineyards and citrus trees. This distinct rural landscape, found mainly in Southern Europe, differentiates the Mediterranean area from Central and Northern Europe. Olive groves constitute an integral and significant part of the Mediterranean environment and culture; however, their ecological importance has only recently been acknowledged (IOOC 2015a; Loumou and Giourga 2003).

The Portuguese and the Spanish transported olive trees to the American continent, expanding the olive culture to Brazil, California and Chile, where important olive groves still exist. These days, Australia and New Zealand are also important olive oil and table olive producers and consumers. The culture is still expanding, with new olive groves being planted in Japan, certain regions of Africa and China. In particular, China is steadily increasing olive grove plantations and olive oil production capacity, primarily for export. Global production increased by an annual rate of 1 % between 1990 and 2005 (IOOC 2015b).

There are hundreds of cultivars of *O. europaea*, the fruits of which vary widely in size and oil content, as well as resulting table olive and olive oil characteristics. Larger fruits with lower lipid contents are normally used to produce table olives, whereas fruits with higher lipid content and generally smaller size are directed towards olive oil production.

Olive Oil. *Olive oil, the juice of the fruit of Olea europaea, is one of the pillars of the Mediterranean food pattern. Pressure-cold-extracted olive oil shows a balanced combination of polyunsaturated fatty acids (n-3, n-6 and n-9), including essential fatty acids, as well as vitamins, antioxidants and many health-promoting minor components. For this reason, health claims, mainly related to the maintenance of normal blood cholesterol levels, have been approved by the European Food Safety Authority. Photo reprinted with kind permission from T. N. Wassermann*

4.2 Olive Oil

Lipids contained in foods can be roughly divided into saturated fatty acids (SFA), carrying only single bonds in the carbon chain; monounsaturated fatty acids (MUFA), containing a single double bond; and polyunsaturated fatty acids (PUFA), containing at least two double bonds. The number of double bonds in the carbon chain, also called unsaturations, affects the physical and chemical properties of fats. When at least two double bonds are present, *cis*- and *trans*-isomers may exist; these last ones are commonly known as trans-fatty acids (TFA), or 'trans-fats'. In other words, TFA in foods are mixtures of several MUFA and PUFA with different length carbon chains. The fatty acid chain length and the location of double bonds affect human health.

Fatty acids may be found free or in triacylglycerides. Another important lipid, besides free fatty acids and triacylglycerides is cholesterol, a primary constituent of cell membranes only found in animal fats.

According to the United Nations (UN) Food and Agriculture Organization (FAO), vegetable oils should be preferred dietary sources of lipids used as an energy source because they include n-3 and n-6 and essential PUFA (FAO 2010). The presence of lipids in the diet increases the absorption of fat-soluble vitamins (e.g. A, D, E and K). In addition to the well-known n-3, n-6 PUFA are also important: linoleic acid (n-6) is an essential component of structural membrane lipids and is involved in cell signalling. It is also a precursor of eicosanoids, a topic covered in more detail in Chap. 7. Linoleic acid is required for normal skin function. On the other hand, the n-3 α-linolenic acid (ALA) is involved in neurological development and growth and is also a precursor of eicosanoids. An appropriate balance of n-3 and n-6 fatty acids should be maintained to limit lipid peroxidation, which is thought to be a component in the development of atherosclerotic plaques.

No essential role, other than as an energy source, has been found for SFA and TFA or cholesterol in the diet, as humans can synthesise these compounds according to their needs. An increased intake of SFA and TFA, particularly from industrially processed food results in the incremental increase in concentrations of plasma total cholesterol and low-density lipoprotein (LDL) cholesterol. The same FAO's report recommends that the intake of each of the above compounds should be minimised, while consuming a nutritionally adequate diet (FAO 2010).

At first glance, edible oils seem much the same, but a few minor differences have important effects on the characteristics of each one. According to the Portuguese National Institute of Health (Instituto Nacional de Saúde Doutor Ricardo Jorge [INSA]) and the 'Portuguese Food Information Resource' (PortFIR), virgin olive oil has no TFA, while refined and hydrogenated vegetable seed oils and margarines are among the major sources of TFA in the diet (INSA 2015). Of those compounds, Δ9 trans (t9) 18:1 (of seed oil origin) has been referred to as particularly deleterious to human health, promoting insulin resistance and increasing the risk of coronary heart disease (Kühlsen et al. 2005).

Moreover, a farmer can do little to alter the quality of seed oils such as canola and sunflower. Conversely, olive oil is a more complex but added-value product, and handling by the grower will strongly affect the quality of the oil. Therefore, olive growers must have a basic understanding of oil quality and how to preserve it.

Traditionally, extreme care is taken—from harvesting to bottling—in order to obtain the best olive oil. Until the mid-twentieth century, fruits were manually harvested at full maturation, when olives turn black in colour. The olive branches were tapped with long sticks to force the ripe olives to fall onto nets or fabrics usually hung a few centimetres above the ground. Workers removed twigs and leaves while carefully placing olives into baskets, from where the olives were transported to the mill and processed within 24 h. Olives were washed at the mill before being ground into a paste using huge mill stones. The compression force was regulated to avoid smashing the pits, which were then separated from the paste before pressing. Olive paste was spread on circular mats that were mounted on top of each other and cold pressed to extract the juice, leaving the 'pomace' behind. The juice (a mixture of the olive's aqueous content and oil) was left to naturally decant for about 1 h. After phase separation, the oil was siphoned off to a special container designed to encourage any particles to quickly sediment at the bottom, enabling the extra-virgin olive oil to be decanted into dark glass bottles without the need for filtration.

The extraction method, as well as the integrity of the fruit, strongly affects the quality of the oil obtained. In current industrial mills, olive oil is first extracted from olives via cold pressing. Sometimes, further extraction with hot water is conducted, resulting in more than one final product. Operations such as decantation, centrifugation and filtration are part of the process.

Given these differences in quality and composition, the IOOC proposed a classification for olive oil, which is not strictly followed in Australia or in the USA. The main classes of olive oil for human consumption are as follows and as shown in Fiorillo and Vercueil (2003) and IOOC (2015c):

(a) Extra virgin olive oil (EVOO). Virgin olive oil that has an organoleptic rating of at least 6.5 and a free acidity, expressed as oleic acid, of not more than 0.8 g/100 g, with due regard for the other criteria laid down in this standard.
(b) Virgin olive oil (VOO). Virgin olive oil that has an organoleptic rating of at least 5.5 and a free acidity, expressed as oleic acid, of not more than 2.0 g/100 g, with due regard for the other criteria laid down in this standard.
(c) Ordinary virgin olive oil (OVOO). Virgin olive oil that has an organoleptic rating of at least 3.3 and a free acidity, expressed as oleic acid, of not more than 3.3 g/100 g, with due regard for the other criteria laid down in this standard.
(d) Lampante virgin olive oil. Virgin olive oil that has an organoleptic rating of less than 3.3 and/or a free acidity, expressed as oleic acid, of more than 3.3 g/100 g, with due regard for the other criteria laid down in this standard.
(e) Refined Olive Oil. Olive oil obtained from virgin olive oil by refining methods, which do not lead to alterations in the initial glyceridic structure.
(f) Olive Oil. A blend of refined olive oil and virgin olive oil fit for consumption as it is.

Olive oil is a complex mixture of triacylglycerides, chlorophylls (accounting for its colour) and minor components responsible for the aroma and other properties, mainly hydrocarbons, phenolic compounds and tocopherols. Triacylglycerides, also known as 'saponifiable fraction', account for about 98–99 % of EVOO and are predominantly esterified with monounsaturated oleic acid, and to a lesser extent palmitic, linoleic and linolenic acids (Bendini et al. 2010; Boskou et al. 2006a; Fazzari et al. 2014; Mailer 2006; Velasco and Dobarganes 2002).

The remainder fraction includes a wide range of minor components, some of which are water soluble (such as phenols and sterols) and are thus called the 'non-saponifiable fraction'. Non-saponifiable compounds give olive oil its unique flavour. Slightly bitter and pungent tastes are classified as positive attributes by panel tests and result from the presence of high levels of polyphenols, greatly contributing to the nutritional benefits and health claims (EFSA 2011a; Visioli et al. 2006). In refined oil, many of these minor components are removed. Therefore, given the considerable differences between refined olive oil and VOO, the discussion below refers solely to virgin olive oil, that is EVOO, VOO or OVOO.

Table 4.1 Allowable ranges of several fatty acid concentrations in extra virgin olive oil

Fatty acid	Number of carbons and unsaturation degree	Allowable range %
Myristic	C14:0	0.0–0.05
Palmitic	C16:0	7.5–20.0
Palmitoleic	C16:1	0.3–3.5
Stearic	C18:0	0.5–5.0
Oleic	C18:1	55.0–83.0
Linoleic	C18:2	3.5–21.0
Linolenic	C18:3	<1.0
Arachidic	C20:0	<0.6–0.8
Gadoleic (eicosenoic)	C20:1	<0.4
Lignoceric	C24:0	0.0–0.2

The second column refers to the length of each listed fatty acid molecule (first figure) and the respective number of double bonds (C=C, second figure). Allowable values are reported from different sources with the preference for broader ranges (Boskou et al. 2006a, b; IOOC 2015e)

The presence and range of fatty acids depends on the cultivar and climate. The saponifiable fraction is responsible for the unique physical properties of olive oils, namely the much higher boiling point and viscosity values than other edible oils.

The triacylglyceride fraction typically includes unsaturated fatty acids (MUFA and PUFA): oleic (C18:1, n-9), linoleic (C18:2, n-6) and α and γ-linolenic (C18:3, respectively n-3 and n-6) acids, as shown in Table 4.1, which summarises the general composition in triacylglycerides of olive oil.

Oleic acid is predominant in olive oil, accounting for about 70 % (INSA 2015), and is regarded as beneficial to health. The positive properties of olive oil have long been observed, and scientific evidence has been collected in support of them. In the early 1990s, Nestel et al. (1994) reported significant differences in human LDL cholesterol when comparing the consumption of oleic acid (lowest LDL cholesterol) with that of palmitoleic acid, which increases LDL levels in blood, even when consumed in small amounts.

The major PUFA in olive oil is linoleic acid (18:2, n-6): 6.2 %, (INSA 2015). On its turn, alfa-linolenic acid (C18:3,n-3), or ALA, is an essential fatty acid required for the biosynthesis of key eicosanoids. Also relevant is the mixture of linoleic acid isomers, also known as conjugated linoleic acid (CLA), consisting of positional (e.g. 7,9; 9,11; 10,12; 11,13) and geometric (*cis* or *trans*) isomers of octadecadienoic acid. Some of these isomers have been associated with health benefits: *cis*-9, *trans*-11-, and *trans*-10, *cis*-12-octadecadienoic acids (Fazzari et al. 2014; Hodge et al. 2007).

Hodge et al. (2007) investigated the associations of fatty acids in plasma and diet with incidence of diabetes. Positive associations with diabetes were seen for total SFA, whereas an inverse association was observed for linoleic acid.

ALA is the precursor of eicosanoids, three important longer-chain n-3 fatty acids (Stark et al. 2008): eicosapentaenoic acid (EPA 20:5, n-3), docosapentaenoic acid (DPA 22:5, n-3) and docosahexaenoic acid (DHA 22:6 n-3). Although there is some interconversion of the omega-3 fatty acids, each fatty acid has its own place in human biology. CLA levels in olive oils can reach 0.2 mg/g of total fat (Chin et al. 1992).

It is important to remember that ALA is the parent molecule of the omega-3 fatty acids, and greater attention should be paid to its independent physiological function. It is of utmost importance in the diet as it cannot be synthesised by humans. Potential benefits of ALA include cardioprotective effects, modulation of the inflammatory response, and positive impacts on both central nervous system function and behaviour (Stark et al. 2008).

In 2009, the European Food Safety Authority (EFSA) approved the health claim that ALA contributes to the maintenance of normal blood cholesterol concentrations, based on proven cause and effect relationships. The reference intake value is 2 g ALA per day. For a food to display this claim on the label, it should provide at least 15 % of that value (EFSA 2009). According to PortFIR data[1] (INSA 2015), VOO contains 0.7 % of linolenic acid on average, which means that 100 g of olive oil (about 109 ml) contains about 35 % of the EFSA reference intake value. Thus, the regular intake of olive oil, as the main dietary fat, may effectively contribute to the improvement of human health.

The VOO non-saponifiable fraction contains minor components such as polar phenolic compounds (more commonly known as polyphenols), sterols, hydrocarbons (such as squalene and β-carotene), tocopherols, fatty alcohols, waxes, chlorophylls and related pigments, as well as a wide range of volatile compounds that contribute to odour and flavour, such as aldehydes, ketones, thiols, alcohols and acids (Boskou et al. 2006a, b; Owen et al. 2000a, b).

4.2.1 Polyphenols

According to different authors (Boskou et al. 2006b; Owen et al. 2000a, b; Velasco and Dobarganes 2002), the major phenolic compounds identified and quantified in VOO belong to three classes (Fig. 4.1):

(a) Simple phenols (hydroxytyrosol, tyrosol)
(b) Secoiridoids (oleuropein, the aglycone of ligstroside, and their respective decarboxylated dialdehyde derivatives: 10-hydroxy-oleuropein and 10-hydroxy-ligstroside)
(c) Lignans, as acetoxypinoresinol and pinoresinol

All three classes have potent antioxidant properties. Structures of the most representative compounds of this class are shown in Fig. 4.1.[2] Polyphenols are present in much higher amounts in table olives (except in Californian-style black

[1] Assumed as the difference between total PUFA and linolenic acid content.

[2] Additional molecular structures have been published by Owen et al. (2000a).

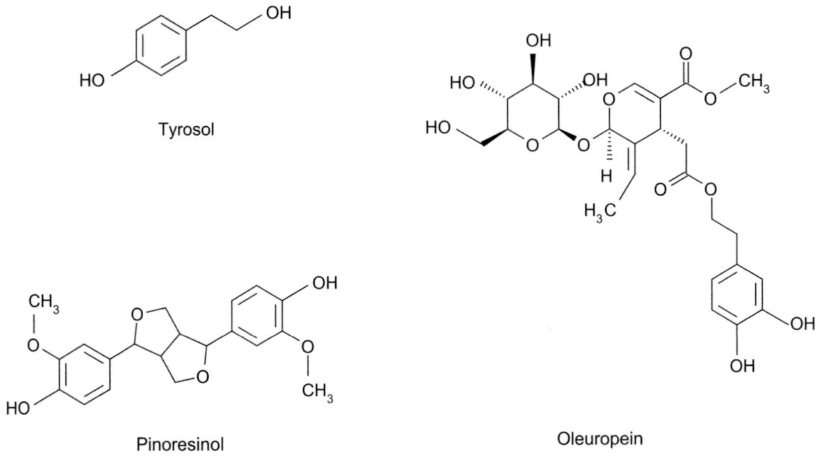

Fig. 4.1 The most representative compounds of the three classes of phenolic compounds found in virgin olive oil (generally obtained from physically pressing the fruit at ambient temperature) and in table olives (generally obtained after debittering and fermentation of fruits): **tyrosol** (simple phenol), **oleuropein** (secoiridoid) and **pinoresinol** (lignan). BKchem version 0.13.0, 2009 (http://bkchem.zirael.org/index.html) was used to draw these structures

olives) than in VOO. Simple phenols are colourless and odourless not transmitting any particular taste to VOO. On the other hand, secoiridoids, particularly oleuropein, confer a characteristic bitter stringent note to VOO.

In terms of the effect of phenolic compounds on human health, Owen et al. (2000a, c) observed that extracts of olive oil containing a mixture of known and unknown phenolics could effectively inactivate reactive oxygen species (ROS) at far lower concentrations than the compounds tested individually.

Both table olives and EVOO contain substantial amounts of compounds deemed to be anti-cancer agents (e.g. squalene and terpenoids). Daily intake may play an important role in the beneficial health effects attributed to the Mediterranean diet (MD) (Visioli et al. 2006). Moreover, under identical experimental conditions, EVOO showed significant ROS-inhibition activity, while extracts of seed oils and refined olive oil were virtually ineffective.

As a result of such studies and collected evidence, the EFSA approved the following health claim in relation to polyphenols in olive oil, given that standardised contents of hydroxytyrosol and its derivatives (e.g. oleuropein complex) are present (EFSA 2011b): 'Protection of LDL particles from oxidative damage (...). A cause-and-effect relationship has been established between the consumption of olive oil polyphenols and the protection of LDL particles from oxidative damage (normally referred to as antioxidant properties). The EFSA considers that, to achieve the benefits of the above claim, 5 mg of hydroxytyrosol and its derivatives (e.g. oleuropein complex and tyrosol) should be consumed daily.'

Comparisons between olive and seed oils support the superiority of olive oil. High levels of 'platelet-activating factor' antagonists have been detected, mainly in total polar lipid fraction, thus explaining the observed anti-thrombotic effect (Karantonis et al. 2002). According to Owen et al. (2000b, c), the reduction of oxidative stress should also afford considerable protection against cancer (colon, breast, skin), coronary heart disease and ageing, and can be achieved by the regular consumption of VOO. Lignans (Fig. 4.1) have been found to be the active molecules that provide protection against breast, colon and prostate cancer (Owen et al. 2000b, c; Visioli et al. 2006). These include (+)-1-acet-oxypinoresinol and (+)-pinoresinol (Owen et al. 2000b).

As with simple phenols and secoiridoids, lignan concentrations vary considerably between olive oils; consequently, the above-mentioned effects on human health are also expected to vary accordingly. Lignans were found to be very abundant in Italian VOO (Boskou et al. 2006b) and virtually absent in refined olive oils, while maximum concentrations are achieved in EVOO (Owen et al. 2000b), despite the expected variation due to geographical origin, variety, etc.

4.2.2 Squalene

Squalene[3] (Fig. 4.2) is an alkane triterpene that constitutes up to 40 % of the VOO unsaponifiable fraction (Velasco and Dobarganes 2002). Squalene is colourless to slightly yellow and absorbs oxygen, becoming viscous. It has been found to quench singlet oxygen (Owen et al. 2000b), but its structure is not amenable to reaction with hydroxyl ion OH^-. Squalene is an intermediate in the biosynthesis of cholesterol (Mazein et al. 2013).

Due to refining processes, squalene is found in only very small amounts in seed oils (Owen et al. 2000a; Visioli et al. 2006). It is noteworthy that fair amounts of squalene are also present in pumpkin seeds and in nuts, such as hazelnuts, walnuts and almonds (Kalogeropoulos et al. 2013; Maguire et al. 2004). The composition of nuts and their impact in the MD and on human health are discussed in Sect. 5.8.

Elenolic[4] and cinnamic[5] acid (the latter is represented in Fig. 4.3) are often associated with this polar phenolic fraction, contributing to the aroma of VOO (Boskou et al. 2006b).

4.2.3 Sterols

Phytosterols are hydrophobic compounds that are similar in structure and function to cholesterol and have long been known to positively impact lipid metabolism

[3] The International Union of Pure and Applied Chemistry (IUPAC) name for squalene is (6Z,10E,14E,18E)-2,6,10,15,19,23-hexamethyltetracosa-2,6,10,14,18,22-hexaene.

[4] IUPAC name for elenolic acid: 2-[(2S,3S,4S)-3-formyl-5-methoxycarbonyl-2-methyl-3,4-dihydro-2H-pyran-4-yl]acetic acid.

[5] IUPAC name for cinnamic acid: (E)-3-phenylprop-2-enoic acid.

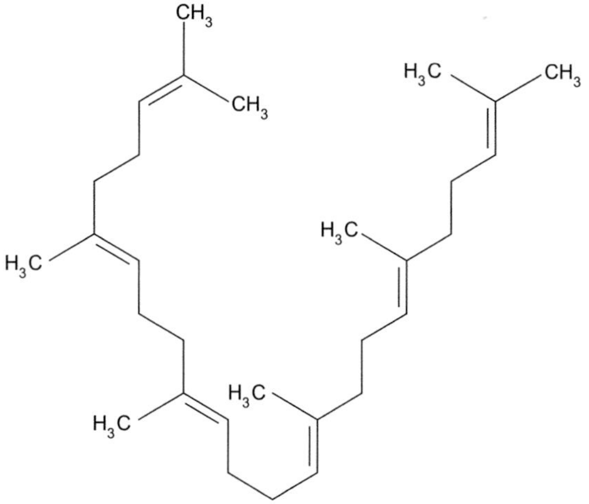

Fig. 4.2 Squalene (molecular structure). BKchem version 0.13.0, 2009 (http://bkchem.zirael.org/index.html) was used to draw this structure

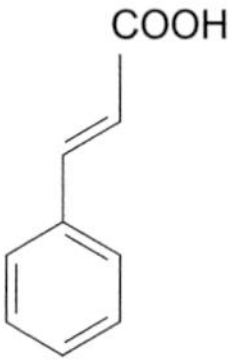

Fig. 4.3 Cinnamic acid (molecular structure). BKchem version 0.13.0, 2009 (http://bkchem.zirael.org/index.html) was used to draw this structure

(Gupta et al. 2011). Sterols are the major constituents of the unsaponifiable fraction of olive oil (around 20 %) and play an important role in the stability of olive oils by acting as inhibitors of polymerisation reactions (see below). Phytosterols are specific to each vegetable species, varying in terms of carbon side chains and/or the presence or absence of a double bond. VOO phytosterols are often used as traceable markers to prove a given protected geographical designation of origin[6] and to detect adulterations (Montealegre et al. 2010).

4.2.4 β-Carotene

β-carotene[7] is a carotenoid and a precursor of vitamin A (Fig. 4.4). It is a yellow pigment that confers a yellowish colour to VOO; it also exhibits antioxidant properties (NCBI 2015a).

[6] More information on protected names of quality agricultural products from Europe can be found at the following web address: http://ec.europa.eu/agriculture/quality/schemes/index_en.htm

[7] IUPAC name for β-carotene: 1,3,3-trimethyl-2-[(1E,3E,5E,7E,9E,11E,13E,15E,17E)-3,7,12,16-tetramethyl-18-(2,6,6-trimethylcyclohexen-1-yl)octadeca-1,3,5,7,9,11,13,15,17-nonaenyl] cyclohexene.

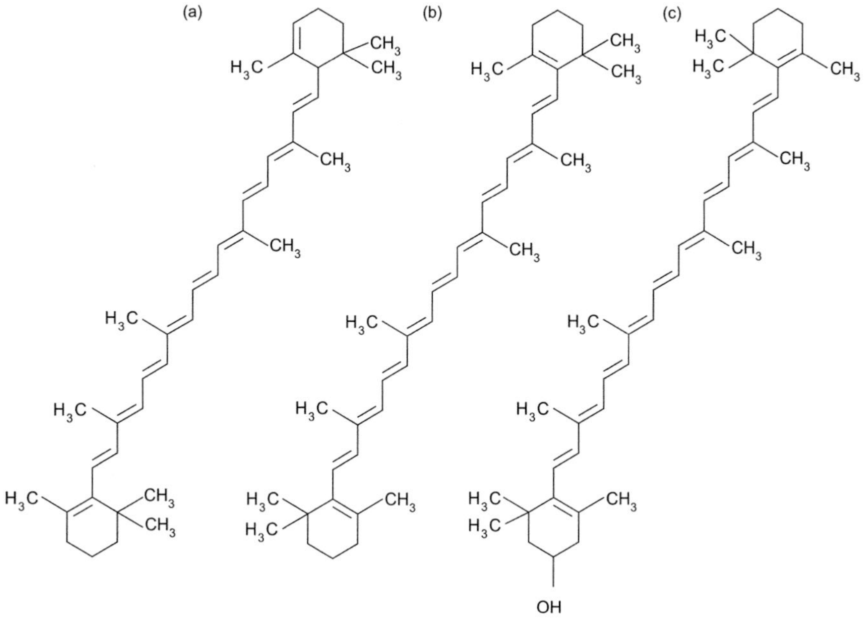

Fig. 4.4 Molecular structures of pro-vitamin A carotenoids: α-carotene (**a**), β-carotene (**b**) and β-cryptoxanthin (**c**). BKchem version 0.13.0, 2009 (http://bkchem.zirael.org/index.html) was used to draw these structures

4.2.5 α-Tocopherol

α-Tocopherol[8] (Fig. 4.5) naturally occurs in VOO as a single stereoisomer. It is traditionally recognised as the most active form of vitamin E in humans and is a powerful biological antioxidant (NCBI 2015b). For this reason, tocopherols are currently added to other oils and fats to prevent rancidity (Boskou et al. 2006b; Velasco and Dobarganes 2002).

4.2.6 Waxes

Waxes are highly viscous secretions that occur naturally in plants and aim to control fruit evaporation and hydration. Waxes are mixtures of substituted long-chain aliphatic hydrocarbons containing alkanes, fatty acids, primary and secondary alcohols, diols, ketones and aldehydes (Nota et al. 1999; Perez-Camino et al. 2003).

[8] IUPAC name: (2R)-2,5,7,8-tetramethyl-2-[(4R,8R)-4,8,12-trimethyltridecyl]-3,4-dihydrochromen-6-ol.

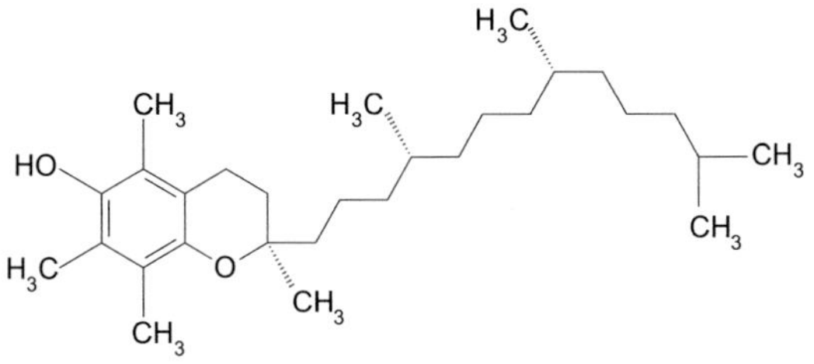

Fig. 4.5 α-tocopherol or vitamin E (molecular structure). BKchem version 0.13.0, 2009 (http://bkchem.zirael.org/index.html) was used to draw this structure

4.2.7 Chlorophylls and Related Pigments

Chlorophylls and related pigments are responsible for the green colour of VOO and were previously involved in photosynthesis. Chlorophylls are porphyrins with a magnesium ion at the centre and generally bonded to a protein moiety. The molecule is hydrophobic and sensitive to light. There are several types of chlorophylls, differing in some radical groups, with the most common types being chlorophylls a and b. Chlorophylls, present in olive oil, may act as photosensitisers (Boskou et al. 2006a; NCBI 2015c). The role of chlorophylls in protecting against or activating VOO oxidation reactions depends on the environmental conditions, such as the presence of acids or light. When exposed to light, chlorophyll molecules can originate free radicals through light activation reactions.

Other minor components in EVOO include volatile compounds (such as hexanol and hexanal), acids and traces of other fruit constituents. The composition of volatile and phenolic fractions as well as the presence of certain sterols and pigments when analysed in combination with fatty acid profiles provides a tool to discriminate between olive oils. Like wines, olive oils are unique harvests of strong regional character. As valuable products, they can be subjected to adulteration, and compositional markers are used to certify their uniqueness (Angerosa et al. 2006; Martins-Lopes et al. 2008; Montealegre et al. 2010).

According to IOOC standards (2015c), the quality of olive oil is defined by some chemical parameters and assays:

- Acidity. This is a measurement of free fatty acids; quality is usually associated with low acidity values.
- Peroxide value. This number gives a measure of the extent to which an oil sample has undergone primary oxidation or the concentration of hydroperoxides. Oils with a high degree of unsaturations (or with a high level of unsaturated fatty acids) are most susceptible to autoxidation (oxidative rancidity).
- Total phenols. These include mainly polar compounds, a complex mixture of compounds with different chemical structures obtained from the oil by extraction with methanol-water (Boskou et al. 2006a). These molecules are divided into simple phenols, lignans and flavonoids (Boskou et al. 2006b; Owen

et al. 2000a, c). Generally, olive oils contain 196 ± 19 mg/kg total phenolics (Owen et al. 2000a).

- Traceability tests. These tests aim to determine the origin and probable mixtures or adulterations of a given sample of olive oil.

Composition-based methods rely on the profile of fatty acids, sterols and/or phenols and the presence and range of certain minor components, such as squalene (Angerosa et al. 2006; Montealegre et al. 2010). As an alternative, deoxyribonucleic acid (DNA)-fingerprint methods with primers selected for genotypic botanical origin may be used. Data are processed by specific statistical software to reveal patterns of similarity (Martins-Lopes et al. 2008).

Variables influencing the composition of olive oil result from a high number of factors taking effect—from the oil formation in the olive tree to the status of the oil at consumption (Velasco and Dobarganes 2002)—and may be divided into three groups: factors acting before oil extraction, those acting during oil extraction and, finally, those acting after oil extraction. The first group includes environmental and agronomic conditions, such as the olive variety and the degree of ripeness at harvest. They mostly account for fixed parameters that cause unavoidable differences in composition. Nevertheless, certain agronomic techniques may have an impact on oil composition, for example, irrigation deficits increase polyphenol content, while ripeness increases the levels of PUFA and decreases polyphenol and pigment contents (Velasco and Dobarganes 2002). Factors acting during oil extraction have the highest impact on the composition, quality and stability of olive oil. Extraction technologies should therefore be carefully chosen; filtration operations and contact with certain materials that may lead to metal contamination should be avoided (Velasco and Dobarganes 2002). Addition of water should be minimised. It has been shown that VOO is less stable after storage in contact with carbon steel than when stored in the absence of metals. Finally, olive oil should be stored in the dark for maximum stability, in impermeable containers with minimum headspace (Velasco and Dobarganes 2002).

The most common degradation reaction affecting VOO is auto-oxidation, commonly known as rancidity. It results from lipid oxidation and causes important deteriorative changes in VOO chemical, sensory and nutritional properties (Velasco and Dobarganes 2002).

The overall mechanism of lipid oxidation consists of three stages that include the following reactions:

(a) Initiation (the formation of free radicals):

$$RH \overset{\text{initiator}}{\rightarrow} R^{\bullet} \tag{4.1}$$

(b) Propagation (the production of hydroperoxides):

$$R^{\bullet} + O_2 \leftrightarrows ROO^{\bullet} \tag{4.2}$$

$$ROO^{\bullet} + RH \leftrightarrows R^{\bullet} + ROOH \tag{4.3}$$

$$ROOH \leftrightarrows RO^{\bullet} + HO^{\bullet} \tag{4.4}$$

(c) Termination or secondary oxidation (the production of non-radical molecules):

$$R^{\bullet} + R^{\bullet} \leftrightarrows RR \tag{4.5}$$

$$R^{\bullet} + ROO^{\bullet} \leftrightarrows ROOR \tag{4.6}$$

$$ROO^{\bullet} + ROO^{\bullet} \leftrightarrows ROOR + O_2 \tag{4.7}$$

where RH represents an oxidation site adjacent to a double bond in any unsaturated fatty acid; $R^{\bullet}$ is the free radical formed; ROOH is a hydroperoxide; the dot refers to free radicals. Hydroxyl radical ($HO^{\bullet}$), RR and ROOR are secondary products.

Normally, oxidation proceeds slowly at the initial stage and then a sudden rise occurs in the oxidation as chain reactions take place.

The initiation step has high activation energy, and thus the production of the first free radicals occurs in the presence of a metal catalyst (such as copper), a hydroperoxide or via the influence of light or high temperatures.

The rate of oxidation of fatty acids increases with the degree of unsaturation because PUFA have more reaction sites (C atom in position α to double bonds), and the position between double bonds is very reactive. For linoleic acid, this position corresponds to C11. Due to the resonance structure, the formed radical may easily shift positions, facilitating the chain reactions. Thus, linoleic acid is ten times more susceptible to oxidation than oleic acid, and the rate of linolenic acid oxidation is 100 times higher because of the conjugated C=C bonds.

Secondary products formed from hydroperoxide dismutation are aldehydes, ketones, alcohols and hydrocarbons, including hexanal, pentanal and malonaldehyde, responsible for off-flavours. Hydroperoxides can oxidise chlorophylls or polymerise to dark (and sometimes toxic) compounds.

For many years, there was little interest in ALA, CLA and other PUFA. Issues were raised concerning the danger of consuming highly unsaturated fatty acids that were susceptible to peroxidation (and consequently were expected to accumulate harmful compounds). Fortunately, the reaction between lipid peroxyl radicals and unoxidised lipids [Eq. (4.3)] is relatively slow, affording phenolic antioxidants the opportunity to intercept peroxyl radicals and thus interrupt chain propagation reactions (Lambert and Elias 2010).

Moreover, ALA, CLA, EPA, DPA and DHA together have recently been found to play vital roles in brain development and function, cardiovascular health, inflammatory response and foetal and infant development (Madden et al. 2009; Stark et al. 2008).

From a technological point of view, the acidity values (free fatty acid levels) and peroxide index values determine the shelf life and quality of newly produced VOO.

EVOO and VOO have a longer shelf life than other edible vegetable oils. Since the main fatty acid in VOO is oleic acid, the shelf life of olive oil may not be the major concern, but quality certainly is. Alterations during storage may occur and involve changes to both major and minor components. Minor components protect against lipid oxidation (rancidity), but by reacting first, they are destroyed in the process, thus decreasing the nutritional and nutraceutical properties of VOO.

Phenolic compounds can inhibit oxidation via a variety of mechanisms based on radical scavenging, hydrogen atom transfer and metal-chelating attributes (Bendini et al. 2006). Phenolic compounds of the *o*-diphenolic category, such as oleuropein aglycon, decarboxymethyl-oleuropein aglycon and hydroxytyrosol, are believed to be the chemicals primarily responsible for the oxidative resistance of EVOO.

To slow the rate of oxidation during storage, certain factors—such as the presence of oxygen, traces of metals and exposure to light—must be controlled (Bendini et al. 2010).

Oxygen availability can be restricted by decreasing both the headspace in the container and the oxygen permeability of the packaging material (Bendini et al. 2010).

Exposure to light should be avoided, as VOO contains chlorophylls that, as mentioned above, may act as photosensitisers, becoming excited by light absorption and initiating auto-oxidation (Bendini et al. 2010; Velasco and Dobarganes 2002). In the dark, chlorophylls may act as antioxidants.

Thus, the opacity to light of the packaging material is important for the preservation of VOO. Another requirement for packaging materials is the absence of trace metals, which may act as catalysts for auto-oxidation reactions. Transition metals, as copper and iron, catalyse the decomposition of hydroperoxides and propagate the free radical chain reactions. Under particular conditions, phenolic compounds become pro-oxidant agents, reducing metal ions to their lowest oxidation state, where they become more active and catalyse hydroperoxide decomposition (Bendini et al. 2006; Lambert and Elias 2010).

Preserving top-quality VOO means preserving the integrity of the minor beneficial components for which health claims are recognised (EFSA 2011a, b). In the case of these added-value products, processing steps and storage and packaging conditions are carefully chosen. Some tools can be useful, such as predictive models. In this regard, a mathematical model, based on the degradation kinetics of chlorophylls, was developed to assess the stability and loss of freshness of VOO and EVOO in different conditions: in the dark, at room temperature and with a limited supply of oxygen (Aparicio-Ruiz et al. 2012). In short, EVOO retains its organoleptic and health beneficial properties over a long shelf-life period at room temperature and under suitable storage conditions.

When considering cooking and food processing, fried foods cannot be forgotten, as they are common in the MD. According to the IOOC (2015d), these foods are not necessarily unhealthy.

Frying can be viewed as a combination of cooking and drying of foods by immersing them in a fat at high temperatures—generally between 160 and 240 °C, ideally at 160 °C. This process is used both at home and in industries using vegetable oils, animal fats or a combination of both. The oil functions as a heat-transfer medium and contributes to the flavour and texture of the food.

During frying, fats and oils are subject to maximum oxidative and thermal abuse if they are repeatedly used at high temperatures in the presence of atmospheric oxygen. Thus, an important requirement of a cooking oil is its chemical stability under high temperatures and moisture (Singh and Debnath 2011). Water released by foods fried in oil enhances the breakdown of fatty acids during heating. Degradation products are generally mono- and diglycerides, dimeric and polymeric triglycerides, free fatty acids and oxidised triglyceride monomers (e.g. triglyceride monohydroperoxides). Further degradation of these products may lead to the formation of saturated and unsaturated aldehydes, ketones, oxidised cyclic fatty acids and 'aldehydic' diglyceride by-products (containing oxo, hydroxy and epoxy groups), as well as other as yet unidentified polar compounds (Kalogeropoulos et al. 2007).

It is noteworthy that cooking oils should be kept at a maximum temperature of 180 °C during frying and that low temperatures (below 120–130 °C) result in increased fat uptake by the food. Higher temperatures, especially if approaching the smoking point of oils, promote the aforementioned chemical changes, which will affect taste and usually produce polar toxic compounds (Bastida and Sánchez-Muniz 2001; Kalogeropoulos et al. 2007).

Thus, in the absence of other variables, the more saturated the oil, the more stable it is expected to be in terms of oxidative and hydrolytic breakdown, and the less likely it is to polymerise (Singh and Debnath 2011). In this regard, oils rich in linoleic acid are expected to become more polymerised upon frying than those richer in oleic acid (Bastida and Sánchez-Muniz 2001).

Nevertheless, the process of frying is far from simple, and olive oil is distinct from seed oils in its reactions. The alterations that occur when vegetable oils are heated for frying take place more quickly in seed oils than in olive oil and depend on the initial acidity of the oil (free fatty acids) and on the presence of antioxidant compounds (IOOC 2015d). These alterations also vary according to the temperature and the duration of heating time, the number of times the oil is used, the manner of frying, and the type of food being fried. As an example, frying fish—especially oily fish—increases the polyunsaturated acid content of the oil, facilitating its decomposition.

In conclusion, and still according to the IOOC (2015d), olive oil is ideal for frying. At proper temperatures and without over-heating, it undergoes no substantial structural changes and keeps its nutritional value better than seed oils, not only because of the presence of antioxidants but also due to its high levels of oleic acid (Bastida and Sánchez-Muniz 2001). In addition, the smoking point of olive oil

(210 °C) is substantially higher than the ideal temperature for frying food (160–180 °C). Another advantage of using olive oil for frying is that it forms a crust around the food (when heated at 160–180 °C) that decreases the penetration of oil and improves its flavour. Food fried in olive oil may therefore have a lower fat content than food fried in other oils. Olive oil, therefore, is the safer, lightest and tastiest medium for frying (IOOC 2015d).

The current trend of food industries is the continuous development of new products as ingredient blends (to facilitate meal preparation) and pre-cooked frozen meals. With respect to the preservation of olive oil qualities, food technologists should pay additional attention to products that include olive oil as an ingredient. Since the inclusion of EVOO can attract consumers because of its health benefits, processing should ensure the preservation of EVOO properties during the product shelf-life.

Some innovative olive oil-based foods and concomitant processing may cause undesirable effects in olive oil and may even raise safety issues. Examples of such foods are garlic in olive oil, or herbs in oil (these products do not fit the IOOC classification for olive oil). The addition of these ingredients may result in significant changes in flavour, stability or even safety of the final product. It presents risks in terms of introducing spores or other dormant forms of microorganisms that do not exist in VOO, impacting preservation conditions and the shelf-life of such products.

Another unexpected example applies to frozen and refrigerated foods. Calligaris et al. (2006) observed that, surprisingly, below the melting point, the oxidation rate of EVOO was higher than that expected based on the Arrhenius equation. The observed deviation was attributed to the physicochemical changes that occur as a consequence of cooling. That deviation is mainly due to the establishment of a concentration gradient in the liquid phase that surrounds fat crystals, in particular because the increase of unsaturated triacylglycerides and the decrease of polyphenols (Calligaris et al. 2006). These findings should be considered by food technologists when formulating pre-cooked frozen meals and taken into consideration in the methodology and analysis of shelf-life tests.

4.3 Table Olives

As with other drupes, olives consist of a waxed epicarp, a mesocarp that is rich in nutrients and an endocarp that is formed by a woody shell containing the kernel seed. The composition of the fruit depends on a great number of factors, from agronomic to environmental.

Table olives. *Many varieties of table olives and forms of preparation can be found in the Mediterranean area. Table olive varieties are chosen for their larger size and lower oil content. All contain some olive oil and dietary fibres and generally high levels of flavonoids and other phenolic compounds. Fermented olives may contain probiotic bacterial strains. Photo reprinted with kind permission from T. N. Wassermann*

Fruit varieties consumed as table olives are chosen for their texture and generally have higher fibre levels (about 4 g/100 g) and lower lipid content than the fruit varieties grown for olive oil production (INSA 2015; USDA 2015). Fats occur in the pulp as triglyceride droplets or in phospholipids, lipoproteins, glycolipids and as some free fatty acids. The oil content in the mesocarp increases during maturation since triglyceride synthesis proceeds until a certain ripening stage (Garrido-Fernandez et al. 1997).

Even after full maturation, olives are inedible, astringent and bitter. Many processing methods currently available usually involve debittering and fermentation stages. The composition of the triglycerides generally remains unaffected by technological procedures (Bianchi 2003). The most important industrial preparations for table olives are the Spanish (or Sevillian) style, the Californian style and the naturally black olive processes. For each type of preparation, the fruits are harvested at different maturation stages, respectively 'green olives' 'semi-ripe olives' (changing colour) and 'ripe olives' (black), to use the designations approved by the IOOC (2006). 'Sevillian-style green olives' or 'treated green olives in brine' refers to a process that involves an alkaline treatment, a washing step and a fermentation stage. Fruits are harvested at their maximum size when their colour is turning to yellow, before full maturity. In the debittering stage, the oleuropein molecule and other polar phenolic compounds undergo an alkaline hydrolysis; reaction products (mainly oleoside-11-methylester and hydroxytyrosol) are washed away (Garrido-Fernández et al. 1997). Fruits are then placed in a brine solution, where nutrients diffuse from the fruit pulp to the brine and a spontaneous microbial fermentation (mainly carried out by lactic acid bacteria) occurs.

The composition data below correspond to Sevillian-style green fermented olives (designated in food databases as 'olives, pickled, canned or bottled, green'). Thus, according to several authors (Bianchi 2003; Garrido-Fernandez et al. 1997; Sánchez et al. 2000) and institutional databases (INSA 2015; USDA 2015), the major constituents of tables olives are primarily water (70–75 %) and olive oil (about 15 %) in a percentage of pulp wet weight (or edible portion).

As with olive oil, a small portion of SFA 16:0 and 18:0 is present (2.03–2.9 g/ 100 g of edible portion), although most of the lipid fraction is MUFA: 9.6–11.3 %, mainly 18:1 (linoleic acid) and some 16:1 (palmitoleic acid). PUFA account for 1.3–2.2 %, mainly 18:2 ALA (1.2–2.0 %).

Polysaccharides such as cellulose, hemicellulose, gums, pectin and lignin are constituents of intercellular lamellae, determining the fruit texture and accounting for fibre content. Pectin and hemicellulose degradation takes place during ripening and further olive processing, resulting in the softening of the pulp. The sugar content decreases with fruit maturation, as sugars are directed to triglyceride synthesis. Moreover, free sugars are the first substrates to be depleted in fermented fruits. The remaining sugars (<1 %) are mainly glucose, fructose and mannitol.

Common organic acids after fermentation (0.1–2.1 %) are lactic, malic, citric and oxalic acids (Sánchez et al. 2000), and the protein content is 1.0–1.5 % (Garrido-Fernandez et al. 1997; INSA 2015; USDA 2015).

Fat-soluble and water-soluble vitamins are commonly present in fermented table olives. According to the most recent INSA and US Department of Agriculture (USDA) databases, the primary vitamins found in olive pulp are the water-soluble β-carotene (231–236 mg/100 g) and the oil-soluble α-tocopherol (3.8 mg/100 g). Other carotenoids are present, as β-cryptoxanthin[9] (9 μg/100 g), and, lutein and its analogue zeaxanthin[10], which are present at a combined concentration of

[9] IUPAC name: (1R)-3,5,5-trimethyl-4-[3,7,12,16-tetramethyl-18-[(1R)-2,6,6-trimethylcyclohex-2-en-1-yl]octadeca-1,3,5,7,9,11,13,15,17-nonaenyl]cyclohex-3-en-1-ol.

[10] IUPAC name: (1R)-4-[(1E,3E,5E,7E,9E,11E,13E,15E,17E)-18-[(4R)-4-hydroxy-2,6,6-trimethyl-cyclohexen-1-yl]-3,7,12,16-tetramethyloctadeca-1,3,5,7,9,11,13,15,17-nonaenyl]-3,5,5-trimethyl-cyclohex-3-en-1-ol.

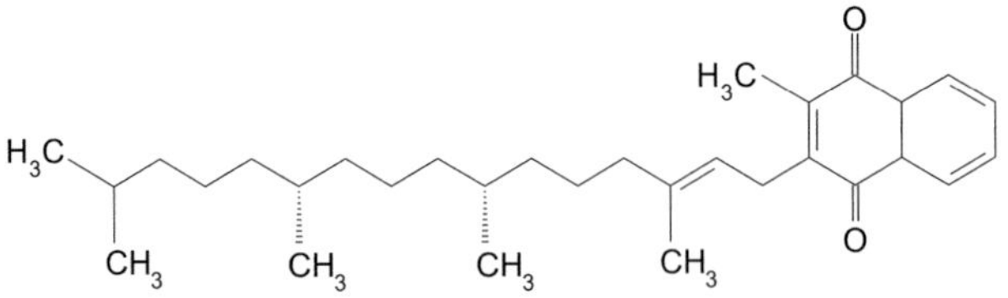

Fig. 4.6 Chemical structure of phylloquinone (also known as vitamin K_1, widespread in vegetables). BKchem version 0.13.0, 2009 (http://bkchem.zirael.org/index.html) was used to draw this structure

510 μg/100 g (USDA 2015). Carotenoids are important components of the retina and are effective for the treatment of age-related macular degeneration (Chew et al. 2014).

Fermented table olives contain the B vitamins thiamine (0.02 mg/100 g), riboflavin (0.070 mg/100 g), niacin (0.24–0.5 mg/100 g), pantothenic acid (0.02 mg/100 g) and B_6 (0.02–0.03 mg/100 g), as well as phylloquinone[11] (or vitamin K_1: 1.4 μg/100 g).

Members of the vitamin K group are phylloquinone (K_1, Fig. 4.6), menaquinone (K_2) and vitamin K_3 (a synthetic form). All these compounds have only one double bond on the proximal isoprene unit and possess anti-haemorrhagic and pro-thrombogenic activities (NCBI 2015d). Vitamin K_1 is widely found in vegetables, whereas vitamin K_2 is the main storage form in animals. The molecules differ by the number of isoprenoid residues in the side chains and are interconvertible.

As a result of processing, table olives contain a high concentration of sodium (Na, 1.6–2.1 %), along with beneficial relevant concentrations of calcium (Ca, 52–54 mg/100 g), phosphorus (P, 4–14 mg/100 g), magnesium (Mg, 11–22 mg/100 g), zinc (Zn, 0.04–0.2 mg/100 g) and selenium (Se, 0.9 μg/100 g) (INSA 2015; USDA 2015).

Free phenols and their glycosides (1–3 %) are a complex mixture of compounds, as mentioned above. Besides a relevant concentration of luteolin (0.6 mg/100 g), table olives also contain many other phenolic compounds, all responsible for the characteristic bitterness. The chemical structures of the most representative types are shown in Fig. 4.1. The major water-soluble phenol is oleuropein; this compound and mannitol are distinctive of olives (Bianchi 2003). Oleuropein is accountable for the bitterness of fruits and decreases with maturation. Small-fruit cultivars are generally characterized by a higher oleuropein content than are large-fruit cultivars (Bianchi 2003). Luteolin,[12] an important flavone with antioxidant and anti-

[11] IUPAC name: 2-methyl-3-[(E,7R,11R)-3,7,11,15-tetramethylhexadec-2-enyl]naphthalene-1,4-dione.

[12] IUPAC name: 2-(3,4-dihydroxyphenyl)-5-hydroxy-7-[(2S,3R,4S,5S,6R)-3,4,5-trihydroxy-6-(hydroxymethyl)oxan-2-yl]oxychromen-4-one.

inflammatory activities (NCBI 2015e), has been found at an average concentration of 0.6 mg/100 mg (USDA 2015).

A certain level of phenols is expected to ensure the desired organoleptic properties in brined olives and the antioxidant capacity of the end product (Garrido-Fernández et al. 1997). Any bacterial population attempting to become established must be resistant to high levels of polyphenols. Besides fibres, antioxidants and vitamins, lactic acid bacteria may also produce in situ bioactive compounds, particularly bacteriocins (Delgado et al. 2005, 2007; Hurtado et al. 2011a, b; Maldonado et al. 2002). From a technological point of view, bacteriocins play a relevant role in establishing the right conditions for spontaneous fermentation; desirable microbial populations must be selected to impair life forms that affect fruit structure and colour and that may cause off-flavours or even pose safety issues (Bautista-Gallego et al. 2013; Delgado et al. 2005, 2007).

In addition, some of these bacteria may show probiotic properties, potentially contributing to a healthy gut (Bautista-Gallego et al. 2013). Probiotics are well accepted in dairy products, but its presence is less known in table olives, although they can have extra health benefits, as described in a patent by Lavermicocca et al. (2004).

A common type of table olive is obtained via the Californian preparation method: olives undergo successive lye treatments and aeration until they acquire an homogenous black colour and soft flavour. Darkening of olive flesh is due to the oxidation and polymerisation of phenolic compounds (Bianchi 2003), particularly anthocyanins (Romero et al. 2004). The black colour is then usually fixed with ferrous gluconate. Finally, olives are canned in brine and subjected to a heat treatment to extend their preservation period. From a nutritional point of view, this type of processing is not very interesting, as most beneficial compounds are degraded.

The natural preparation styles of black olives are among many traditional olive brining methods. They generally involve several washing steps followed by fermentation in brine (to which herbs, garlic, lemon, etc. are added according to traditional recipes). Sodium chloride (NaCl) is always used, in concentrated solutions or in dry forms. As an alternative to brine fermentation, olives are submitted to dehydration processes, whereby the mature fruit are covered with NaCl crystals, causing osmotic stress to the still living cells. Hydrophilic compounds, such as oleuropein, are eliminated in the released water phase, thus decreasing the bitterness and astringency of fruits. This type of preparation does not involve a fermentation process and is characteristic of Greece (Laurent and Barnouin 2000). Fruits lose about 50 % of their water content, and oil and protein levels increase proportionally.

In short, table olives are produced in many regions of the world, although they are consumed in different proportions according to local availability, cultural aspects, diet and health-related beliefs. Except in Spain, the consumption of table olives seems to be generally decreasing, probably due to the lack of nutritional information.

Table olives are more than valuable nutritive appetisers; they have a place in meal composition. As table olives are prepared from fruit varieties with lower oil content (although this varies according to fruit variety and preparation method), their average energy content ranges from 145 to 172 kcal/100 g (INSA 2015; USDA 2015), up to a maximum of 220 kcal/100 g for certain varieties (Garrido-Fernández et al. 1997; Laurent and Barnouin 2000). Thus, table olives certainly have a lower energy content and higher nutritional value than such common snacks as potato chips (526 kcal/100 g), average cheese bites (316 kcal/100 g) or streaky pork cracklings (810 kcal/100 g) (INSA 2015; USDA 2015). On the other hand, olives are a valuable source of bioactive phytochemicals, including a wide range of anti-oxidant phenolic compounds. However, moderate consumption is recommended because of the (high) sodium content.

References

Angerosa F, Campestre C, Giansante L (2006) Chemistry properties, health effects. 7: Analysis and authentication. In: Boskou D (ed) Olive oil chemistry and technology, 2nd edn. AOCS, Champaign, IL

Aparicio-Ruiz R, Roca M, Gandul-Rojas B (2012) Mathematical model to predict the formation of pyropheophytin a in virgin olive oil during storage. J Agric Food Chem 60(28):7040–7049. doi:10.1021/jf3010965

Bastida S, Sánchez-Muniz FJ (2001) Thermal oxidation of olive oil, sunflower oil and a mix of both oils during forty discontinuous domestic fryings of different foods. Food Sci Technol Int 7(1):15–21. doi:10.1106/1898-PLW3-6Y6H-8K22

Bautista-Gallego J, Arroyo-López FN, Rantsiou K, Jiménez-Díaz R, Garrido-Fernández A, Cocolin L (2013) Screening of lactic acid bacteria isolated from fermented table olives with probiotic potential. Food Res Intern 50(1):135–142. doi:10.1016/j.foodres.2012.10.004

Bendini A, Carretani L, Vecchi S, Carrasco-Pancorbo A, Lercker G (2006) Protective effects of extra virgin olive oil phenolics on oxidative stability in the presence or absence of copper ions. J Agric Food Chem 54(13):4880–4887. doi:10.1021/jf060481r

Bendini A, Cerretani L, Salvador MD (2010) Olive oil, stability of the sensory quality of virgin olive oil during storage. Ital J Food Sci 60:5–18

Bianchi G (2003) Lipids and phenols in table olives. Eur J Lipid Sci Technol 105(5):229–242. doi:10.1002/ejlt.200390046

Boskou D, Blekas G, Tsimidou M (2006a) Part 2: Chemistry properties, health effects. 4: Olive oil composition. In: Boskou D (ed) Olive oil chemistry and technology, 2nd edn. AOCS, Champaign, IL

Boskou D, Tsimidou M, Blekas G (2006b) Part 2: Chemistry properties, health effects. 5: Polar phenolic compounds. In: Boskou D (ed) Olive oil chemistry and technology, 2nd edn. AOCS, Champaign, IL

Calligaris S, Sovrano S, Manzocco L, Nicoli MC (2006) Influence of crystallization on the oxidative stability of extra virgin olive oil. J Agric Food Chem 54(2):529–535. doi:10.1021/jf051808b

Chew EY, Clemons TE, Sangiovanni JP, Danis RP, Ferris FL, Elman MJ, Antoszyk AN, Ruby AJ, Orth D, Bressler SB, Fish GE, Hubbard GB, Klein ML, Chandra SR, Blodi BA, Domalpally A, Friberg T, Wong WT, Rosenfeld PJ, Agrón E, Toth CA, Bernstein PS, Sperduto RD (2014) Secondary analyses of the effects of lutein/zeaxanthin on age-related macular degeneration progression: AREDS2 report No. 3. JAMA Ophthalmol 132(2):142–149. doi:10.1001/jamaophthalmol.2013.7376

Chin SF, Liu W, Storkson JM, Ha YL, Pariza MW (1992) Dietary sources of conjugated dienoic isomers of linoleic acid, a newly recognized class of anticarcinogens. J Food Compost Anal 5(3):185–197. doi:10.1016/0889-1575(92)90037-K

Delgado A, Brito D, Peres C, Noe-Arroyo F, Garrido-Fernández A (2005) Bacteriocin production by Lactobacillus pentosus B96 can be expressed as a function of temperature and NaCl concentration. Food Microbiol 22(6):521–528. doi:10.1016/j.fm.2004.11.015

Delgado A, Arroyo López FN, Brito D, Peres C, Fevereiro P, Garrido-Fernández A (2007) Optimum bacteriocin production by Lactobacillus plantarum 17.2b requires absence of NaCl and apparently follows a mixed metabolite kinetics. J Biotechnol 130(2):193–201. doi:10.1016/j.jbiotec.2007.01.041

EFSA (2009) Opinion on the substantiation of health claims related to alpha-linolenic acid and maintenance of normal blood cholesterol concentrations (ID 493) and maintenance of normal blood pressure (ID 625) pursuant to Article 13(1) of Regulation (EC) No 1924/2006. EFSA J 7(9):1252–1269. doi:10.2903/j.efsa.2009.1252

EFSA (2011a) Scientific opinion on the substantiation of health claims related to polyphenols in olive and protection of LDL particles from oxidative damage (ID 1333, 1638, 1639, 1696, 2865), maintenance of normal blood HDL-cholesterol concentrations (ID 1639), maintenance of normal blood pressure (ID 3781), 'anti-inflammatory properties' (ID 1882), 'contributes to the upper respiratory tract health' (ID 3468), 'can help to maintain a normal function of gastrointestinal tract' (3779), and 'contributes to body defences against external agents' (ID 3467) pursuant to Article 13(1) of Regulation (EC) No 1924/2006. EFSA J 9(4): 2033–2058. doi:10.2903/j.efsa.2011.2033

EFSA (2011b) Scientific opinion on the substantiation of health claims related to walnuts and maintenance of normal blood LDL-cholesterol concentrations (ID 1156, 1158) and improvement of endothelium-dependent vasodilation (ID 1155, 1157) pursuant to Article 13(1) of Regulation (EC) No 1924/2006. EFSA J 9(4):2074–2093. doi:10.2903/j.efsa.2011.2074

FAO (2010) Fats and fatty acids in human nutrition: report of an expert consultation. FAO Food and Nutrition Paper 9. Food and Agricultural Organization of the United Nations, Rome

Fazzari M, Trostchansky A, Schopfer FJ, Salvatore SR, Sánchez-Calvo B, Vitturi D, Valderrama R, Barroso JB, Radi R, Freeman BA (2014) Olives and olive oil are sources of electrophilic fatty acid nitroalkenes. PLoS One 9(1):e84884. doi:10.1371/journal.pone.0084884

Fiorillo C, Vercueil J (eds) (2003) Syrian agriculture at the crossroads. FAO Agricultural Policy and Economic Development Series N° 8. Food and Agriculture Organization of the United Nations, Rome

Garrido-Fernández A, Fernández-Díez MJ, Adams MR (1997) Table Olives. Production and processing. Chapman & Hall, London

Gupta AK, Savopoulos CG, Ahuja J, Hatzitolios AI (2011) Role of phytosterols in lipid-lowering: current perspectives. Q J Med 104(4):301–308. doi:10.1093/qjmed/hcr007, Epub 2011 Feb 15

Hodge AM, English DR, O'Dea K, Sinclair AJ, Makrides M, Gibson RA, Giles GG (2007) Plasma phospholipid and dietary fatty acids as predictors of type 2 diabetes: interpreting the role of linoleic acid. Am J Clin Nutr 86(1):189–197

Hurtado A, Ben Othman N, Chammem N, Hamdi M, Ferrer S, Reguant C, Bordons A, Rozès N (2011a) Characterization of Lactobacillus isolates from fermented olives and their bacteriocin gene profiles. Food Microbiol 28(8):1514–1518. doi:10.1016/j.fm.2011.07.010

Hurtado A, Reguant C, Bordons A, Rozès N (2011b) Expression of Lactobacillus pentosus B96 bacteriocin genes under saline stress. Food Microbiol 28(7):1339–1344. doi:10.1016/j.fm.2011.06.004

INSA (2015) Tabela da Composição de Alimentos (TCA). http://www.insa.pt/sites/INSA/Portugues/AreasCientificas/AlimentNutricao/AplicacoesOnline/TabelaAlimentos/PesquisaOnline/Paginas/PorPalavraChave.aspx. Accessed 5 Nov 2015

IOOC (2015a) The olive tree. International Olive Oil Council (IOOC). Available at http://www.internationaloliveoil.org/estaticos/view/76-the-olive-tree. Accessed 5 May 2015

IOOC (2015b) World olive oil figures. International Olive Oil Council (IOOC). Available at http://www.internationaloliveoil.org/estaticos/view/131-world-olive-oil-figures. Accessed 5 May 2015

IOOC (2015c) Designations and definitions of olive oils. International Olive Oil Council (IOOC). Available at http://www.internationaloliveoil.org/estaticos/view/83-designations-and-definitions-of-olive-oils. Accessed 5 May 2015

IOOC (2015d) The olive world. Olive oil. Frying with olive oil. International Olive Oil Council (IOOC). Available at http://www.internationaloliveoil.org/web/aa-ingles/oliveWorld/aceite3.html. Accessed 12 May 2015

IOOC (2015e) Trade standards applying to olive oil and olive pomace oil. COI/T.15/NC No 3/Rev. 8, February 2015. International Olive Oil Council, Madrid, Spain. Available at http://www.internationaloliveoil.org/documents/viewfile/3615-normaen1. Accessed 9 Nov 2015

Kalogeropoulos N, Salta FN, Chiou A, Andrikopoulos NK (2007) Formation and distribution of oxidized fatty acids during deep and pan frying of potatoes. Eur J Lipid Sci Technol 109(11): 1111–1123. doi:10.1002/ejlt.200700007

Kalogeropoulos N, Chiou A, Ioannou MS, Karathanos VT (2013) Nutritional evaluation and health promoting activities of nuts and seeds cultivated in Greece. Int J Food Sci Nutr 64(6): 757–767. doi:10.3109/09637486.2013.793298, Epub 2013 May 3

Karantonis HC, Antonopoulou S, Demopoulos CA (2002) Antithrombotic lipid minor constituents from vegetable oils. Comparison between olive oils and others. J Agric Food Chem 50(5): 1150–1160. doi:10.1021/jf010923t

Kühlsen N, Pfeuffer M, Soustre Y, MacGibbon A, Lindmark-Mansson H, Schrezenmeir (2005) Trans fatty acids: scientific progress and labelling. In: Bulletin of the International Dairy Federation 393/2005. FIL/IDF, Fédération International de Laiterie/International Dairy Federation, Brussels

Lambert JD, Elias RJ (2010) The antioxidant and pro-oxidant activities of green tea polyphenols: a role in cancer prevention. Arch Biochem Biophys 501(1):65–72. doi:10.1016/j.abb.2010.06.013

Laurent A, Barnouin A (2000) L'olive. Aubanel, Editions Minerva, Genéve

Lavermicocca P, Lonigro SL, Visconti A, De Angelis M, Valerio F, Morelli L (2004) Table olives containing probiotic microorganisms. International Patent PCT/EP2004/013582, 16 June 2005

Loumou A, Giourga C (2003) Olive groves: the life and identity of the Mediterranean. Agric Hum Values 20(1):87–95. doi:10.1023/A:1022444005336

Madden SMM, Garrioch CF, Holub BJ (2009) Direct diet quantification indicates low intakes of (n-3) fatty acids in children 4 to 8 years old. J Nutr 139(3):528–532. doi:10.3945/jn.108.100628

Maguire LS, O'Sullivan SM, Galvin K, O'Connor TP, O'Brien NM (2004) Fatty acid profile, tocopherol, squalene and phytosterol content of walnuts, almonds, peanuts, hazelnuts and the macadamia nut. Int J Food Sci Nutr 55(3):171–178. doi:10.1080/09637480410001725175

Mailer R (2006) Chemistry and quality of olive oil. Primefacts 227. Available at www.dpi.nsw.gov.au. Accessed 6 Nov 2015

Maldonado A, Ruiz-Barba JL, Floriano B, Jiménez-Díaz R (2002) The locus responsible for production of plantaricin S, a class IIb bacteriocin produced by Lactobacillus plantarum LPCO10, is widely distributed among wild-type Lact. plantarum strains isolated from olive fermentations. Int J Food Microbiol 77(1–2):117–124. doi:10.1016/S0168-1605(02)00049-1

Martins-Lopes P, Gomes S, Santos E, Guedes-Pinto H (2008) DNA markers for Portuguese olive oil fingerprinting. J Agric Food Chem 56(24):11786–11791. doi:10.1021/jf801146z

Mazein A, Watterson S, Hsieh W-Y, Griffiths J, Ghazal P (2013) A comprehensive machine-readable view of the mammalian cholesterol biosynthesis pathway. Biochem Pharmacol 86(1): 56–66. doi:10.1016/j.bcp.2013.03.021

Montealegre C, Alegre MLM, Garcia-Ruiz C (2010) Traceability markers to the botanical origin in olive oils. J Agric Food Chem 58(1):28–38. doi:10.1021/jf902619z

NCBI (2015a) PubChem compound Database. CID 5280489. National Center for Biotechnology Information. http://pubchem.ncbi.nlm.nih.gov/compound/5280489. Accessed 4 May 2015

NCBI (2015b) PubChem compound Database. CID 14985. National Center for Biotechnology Information. http://pubchem.ncbi.nlm.nih.gov/compound/14985. Accessed 4 May 2015

NCBI (2015c) PubChem compound Database. CID 6477652. National Center for Biotechnology Information. http://pubchem.ncbi.nlm.nih.gov/compound/6477652. Accessed 4 May 2015

NCBI (2015d) PubChem compound Database. CID 5284607. National Center for Biotechnology Information. http://pubchem.ncbi.nlm.nih.gov/compound/5284607. Accessed 5 May 2015

NCBI (2015e) PubChem compound Database. CID 5280445. National Center for Biotechnology Information. http://pubchem.ncbi.nlm.nih.gov/compound/5280445. Accessed 5 May 2015

Nestel P, Clifton P, Noakes M (1994) Effects of increasing dietary palmitoleic acid compared with palmitic and oleic acids on plasma lipids of hypercholesterolemic men. J Lipid Res 35(4): 656–662

Nota G, Naviglio D, Romano R, Sabia V, Musso SS, Improta C (1999) Determination of the wax ester content in olive oils. Improvement in the method proposed by EEC regulation 183/93. J Agric Food Chem 47(1):202–205. doi:10.1021/jf980648j

Owen RW, Mier W, Giacosa A, Hull WE, Spiegelhalder B, Bartsch H (2000a) Phenolic compounds and squalene in olive oils: the concentration and antioxidant potential of total phenols, simple phenols, secoiridoids, lignans and squalene. Food Chem Toxicol 38(8): 647–659. doi:10.1016/S0278-6915(00)00061-2

Owen RW, Mier W, Giacosa A, Hull WE, Spiegelhalder B, Barst H (2000b) Identification of lignans as major components in the phenolic fraction of olive oil. Clin Chem 46(7):976–988

Owen RW, Giacosa A, Hull WE, Haubner R, Würtele G, Spiegelhalder B, Bartsch H (2000c) Olive-oil consumption and health: the possible role of antioxidants. Lancet Oncol 1(2): 107–112. doi:10.1016/S1470-2045(00)00015-2

Perez-Camino MC, Moreda W, Mateos R, Cert A (2003) Simultaneous determination of long-chain aliphatic aldehydes and waxes in olive oils. J Chromatogr A 983(1–2):283–288. doi:10. 1016/S0021-9673(02)01608-4

Romero C, Brenes M, Garcia P, Garcia A, Garrido A (2004) Polyphenol changes during fermentation of naturally black olives. J Agric Food Chem 52(7):1973–1979. doi:10.1021/jf030726p

Sánchez AH, Castro A, Rejano L, Montaño A (2000) Comparative study on chemical changes in olive juice and brine during green olive fermentation. J Agric Food Chem 48(12):5975–5980. doi:10.1021/jf000563u

Singh RP, Debnath S (2011) Heat and mass transfer in foods during deep-fat frying. In: Kulp K, Lorenz K, Brümmer J (eds) Frozen and refrigerated doughs and batters, 2nd edn. American Association of Cereal Chemists, St. Paul, MN

Stark AH, Crawford MA, Reifen R (2008) Update on alpha-linolenic acid. Nutr Rev 66(6): 326–332. doi:10.1111/j.1753-4887.2008.00040.x

USDA (2015) Agricultural Research Service National Nutrient Database for Standard Reference Release 27, Software v.2.2.4, The National Agricultural Library. http://ndb.nal.usda.gov/ndb/foods. Accessed 5 May 2015

Velasco J, Dobarganes C (2002) Oxidative stability of virgin olive oil. Eur J Lipid Sci Technol 104 (9–10):661–676. doi:10.1002/1438-9312(200210)104:9/10<661::AID-EJLT661>3.0.CO;2-D

Visioli F, Bogani P, Galli C (2006) Chemistry properties, health effects. 8: Healthful properties of olive oil minor components. In: Boskou D (ed) Olive oil chemistry and technology, 2nd edn. AOCS, Champaign, IL

Greens and Other Vegetable Foods

5

Abstract

This food category is predominant in a Mediterranean diet pattern. The term 'vegetable food' includes a highly heterogeneous variety of food items, such as leafy vegetables and analogues, tomato, pumpkin, wild leaf vegetables and weeds, aromatic plants and spices, starchy foods, pulses, fresh fruits and nuts. The nutrient composition taken from two different databases (one in the European Union and the other in the USA) is described and discussed. Some of these foods are consumed as energy sources (such as wheat, rice and potato), as protein sources (pulses) or as a mineral and vitamin supply (greens and fruits). Nevertheless, all have complex compositions supplying a wide variety of nutrients. Given the variety of a large number of factors, from the climate to analytical methodologies, the figures on the composition of food items (obtained from the food databases) should be regarded solely as guidance. Nevertheless, the concentration of the so-called minor components in each food item is highlighted, given the increased awareness of their roles in human health and wellness. In this regard, increasing levels of scientific evidence are being collected on the benefits conveyed by oligosaccharides, dietary fibres, flavonoids (e.g. phytosterols), as well as by many other phytochemicals. In this regard, the European Food Safety Authority granted walnuts a health claim related to cardiovascular health. Many epidemiological studies have correlated a high intake of greens, fruits and other vegetables with lower incidence rates of many non-communicable diseases and lighter body weight. This chapter also makes mention of compounds specific to certain vegetable species and their effects on health. Synergies between several vegetable foods for improved bioavailability and nutraceutical effects of some compounds are also mentioned.

A.M. Delgado et al., *Chemistry of the Mediterranean Diet*,
DOI 10.1007/978-3-319-29370-7_5

5.1 Vegetable Foods – An Introduction

In terms of components of the Mediterranean diet (MD), greens and other vegetables, fruits, nuts and olive oil are mainly responsible for the protection conveyed against hypertension and other health benefits. The high mineral content (including potassium [K], magnesium [Mg] and calcium [Ca]) of plant foods tend to reduce arterial blood pressure and may represent the mediating mechanisms of the protective effects of these foods (Psaltopoulou et al. 2004). Consumption of olive oil may be as important as that of fruit and vegetables in terms of the beneficial effect of the MD on arterial blood pressure control. Greens, fruits, nuts and other vegetables are also a source of many vitamins, flavonoids, (including phytosterols and anthocyanins and their polymers), oligosaccharides and many other compounds with relevance to human health; some mechanisms of action have recently been disclosed. Not all compounds from the same chemical family are relevant, as bioactivity may depend on the presence/absence of a certain radical or only exist for a certain isomer. More than 300 compounds of interest have been identified, and the number is rising. Some compounds are ubiquitous, whereas others are restricted to specific families or species (e.g. isoflavones in legumes). The composition may also vary within the same species, mainly due to genetic, geographic and agronomic factors.

On the other hand, some differences in the chemical form and bioavailability of micronutrients—namely, vitamins—are found between foods of animal and plant origins. Vitamin A, in its active form (all-*trans*-retinol), is only found in animal food sources. Structurally, it is a C20-apocarotenoid derivative (NCBI 2015a). Retinol is a transport and precursor form that is enzymatically activated to retinoic acid via a two-step oxidation process. The primary role of retinol is in the eye, where 11-*cis*-retinal is required for the formation of visual pigment (O'Byrne and Blaner 2013).

Biosynthetic precursors of retinol are provitamin A carotenoids β-carotene, α-carotene and β-cryptoxanthin, which are found in vegetables (Fig. 4.5). All retinoids are then originally derived from these pro-retinoid carotenoids (O'Byrne and Blaner 2013).

Other carotenoids (such as lycopene, lutein and zeaxanthin) are non-pro–vitamin A, despite their health-protective effects (Fitzpatrick et al. 2012). Thus, pro-vitamin A concentrations in foods are expressed below as milligrams of the quantified carotenoid per 100 g of edible food item or as retinol activity equivalents (RAE), where 1 RAE equals 1 μg of retinol, 12 μg of β-carotene, 24 μg of α-carotene or 24 μg of β-cryptoxanthin (Fitzpatrick et al. 2012).

As previously mentioned, vitamin K is essential for the functioning of several proteins, including those involved in blood clotting, and exists in two naturally occurring and inter-convertible forms: vitamin K_1 and vitamin K_2. Vitamin K_1, or phylloquinone, is synthesised by plants and is the predominant form in the human diet (Fig. 4.6). Additional information about vitamins and the molecular structure of the 13 vitamin compounds required in the human diet can be found in a recent review (Fitzpatrick et al. 2012).

Besides many nutrients, vegetables also supply a large number and variety of phytochemicals and many other compounds—from aldehydes to alkaloids—as detailed below. Many of these compounds have been associated with beneficial health effects, namely, chemoprotective properties, although some others may be responsible for unwanted reactions (e.g. allergies). Polyphenols are an important group of compounds that exhibit a wide range of properties, depending on particular structures, ranging from pigments to flavours, primarily bitterness and astringency. Other characteristics include reactive oxygen species (ROS)-scavenging capacity and the ability to interact with dietary proteins, which may affect their bioavailability.

In regard to health, it should be noted that cancer is a disease resulting from deregulated cell growth and is caused by an interaction of dietary, genetic and environmental risk factors. Dietary factors, as well as physical activity, may contribute to approximately one-third of all cancers (Williams and Hord 2005). The strongest evidence linking specific foods to the decreased risk of certain cancers includes high consumption of fruits, vegetables and whole grains. Secondary prevention trials and observational prospective epidemiologic studies have demonstrated the efficacy of a Mediterranean-type dietary pattern to a decreased risk of both cancer and cardiovascular diseases (Williams and Hord 2005; Zamora-Ros et al. 2013). Bentley and co-workers (2012) registered improvements in respiratory function in continuing smokers after higher intake of vitamin C and fruit and vegetables. Their epidemiological study led to the conclusion that the intake of nutrients with antioxidant properties may modulate decline in lung function in older adults exposed to cigarette smoke. These observed chemoprotective and anticancer properties have been attributed primarily to flavonoids (Labbé et al. 2009; Lee et al. 2007; Lu et al. 2006; Torkin et al. 2005).

On the other hand, a diet high in fruits and vegetables is associated with lower body weight than a diet poor in fruits and vegetables (Psaltopoulou et al. 2004; Ziesenitz et al. 2012). In this respect, the role of pulses in a healthy diet and their satiating effect, which contributes to weight control, should also be highlighted (McCrory et al. 2010).

The major edible plant families are Solanaceae (e.g. potato), Poaceae (e.g. rice), Fabaceae (e.g. beans), Brassicaceae (e.g. cabbage), Asteraceae (e.g. sunflower, lettuce) and Rosaceae (e.g. apples). These include crops as well as edible wild plants and weeds.

Every plant cell has a rigid cellulose wall that contributes to cellular mechanical resistance, a phospholipid bilayer cell membrane that also contains proteins, diverse organelle with similar membrane structures (including chloroplasts where chlorophylls and other pigments are found), a large water vacuole and cytoplasmic contents, mostly hydrophilic solutes and proteins. The degree of cell specialisation depends on evolutionary features and determines the function and composition of the plant's organ.

Depending on the species, the edible parts may include one or more organs of the plant: leaves, tubercles, roots, stems, fruits, seeds and even flowers. Leaves and fruits are usually consumed fresh, whereas tubercles and roots are more often cooked in several ways. Infusions can be made of weeds, leaves or flowers.

Many different factors determine the acceptance and choice of a specific plant species as a food. The most prominent are its abundance and availability, cultural preferences, the ability to harvest it in the optimal period and/or the capacity to preserve it or industrially process the food commodity.

The aim of this section is to give an insight into the chemical composition of some representative vegetable products and fruits, given their importance in the MD and according to their composition and dietary function. Plants are divided herein according to their nutritional significance: leaf vegetables (including weeds and aromatic herbs), sources of starch (such as grains and potatoes), pulses, fruits and nuts. The chemical composition of plants, mainly in minor components (highly relevant for human health), depends on many factors, such as the cultivar or variety, agronomic and climate factors, geographical location and degree of maturation at harvest. The food composition databases that are cited within this work (INSA 2015; USDA 2015) detail the composition of foods that are available, respectively, in the USA and in Europe (Portugal). The mass of each nutrient is herein expressed in a basis of 100 g of edible portion of the food in question. Because local variations can be significant, the presented values should be regarded solely as guidance.

5.2 Leafy Vegetables and Analogues

Different types of leaves, depending on the place and the season, have been part of the human diet since pre-historic times, providing minerals and vitamins. Many of those traditional leafy vegetables later became crops; as a consequence, modern cultivars have tender leaves but a more neutral taste. The leaves that were part of ancient traditional diets are still found in the wild, and gathering weeds for cuisine is a tradition in many parts of the world, including certain areas of the Mediterranean basin (Rivera et al. 2006). Emblematic examples can be found in the Italian region of Liguria, where different kinds of edible leaves of wild plants are gathered in spring to prepare 'preboggion', a local green mixture generally used to stuff ravioli. In Greece, the tradition of eating a great variety of different leafy greens gathered in the wild is symbolised by a dish called 'horta'. Carvalho and Morales (2005) reported 57 vascular plants and 16 fungi species gathered from the wild flora in northern Spain and Portugal (Galicia and Trás-os-Montes). These traditional edible plants (Carvalho and Morales 2005) are consumed in many different and creative ways, such as raw in salads, boiled or fried, as seasoning for potato-based, bean-based or chickpea-based soup or as beverages. The impact of plant gathering on biodiversity, on the preservation of traditional knowledge, on the supply of micronutrients to the diet and as a source of nutraceuticals is discussed below.

Many vegetables are a valuable source of phytosterols (Fig. 5.1), which a growing number of studies show as relevant to the diet. Phytosterols are referred to as being able to decrease the risk of coronary diseases as a result of their

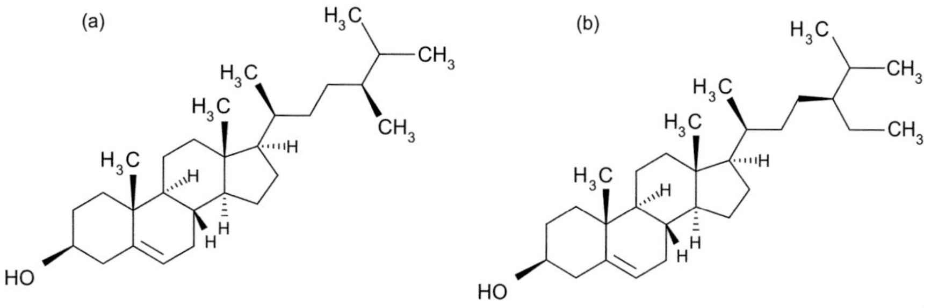

Fig. 5.1 Molecular structure of the phytosterols campesterol (**a**) and γ-sitosterol (**b**), different phytosterols vary in carbon side chains and/or in the presence or absence of a double bond. BKchem version 0.13.0, 2009 (http://bkchem.zirael.org/index.html) was used to draw these structures

similarities to and interference in cholesterol metabolic pathways (Ramprasath et al. 2014; Schonewille et al. 2014). Biosynthesis of cholesterol is directly regulated by the levels of different lipoproteins in the blood, and is closely related to diet. Low-density lipoprotein (LDL) particles are the major carriers of blood cholesterol. When this process becomes unregulated, LDL molecules accumulate in the blood, where they may be oxidised and taken up by macrophages to form foam cells. These foam cells often become trapped in the walls of blood vessels and contribute to the formation of atherosclerotic plaque. The nutritional role of phytosterols is thought to be a consequence of its similarity to the structure of cholesterol, consequently relying on a mechanism of competitive replacement of dietary and biliary cholesterol in mixed micelles. Daily consumption of phytosterols can lower plasma cholesterol (and consequently LDL cholesterol in the bloodstream), thus decreasing the risk of atherosclerotic disease. More information on these mechanisms can be found in selected papers (Lin et al. 2015; Schonewille et al. 2014). The most representative examples of traditional MD leafy crops that are good source of phytosterols are cabbage, lettuce, watercress, cultivated aromatic herbs and turnip leaves.

5.2.1 Cabbage (*Brassica oleracea*)

Cabbage (*Brassica oleracea*) is native to coastal southern and Western Europe, and has become established as an important crop because of its large nutrient reserves, which are stored over the winter in its leaves. Many cultivars of cabbages can be found, from curly to flat-leafed varieties and are generally dark green; they are mainly eaten in soups, stews or sliced in salads.

Cabbage (common, fresh and raw; Galega cultivar), on average, contains 89–92 % (weight/weight) water and supplies 25 kcal/100 g: 2.4 % protein, 3.1 % carbohydrates (of which 2.7 g/100 g are mono- and disaccharides and 0.4 g/100 g are starch), 0.4 % total fat (of which 0.1 % are saturated fatty acids [SFA] and 0.3 %

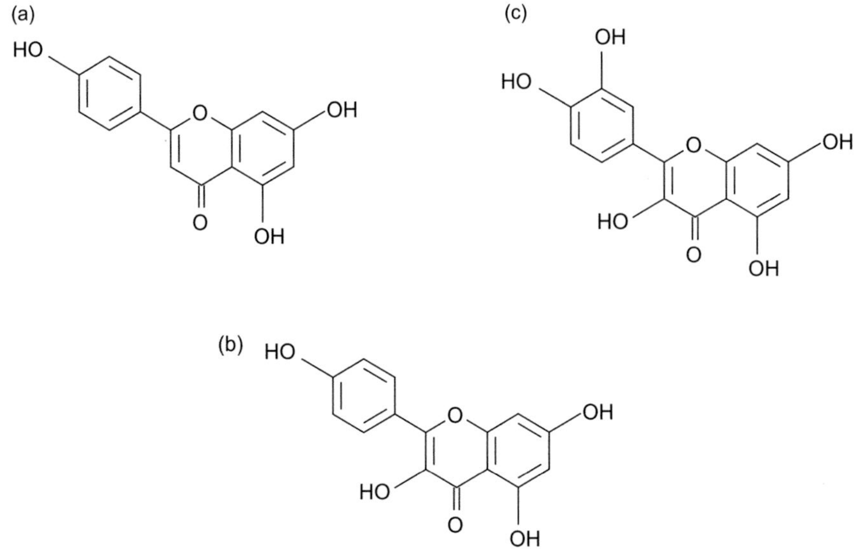

Fig. 5.2 Molecular structure of the flavonoids apigenin, a flavone (**a**), and the flavonols kaempferol (**b**) and quercetin (**c**). BKchem version 0.13.0, 2009 (http://bkchem.zirael.org/index.html) was used to draw these structures

is α-linolenic acid [ALA]); also providing 2.3–3.1 % dietary fibre, 40–286 mg/100 g Ca, 26–40 mg/100 g P, 12–18 mg Mg and the microelements manganese (Mn: 0.16 mg/100 g), selenium (Se: 0.3 μg/100 g) and fluorine (F: 1.0 μg/100 g).

Cabbage is a relevant source of some vitamins, such as ascorbic acid (36.6–51 mg/100 g), folate (43–78 μg/100 g), thiamine (0.061–0.21 mg/100 g), riboflavin (0.040–0.11 mg/100 g), niacin (0.23–1.50 mg/100 g), vit. B_6 (0.12–0.15 mg/100 g), phylloquinone (76 μg/100 g), α-tocopherol (0.15–0.20 mg/100 g) and pro-vitamin A carotenoids: α-carotene (33 μg/100 g) and β-carotene (42 μg/100 g) in a total amount of 5–414 μg/100 g RAE (INSA 2015; USDA 2015). In addition, cabbage contains the flavonoids apigenin (0.1 mg/100 g), luteolin (0.1 mg/100 g), kaempferol[1] (0.2 mg/100 g) and quercetin (0.3 mg/100 g), with molecular structures as shown in Fig. 5.2.

A diet rich in cruciferous vegetables (e.g. cabbage, broccoli, cauliflower) is linked to a reduced risk of several human cancers, probably associated with the presence of ROS scavengers as carotenoids, xanthophylls and flavonoids. Molecules such as quercetin have much greater antioxidant potential in vitro than vitamins A, C and E (Graf et al. 2005), which—along with other bioactive phytochemicals—result in the observed anti-inflammatory and anti-cancer activities (Abu-Yousif et al. 2008; Chiang et al. 2007; Shi et al. 2007; Siegelin et al. 2008; Suzuki et al. 2011; Torkin et al. 2005). Some phytochemicals are specific

[1] International Union of Pure and Applied Chemistry (IUPAC) name: 3,5,7-trihydroxy-2-(4-hydroxyphenyl)-4H-chromen-4-one.

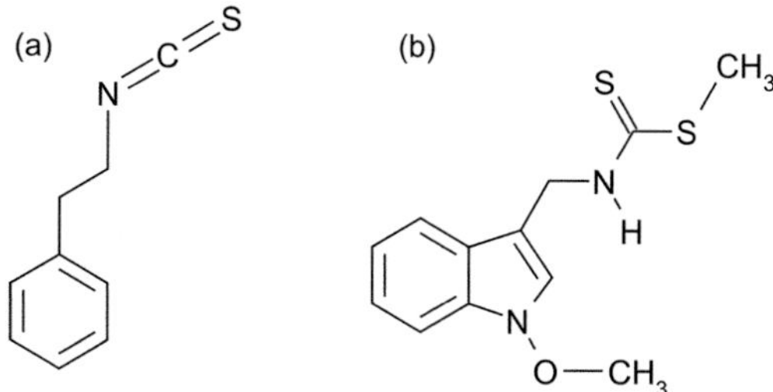

Fig. 5.3 Molecular structure of phytoalexins, sulphur-containing polycyclic compounds found in edible plants of the *Brassicaceae* family: brassinin (**a**) or 2-phenethyl isothiocyanate (PEITC); methoxybrassinin (**b**). BKchem version 0.13.0, 2009 (http://bkchem.zirael.org/index.html) was used to draw these structures

to this botanic family, such as the phytoalexins brassinin[2] and methoxybrassinin[3] (Fig. 5.3), which are sulphur-containing polycyclic compounds with reported anti-fungal activity (Pedras et al. 2004). Synthetic analogues of phytoalexins are being tested to confirm effects on the treatment of human leukaemia and solid tumours; studies to determine concentration–response relationships are also ongoing (NCBI 2015b).

5.2.2 Turnip (*Brassica rapa* subsp. *rapa*)

The common white turnip is a root vegetable generally grown in temperate climates worldwide for its white, bulbous taproot. In the Mediterranean area, both, the bulb and the leaves (turnip greens) are consumed; they have different compositions and thus distinct nutritional values. The bulb is a common ingredient in soups that usually contain a mixture of many other vegetables. It is also consumed in stews or just boiled in water (as is the case in *cozido à portuguesa*, which consists of a mixture of several vegetables and meats, including smoked meats and beans, all boiled together and seasoned with olive oil).

The turnip bulb contains more than 90 % water (INSA 2015; USDA 2015) and supplies moderate amounts of energy (28–29 kcal/100 g) from: about 3 % carbohydrates (2.9 g/100 g of mono- and disaccharides and 0.1 g/100 g starch), 0.4–0.9 % proteins and 0.1–0.4 % lipids (16:0, 18:0, 16:1, 18:1, 18:2, 18:3—the first figure represents the length of the carbon chain (number of C) of a fatty acid, and the second figure refers to the number of double bounds (unsaturations)). PUFA predominate but no ALA has been reported. Turnip bulb also includes dietary fibre (1.8–2.0 %). It is a good source of minerals such as iron (Fe: 0.2–0.3 mg/100 g) and the microelements Mn (0.134 mg/100 g) and Se (0.7 μg/100 g). It also contains ascorbic acid (18–21 mg/100 g), total folates (14–15 μg/100 g), niacin (0.4–0.9 mg/100 g),

[2] IUPAC name: methyl *N*-(1H-indol-3-ylmethyl)carbamodithioate.

[3] IUPAC name: methyl *N*-[(1-methoxyindol-3-yl)methyl]carbamodithioate.

thiamine (0.04–0.05 mg/100 g), riboflavin (0.02–0.03 mg/100 g) and phylloquinone (0.1 μg/100 g) (INSA 2015; USDA 2015).

It is noteworthy that the 2015 edition of the Portuguese National Institute of Health (Instituto Nacional de Saúde Doutor Ricardo Jorge [INSA]) database reports the presence of 20 mg of carotenes/100 g of Portuguese turnip root (or 3.0 μg RAE) but no tocopherols, whereas no carotenes are reported in the equivalent average food item found in the USA. On the other hand, 0.03 mg/100 g of α-tocopherol are reported in American turnips (USDA 2015). According to the 2015 edition of the US Department of Agriculture (USDA) database, turnip root contains flavonoids (7 mg/100 g), which include apigenin,[4] a flavone found in a wide variety of fruits and vegetables, particularly those of the *Brassicaceae* family (Horinaka et al. 2006; Jeyabal et al. 2005). As with the other flavones, apigenin has ROS-scavenging activity and therefore has a relevant role in protecting against oxidative stress probably related to apigenin's anti-cancer activities; it markedly induces the expression of death receptor 5 and synergistically acts with exogenous soluble recombinant human tumour necrosis factor-related apoptosis-inducing ligand to induce apoptosis in malignant tumour cells (Abu-Yousif et al. 2008; Franzen et al. 2009; Horinaka et al. 2006; Jeyabal et al. 2005; Torkin et al. 2005).

Turnip greens are generally consumed boiled and/or slightly fried with garlic in olive oil, as a complement to many dishes. Turnip greens contain more vitamins than the root, including β-carotene (6.9 mg/100 g), lutein and zeaxanthin (12.8 mg/100 g), α-tocopherol (2.9 mg/100 g) and higher levels of phylloquinone than the corresponding root (251 μg/100 g vs. 0.1 μg/100 g). Turnip greens also contain important nutraceutical flavonols: kaempferol (11.9 mg/100 g) and quercetin (0.7 mg/100 g), common to other green vegetables and discussed below (Fig. 5.2).

The health-promoting effects of flavonoids have been attributed to their antioxidant activities; however, the biological functions of flavonoids may also be exerted through their action on enzyme activities and binding to signalling molecules (Labbé et al. 2009; Suzuki et al. 2011).

Quercetin is readily absorbed as quercetin-4′-glucoside, and the main site of quercetin metabolism is the gastrointestinal tract (Graf et al. 2005). Consequently, it may exert a major influence on the gut mucosal epithelium and on colonic disease. Mechanisms for the observed anti-cancer and chemopreventive effects have been proposed (Jung et al. 2010; Lee et al. 2007; Lu et al. 2006).

Kaempferol (Fig. 5.2) was reported to enhance the intestinal barrier, acting at the level of epithelial cell tight junctions (Suzuki et al. 2011). It has also been noted as increasing energy expenditure, possibly influencing metabolic control in humans, with an impact on obesity and diabetes (Silva et al. 2007). Kaempferol may sensitise certain tumour cell lines, thus enhancing the effect of chemotherapy drugs (Siegelin et al. 2008).

[4] IUPAC name: 5,7-dihydroxy-2-(4-hydroxyphenyl)-4H-chromen-4-one.

5.2.3 Lettuce (*Lactuca sativa*)

Lettuce is botanically classified in the family *Asteraceae*, with a worldwide distribution. Lettuce was first cultivated by the ancient Egyptians, who turned it from a weed, the seeds of which were used to produce oil, into a plant grown for its leaves. Lettuce spread to Greece and Rome, acquiring the name *lactuca*, from which the English 'lettuce' is ultimately derived. Lettuce is generally consumed in high amounts in the Mediterranean regions, particularly during the warmer months of the year (from April to October). It is generally combined, in salads, with other leafy vegetables and possibly also sliced tomato and fresh onion as well as grated carrot and aromatic herbs. Minimally processed blends of pre-washed and pre-prepared salad vegetables (preserved in a modified atmosphere) are becoming increasingly popular in Southern Europe. Salad dressings are commonly olive oil, vinegar and salt. According to the 2015 editions of the INSA and USDA databases, the most common variety of lettuce, green leaf (raw), contains about 95 % water and 1.3 % dietary fibre; it is low in calories (15–22 kcal/100 g) but is an important source of vitamins, minerals and phytochemicals. Lettuce contains α-tocopherol (0.60 mg/100 g), thiamine (0.060–0.070 mg/100 g), riboflavin (0.020–0.080 mg/100 g), niacin (0.375–0.40 mg/100 g), folates (38–55 μg/100 g), vitamin B_6 (0.04–0.09 mg/100 g), vitamin C (4.0–9.2 mg/100 g), β-carotene (4.44 mg/100 g), lutein + zeaxanthin (1.73 mg/100 g), α-tocopherol (0.22 mg/100 g), γ-tocopherol (0.41 mg/100 g), δ-tocopherol (0.02 mg/100 g) and phylloquinone (126.3 μg/100 g).

Lettuce is also an important source of Fe (0.86–1.50 mg/100 g), Mg (13–22 mg/100 g) and micronutrients such as Se (0.6 μg/100 g), zinc (Zn, 0.18 mg/100 g), and Mn (0.25 mg/100 g). Lettuce contains the flavonoids apigenin (0.1 mg/100 g) and luteolin (0.3 mg/100 g), for which anti-cancer activities have been reported (Byun et al. 2010; Chiang et al. 2007; Shi et al. 2007; Zhou et al. 2009), as well as myricetin (0.1 mg/100 g) and quercetin (4.2 mg/100 g), with similar structures and properties (Fig. 5.2).

5.2.4 Tomato (*Solanum lycopersicum*)

Botanically, tomato is a fruit, but it is considered a vegetable in terms of its culinary uses. Tomatoes are consumed worldwide in many diverse ways (including raw in salads or juices, in jam, and as an ingredient in many dishes, sauces, soups and drinks). It was probably brought to the Mediterranean from Mexico during the Spanish colonization period. Despite the many varieties of tomato available and consequent differences in chemical composition, the present work is focused on average data from the most common varieties in the MD, having in mind the preference for ripe fruits. Processed tomato is also herein compared with the ripe fresh fruit.

Tomato. *Tomato (Solanum lycopersicum) is widely used in Mediterranean cuisine, particularly fried in olive oil, which favours the release of lycopene, a carotenoid with no vitamin activity but with health-protective effects, mainly associated with the reduction of cancer and heart disease risks. Photo reprinted with kind permission from T. N. Wassermann*

Red, ripe, raw tomatoes contain more than 90 % water and provide 18–19 kcal/100 g from 2.63 to 3.50 % mono- and disaccharides (glucose and fructose), 0.8 % proteins, 0.3 % total fat (0.1 % monounsaturated fatty acids [MUFA], 16:1 and 18:1, and 0.2 % PUFA, mainly ALA), as well as 1.3 % dietary fibre (INSA 2015; USDA 2015). Tomatoes are a good source of vitamins, particularly carotenes, containing about 510 mg/100 g of total carotenes, corresponding to 42–85 μg/100 g RAE. These carotenes are α- and β-carotene (USDA 2015), found in concentrations of 101 and 449 μg/100 g, respectively. Other carotenoids present in tomatoes are lutein + zeaxanthin (123 μg/100 g) and lycopene (2.57 mg/100 g). Tomatoes also supply 0.54–1.2 mg/100 g α-tocopherol, 0.01 mg/100 g β-tocopherol and 0.12 mg/100 g γ-tocopherol. In addition, tomatoes contain the B vitamins thiamine (0.037–0.050 mg/100 g), riboflavin (0.019–0.030 mg/100 g), niacin (0.594–0.060 mg/100 g), pantothenic acid (0.089 mg/100 g), B_6 (0.06–0.14 mg/100 g) and folates (15–17 μg/100 g), as well as vitamin C (13.7–20.0 mg/100 g) and phylloquinone (7.9 μg/100 g).

Besides being so rich in vitamins, tomatoes supply several minerals, including the microelements Mn (0.114 mg/100 g) and F (2.3 μg/100 g), as well as the flavonoids kaempferol (0.1 mg/100 g), myricetin (0.1 mg/100 g) and quercetin (0.6 mg/100 g).

Lycopene is the compound that makes the tomato distinctive. It is a lipid-soluble carotenoid responsible for the red colour. Therefore, it is not found in unripe fruits.

Chemically, lycopene is an acyclic, symmetrical polyene consisting of 13 double bonds, 11 of which are conjugated (Fig. 5.4). Consequently, a large number of

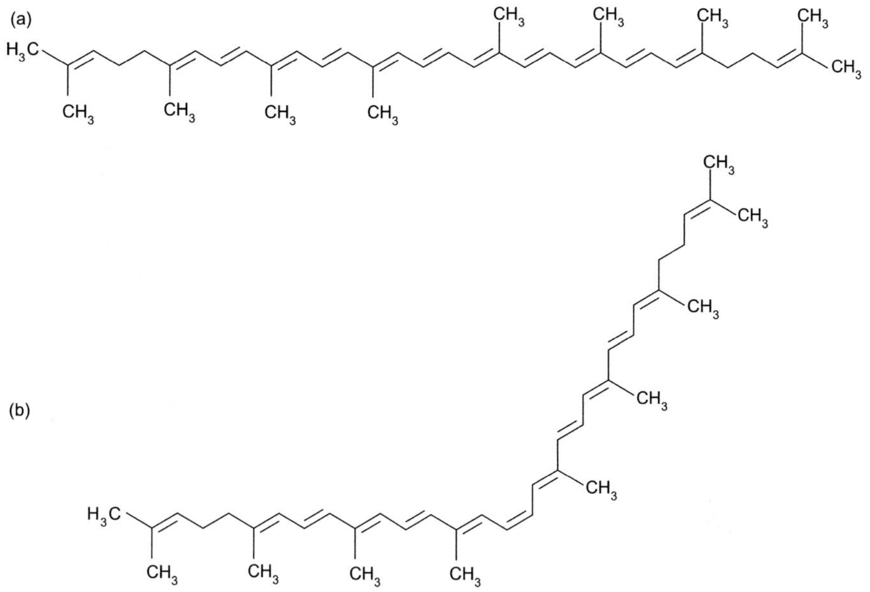

Fig. 5.4 Molecular structure of lycopene: all-*E*, also named all-*trans*, isomer (**a**) and the 15*Z*-isomer, *cis*-form (**b**). The first kind, all-*trans* isomer, is abundant in fresh fruits, while *cis*-forms are mainly found in processed foods, such as tomato paste. Some *Z*-isomers have higher antioxidant activity than natural (all-*trans*) isomers. BKchem version 0.13.0, 2009 (http://bkchem.zirael.org/index.html) was used to draw these structures

isotopes are theoretically possible, but only a few are usually found (NCBI 2015c). All-*E*-lycopene isomers[5] are most common in fresh fruits, while (13Z)-isomers[6] and (15Z)-isomers[7] are predominant in processed tomato matrices (Fig. 5.4). It is noteworthy that 'processed tomato' does not refer exclusively to the commercial end product but also includes home-cooked foodstuffs.

Z-lycopene isomers (*cis*-isomers) have distinct characteristics from their all-*E* counterparts (*trans*-isomers) and are of special interest because some *Z*-isomers have been proven to have higher antioxidant capacities than all-*E*-lycopene (Colle et al. 2013). Although lycopene is a carotenoid, it shows no pro-vitamin A activity. Nevertheless, many positive health properties have been attributed to lycopene, mainly in reducing the risk of some cancers and heart disease (Gajendragadkar et al. 2014; Tang et al. 2005; Wang et al. 2014a).

[5] IUPAC name: (6*E*,8*E*,10*E*,12*E*,14*E*,16*E*,18*E*,20*E*,22*E*,24*E*,26*E*)-2,6,10,14,19,23,27,31-octamethyl-dotriaconta-2,6,8,10,12,14,16,18,20,22,24,26,30-tridecaene.

[6] IUPAC name: (6*E*,8*E*,10*E*,12*E*,14*Z*,16*E*,18*E*,20*E*,22*E*,24*E*,26*E*)-2,6,10,14,19,23,27,31-octamethyl-dotriaconta-2,6,8,10,12,14,16,18,20,22,24,26,30-tridecaene.

[7] IUPAC name: (6*E*,8*E*,10*E*,12*E*,14*E*,16*Z*,18*E*,20*E*,22*E*,24*E*,26*E*)-2,6,10,14,19,23,27,31-octamethyl-dotriaconta-2,6,8,10,12,14,16,18,20,22,24,26,30-tridecaene.

Industrial processing of tomato affects its composition and consequently its nutritional value. Although some reactions can be advantageous, others are not desirable. Some changes are evident by comparing the composition of canned and fresh tomato (detailed composition listed above). Although the concentration of macronutrients seems to be about the same, the chemical reactions that take place during processing and heating are most probably responsible for the presence of SFA in canned tomato (absent in the fresh fruit). The presence of SFA may result from the hydrogenation of MUFA and PUFA, which are present in high quantities in fresh tomatoes. Carotenoids are also affected: α-carotene and β-cryptoxanthin seem to be totally degraded, while some β-carotene remains, and lutein and zeaxanthin are strongly reduced. Flavonoids are mostly destroyed, with the exception of quercetin. Lycopene mainly undergoes isomerisation reactions, although their content may be partially reduced. Thus, in industrial processing, while some components may be degraded, the bioactivity of others may increase, as happens with lycopene (Colle et al. 2013; Schieber and Carle 2005).

During the commercial processing of tomatoes, thermal processes (bleaching, retorting and freezing) generally cause some loss of lycopene in tomato-based foods. However, heat induces isomerisation of the all-*trans*- to *cis*-forms of higher bioavailability, and the content of *cis*-isomers increases with temperature and processing time. Frozen foods and heat-sterilized foods exhibit excellent lycopene stability over the shelf life (Shi and Le Maguer 2000).

The bioavailability of lycopene in processed tomato products is higher than in unprocessed fresh tomatoes. This is because processing operations break down cell walls, which weakens the bonding forces between lycopene and the tissue matrix, thus making lycopene more bioavailable and enhancing the *cis*-isomerisation (Shi and Le Maguer 2000).

On the other hand, the main causes of lycopene degradation in tomatoes are isomerisation to non-bioavailable forms and oxidation. In the presence of lipids, lycopene reacts very easily with free radicals generated from fatty acid oxidation (Colle et al. 2013). It is well known that the oxidative stability of fatty acids depends on their degree of saturation and on the presence of protective compounds.

Cooking tomatoes in olive oil is a characteristic feature of the Mediterranean cuisine. This cooking process breaks down the tomato cell matrix and increases lycopene bioavailability without the many deleterious effects observed in industrial processing (Fielding et al. 2005). The desirable *trans–cis* isomerisation that increases the bioavailability of lycopene seems to be favoured by slightly frying fresh tomato in olive oil but not in other oils (Colle et al. 2013; Fielding et al. 2005). Therefore, since olive oil has a more complex composition than seed oils (which are generally simple mixtures of fatty acids), it is likely that olive oil might influence the kinetics of degradation and isomerisation of lycopene. This might explain the differences in serum lycopene concentration registered in individuals who regularly consume tomatoes cooked in olive oil compared with raw tomatoes (Fielding et al. 2005) or tomatoes cooked in any other oil (Colle et al. 2013).

5.2.5 Pumpkin (*Cucurbita* spp.)

The designation "Pumpkin" encompasses many species and cultivars with different colours, shapes and compositions. Pumpkins were introduced to Europe by Portuguese and Spanish navigators when settling in the 'New World'. The most widely distributed pumpkin variety seems to be a large fruit with thick orange flesh that is generally cooked before eating. Its edible seeds are consumed as dried and salted appetisers. Oil extracted from pumpkin seeds has been used in folk medicine primarily for their alleged pharmacological properties. Some authors relate these alleged healing properties to the presence of certain non-polar bioactive compounds (Gossell-Williams et al. 2006; Nishimura et al. 2014).

Herein we chose the more general description (pumpkin, raw) to search for the 'average composition' in both databases (INSA 2015; USDA 2015). Common raw pumpkin provides 9–26 kcal/100 g from: 1.4–2.76 % mono- and disaccharides (glucose and fructose), 0.3 % starch, 0.3–1.0 % proteins, and 0.1–0.2 % total fat as well as 0.5–0.7 % dietary fibre. An apparent divergence in the compositional data for pumpkins from South Europe (Portugal) versus the USA concerns its lipid profile. The 2015 edition of the INSA database reports the lipid fraction to contain only SFA, while the USDA database reports (in the same year) a more complex lipid fraction containing SFA (mainly 16:0), MUFA (16:1 and 18:1) and PUFA (18:2 and 18:3).

Pumpkin is a good source of vitamins, particularly carotenes, containing 962 mg/100 g of total carotenes, corresponding to 160–426 μg/100 g RAE. Carotenes are α- and β-carotene (4.02 and 3.10 mg/100 g, respectively) (USDA 2015). Other carotenoids present in pumpkins are lutein + zeaxanthin (1.50 mg/100 g). Pumpkins also supply 1.0 mg/100 g α-tocopherol and 1.06 mg/100 g β-tocopherol. B vitamins are also reported: thiamine (0.05–0.010 mg/100 g), riboflavin (0.010–0.110 mg/100 g), niacin (0.60 mg/100 g), pantothenic acid (0.30 mg/100 g), folates (8.0–16.0 μg/100 g) and B_6 (0.040–0.061 mg/100 g), as well as vitamin C (9–12 mg/100 g) and phylloquinone (1.1 μg/100 g).

As with other vegetables, pumpkin is a good source of several minerals, including the microelements Mn (0.125 mg/100 g) and Se (0.3 μg/100 g). Also relevant is the reported presence of the flavone luteolin (1.6 mg/100 g), with a similar structure of those represented in Fig. 5.2. Luteolin is present in other vegetables such as lettuce, cabbage and edible weeds, although in lower concentrations than in pumpkin. As mentioned above, luteolin is described as a potent chemoprotective and anti-cancer agent (Byun et al. 2010; Chiang et al. 2007; Shi et al. 2007; Zhou et al. 2009).

Relatively high amounts—12 mg/100 g—of (non-identified) total phytosterols have been reported in the flesh of pumpkins (USDA 2015). As previously mentioned, these compounds have structures and functions analogous to those presented in Fig. 5.1.

5.3 Wild Leafy Vegetables and Weeds

In the Mediterranean basin, gathering plants for consumption or even medicinal uses is still strongly correlated with hunger, insufficient medical care, lack of formal education, and geographic isolation, as shown in ethnobotanical studies.[8] Gathering techniques and herbal preparation is not currently a valorised knowledge in rural populations of the area (Carvalho and Morales 2005). On the other hand, in northern Europe, there is no such stigma, and the gathering of wild plants has evolved from being a necessity in the past to a pleasurable activity more recently. The gathering of wild plants should receive renewed attention in order to favour the preservation of biodiversity and local empirical knowledge that expresses the intangible cultural heritage of local preferences, habits and human's relationship with nature (Rivera et al. 2006).

Examples of valuable wild leafy vegetables and weeds include watercress (*Nasturtium officinale*), purslane (*Portulaca oleracea*) and borage (*Borago officinalis*), all of which are still valued and consequently often cultivated. Many other edible weeds that are only found locally do not have a common name in English. Sometimes it is difficult to categorise them as a food ingredient versus as a seasoning, since the portions consumed vary widely. Some coexist in the cultivated and in the wild form, further complicating categorisation. Thus, the categorisation below is not based on strict criteria.

5.3.1 Watercress (*Nasturtium officinale*)

Watercress is a fast-growing, aquatic or semi-aquatic perennial plant native to Europe and Asia, and one of the oldest known leafy vegetables to be consumed by humans. It is a member of the family *Brassicaceae*, botanically related to garden cress, mustard and radish—all noteworthy for a peppery, tangy flavour. Watercress is rich in water, as with other leaf vegetables. Watercress supplies 11–23 kcal/100 g from: 0.2–0.4 % mono- and disaccharides, 0.1–0.9 % total fat, 2.3–3.4 % proteins and 0.5–3 % dietary fibre (INSA 2015; USDA 2015).

The lipid profile is complex and includes 0.3 % SFA (mainly 16:0 and some 18:0), 0.1 % MUFA (16:1 and 18:1) and 0.4 % PUFA (18:2 and 18:3), of which 0.1 g/100 g is ALA (INSA 2015). Watercress is a good source of vitamins, supplying high amounts of β-carotene (1.91–1.95 mg/100 g) equivalent to about 160 μg of RAE. Other carotenoids, such as lutein + zeaxanthin are found in abundant concentrations (5.77 mg/100 g); α-tocopherol (1.0–1.5 mg/100 g) is also found. Watercress contains the B vitamins thiamine (0.090 mg/100 g), riboflavin (0.07–0.12 mg/100 g), niacin (0.20–0.60 mg/100 g), pantothenic acid (0.310 mg/100 g), folates (9–200 μg/100 g) and B_6 (0.13–0.23 mg/100 g), as well

[8] Ethnobotany is the interdisciplinary study of plant–human relationships embedded in a complex and dynamic system of natural and social components.

as vitamin C (43–77 mg/100 g) and phylloquinone (250 μg/100 g). It is also rich in minerals, namely Ca (120–198 mg/100 g) and P (56–60 mg/100 g) at a rate that favours Ca absorption; it also contains relevant concentrations of Mg (15–21 mg/100 g) and the microelements Mn (0.244 mg/100 g) and Se (0.9 μg/100 g), among others. Like other *Brassicaceae*, watercress is an important source of bioactive phytochemicals such as the natural flavonols: kaempferol (23 mg/100 g), myricetin[9] (0.2 mg/100 g) and quercetin[10] (30 mg/100 g). The general chemical structure of this class of phenolic compounds is shown in Fig. 5.2.

In addition, watercress is rich in gluconasturtiin,[11] which has a pungent taste and, after enzymatic hydrolysis, produces 2-phenethyl isothiocyanate (PEITC), a small aromatic molecule with the characteristic chemical group –N=C=S of phytoalexins (Fig. 5.3) (Palaniswamy et al. 2003). Phytoalexins are found in several edible cruciferous vegetables. The anti-carcinogenic mechanism of action of these sulphur-containing aromatic compounds has been recently elucidated (Jutooru et al. 2014).

PEITC induces apoptosis in certain cancer cell lines, including cells that are resistant to some current chemotherapeutic drugs. Isothiocyanates (PEITC and other phytoalexins) have been observed to inhibit the development of adenomas to adenocarcinomas in the lung, in animal models (Conaway et al. 2005); moreover, PEITC can induce apoptosis in human prostate cancer cells (Tang et al. 2011), and is also involved in the inhibition of migration and in preventing invasion of human gastric cancer cells (Yang et al. 2010). Epidemiologic studies have provided evidence for the protective role of dietary isothiocyanate compounds in reducing the risk of several types of cancer (Conaway et al. 2005).

5.3.2 Purslane (*Portulaca oleracea*)

Purslane has a wide Mediterranean distribution, from North Africa through the Middle East, and tolerates poor, compacted soils and drought.

According to the USDA database, purslane supplies 20 kcal/100 g from: 3.39 % carbohydrates, 0.36 % total fat and 2.03 % proteins. Moreover, purslane is a source of the B vitamins thiamine (0.047 mg/100 g), riboflavin (0.112 mg/100 g), niacin (0.48 mg/100 g), pantothenic acid (0.036 mg/100 g), folates (12 μg/100 g) and B_6 (0.073 mg/100 g) as well as some vitamin C (21.0 mg/100 g). As with other leafy vegetables, it contains minerals such as Ca (65 mg/100 g) and P (44 mg/100 g) at a rate that facilitates Ca absorption, and the microelements Mn (0.303 mg/100 g) and Se (0.9 μg/100 g), among others (USDA 2015).

[9] IUPAC name: 3,5,7-trihydroxy-2-(3,4,5-trihydroxyphenyl)chromen-4-one.

[10] IUPAC name: 2-(3,4-dihydroxyphenyl)-3,5,7-trihydroxychromen-4-one.

[11] IUPAC name: [(2*S*,3*R*,4*S*,5*S*,6*R*)-3,4,5-trihydroxy-6-(hydroxymethyl)oxan-2-yl] (1*E*)-3-phenyl-*N*-sulfooxypropanimidothioate.

In addition, purslane contains the flavonols isorhamnetin,[12] with an average concentration of 2.8 mg/100 g, kaempferol (0.7 mg/100 g) and quercetin (0.8 mg/100 g).

5.3.3 Borage (*Borago officinalis*)

Borage is an annual herb native to the Mediterranean region. It grows satisfactorily in many other regions, remaining in gardens year after year by self-seeding. In mild climates, borage will bloom continuously all year round. The leaves are edible, and the plant is grown in gardens to be used as a seasoning.

Borage supplies 21 kcal/100 g from: 3.06 % carbohydrates, 0.7 % total fat, of which 0.17 % are SFA (16:0 and 18:0), 0.211 % MUFA (16:1 and 22:1) and 0.109 % PUFA (18:2), and 1.8 % proteins. Moreover, borage is a source of the B vitamins thiamine (0.060 mg/100 g), riboflavin (0.15 mg/100 g), niacin (0.90 mg/100 g), pantothenic acid (0.041 mg/100 g), B_6 (0.084 mg/100 g) and folates (13 μg/100 g) as well as vitamin C (35 mg/100 g). As with other leafy vegetables, it contains minerals such as Ca (93 mg/100 g), Fe (3.3 mg/100 g), Mg (52 mg/100 g) and P (53 mg/100 g). In addition, borage is a source of the microelements Mn (0.349 mg/100 g) and Se (0.9 μg/100 g), among others (USDA 2015).

The plant is commercially cultivated for borage seed-oil extraction. The seed oil is a source of *cis*-γ-linolenic acid (GLA), for which borage is the best-known plant-based source (17–28 %). The seed oil contains between 26 and 38 % GLA, which has been used in clinical trials and as a food supplement (despite that healthy adults may easily synthesise GLA from dietary linolenic acid); GLA is a folk medicine ingredient; however, somewhat conflicting results have been reported (Foster et al. 2010; Herrmann et al. 2002). In this sense, Tasset-Cuevas and colleagues (2013) confirmed the low toxicity and antioxidative capacity of borage seed oil, while Vacillotto and co-workers (2013) mentioned the presence of pyrrolizidine-derived alkaloids (also found in other weeds),which may be responsible for clinical side effects.

5.4 Aromatic Plants and Spices

Aromatic herbs and dry spices play a key role in the Mediterranean cuisine. Although they vary from region to region, the abundant use of fresh and dry herbs or their seeds is a common feature in the MD. The used herbs include parsley (*Petroselinum crispum*), oregano (*Origanum vulgare*), coriander (*Coriandrum sativum*), cumin (*Cuminum cyminum*), saffron (*Crocus sativus*) and rosemary (the general English designation for several botanically related species, of which *Rosmarinus officinalis* is herein mentioned). A wild weed, fennel (*Foeniculum vulgare*) is also described, and finally garlic (*Allium sativum*) and onion (*Allium cepa*) are mentioned in this section for their seasoning applications.

[12] IUPAC name: 5,7-trihydroxy-2-(4-hydroxy-3-methoxyphenyl)chromen-4-one.

The preparation of many traditional Mediterranean dishes (such as stews and rice-based dishes) commonly starts by frying onion and/or garlic in olive oil, often adding tomato. Oleic acid (from olive oil) combined with allicin and diallyl thiosulfinate (from garlic and onion) form a potent anti-microbial mixture, which activity spectrum and potency have been the subject of several studies (Bjarnsholt et al. 2005; Borlinghaus et al. 2014; Kohler et al. 2009). Coriander and chilli pepper are also included in this group, showing similar properties (Silva et al. 2013).

Different combinations of spices have different effects on microbial growth. Thus, although seasoning ingredients are selected mainly according to sensorial aspects, herbs play an important role in food preservation. Most often, herbs (particularly when fresh) are added by the end of cooking to preserve flavour, also allowing the preservation of potentially beneficial chemical entities, some of which may still be unknown.

Aromatic plants are rich in vitamins but most particularly in bioactive flavonoids. Their vitamin content plays a secondary role in the diet, because of the low intakes of these food items but many associated phytochemicals are likely to play a relevant role in health and wellness.

5.4.1 Parsley (*Petroselinum crispum*)

Parsley, native to the central Mediterranean region (southern Italy, Algeria and Tunisia), is now used worldwide and is one of the most widely used aromatic herbs for its mild flavour. Again, we found a discrepancy between the compositional data from the 'Portuguese Food Information Resource' (PortFIR) and USDA databases. The absence of fat is noticeable in the record corresponding to the average sample of fresh parsley collected in the Portuguese market (INSA 2015), whereas the USDA database reports a lipid fraction of 0.79 %, of which 0.132 g/100 g of SFA (16:0 and 18:0), 0.295 g MUFA and 0.124 g PUFA (USDA 2015).

Parsley is rich in the carotenoids β-carotene (5.05 mg/100 g) and lutein + zeaxanthin (5.56 mg/100 g), folates (152 μg/100 g) and phylloquinone (1.64 mg/100 g) (USDA 2015). It also contains vitamin C (133–220 mg/100 g), riboflavin (0.060–0.098 mg/100 g), pantothenic acid (0.40 mg/100 g), thiamine (0.28–0.086 mg/100 g) and vitamin B_6 (0.090 mg/100 g) as well as α-tocopherol (0.75 mg/100 g) and γ-tocopherol (0.53 mg/100 g) (INSA 2015; USDA 2015). Moreover, the following data have been reported (USDA 2015): non-specified phytosterols (5 mg/100 g) (Fig. 5.1), in addition to the flavones apigenin (215.4 mg) and luteolin (1.1 mg), and the flavonols kaempferol (1.5 mg/100 g), myricetin (14.8 mg/100 g) and quercetin (0.3 mg/100 g) (Fig. 5.2). In addition to the above-mentioned nutraceutical properties of flavonoids, luteolin has been suggested as a potential chemosensitiser in cancer therapy for a wide range of solid tumours (Shi et al. 2007). The therapeutic use of apigenin, combined with other agents, has also been mentioned for the treatment of malignant tumours (Horinaka et al. 2006).

5.4.2 Oregano (*Origanum vulgare*)

Oregano is native to warm-temperate western and south-western Eurasia and the Mediterranean region. It is an important culinary herb, used for the flavour of its leaves, which have a warm and slightly bitter taste that is generally more intense when dried, which is the most used form.

According to the USDA database, oregano is rich in the B vitamins thiamine (0.177 mg/100 g), riboflavin (0.528 mg/100 g), niacin (4.64 mg/100 g), pantothenic acid (0.921 mg/100 g) and B_6 (1.044 mg/100 g). Oregano contains a relevant amount of phylloquinone (621.7 μg/100 g) as well as the carotenoids β-carotene (1.0 mg/100 g), α-carotene (20 μg/100 g), β-cryptoxanthin (7 μg/100 g) and lutein + zeaxanthin (1.9 mg/100 g) (USDA 2015). Associated to the lipid fraction, oregano contains γ-tocopherol (24.42 mg/100 g) and δ-tocopherol (0.92 mg/100 g), which have antioxidant and vitamin activity, as well as essential oils. This lipid fraction accounts for 4.28 % of the edible portion and contains 1.55 % SFA (10:0, 12:0, 14:0, 16:0, 18:0), 0.72 % MUFA (16:1, 18:1c), 1.37 % PUFA (18:2 and ALA), expressed in total edible portion. In addition, 203 mg/100 g of non-specified phytosterols are also associated with oregano's lipid fraction.

The typical oregano's aroma is attributed to the richness of terpenes and sesquiterpenes in its essential oil (Crocoll et al. 2010), which includes *α*-pinene,[13] limonene,[14] camphor,[15] citronellol and carvacrol (Yang et al. 2007), among many others, some of which are yet to be isolated and identified.

Oregano has shown in vitro anti-microbial activity, namely against strains of several food borne pathogens including *Listeria monocytogenes* (Firouzi et al. 2007; Silva et al. 2013).

5.4.3 Coriander (*Coriandrum sativum*)

Coriander is an annual herb of the *Apiaceae* family. It is native to regions spanning from southern Europe and North Africa to south-western Asia, and its fresh leaves and seeds are both consumed. Fresh leaves have a stronger flavour than seeds, and a slightly different composition. Coriander contains a fair amount of vitamin C (27 mg/100 g) and—like other herbs—is rich in B vitamins (except B_{12}) and in carotenoids (β-carotene [3.93 mg/100 g], β-cryptoxanthin [202 μg/100 g], lutein + zeaxanthin [865 μg/100 g]) as well as in phylloquinone (310.0 μg/100 g) (Fig. 4.6) and quercetin (52.9 mg/100 g) (Fig. 5.2).

[13] IUPAC name: 4,6,6-trimethylbicyclo[3.1.1]hept-3-ene.

[14] IUPAC name: (4*R*)-1-methyl-4-prop-1-en-2-ylcyclohexene.

[15] IUPAC name: 1,7,7-trimethylbicyclo[2.2.1]heptan-2-one.

Coriander. *Coriander (Coriandrum sativum) is an aromatic herb; both the fresh leaves and dry seeds are used to season many dishes and, as with many other aromatic herbs, it is rich in phytosterols that are thought to positively interfere with cholesterol levels in the blood. Photo reprinted with kind permission from T. N. Wassermann*

The lipid fraction of coriander contains 0.014 g/100 g SFA (16:0 and 18:0), 0.275 g MUFA (18:1 and traces of 16:1) and 0.04 g/100 g PUFA (18:2), as well as the oil soluble γ-tocopherol (24.41 mg/100 g), δ-tocopherol (0.93 mg) and the phytosterols stigmasterol (3 mg) and β-sitosterol (2 mg/100 g) (INSA 2015; USDA 2015). The role of these two last compounds in the control of lipid transport in the blood stream is explained above. Several essential oils are associated with the lipid fraction of coriander, as is the case of linalool,[16] a terpene alcohol found in many flowers and aromatic herbs, with many commercial applications primarily based on its pleasant scent. Linalool accounts for about 70 % of coriander's essential oil fraction; despite this, it can be safely used as a food ingredient (Burdock and Carabin 2009).

Coriander, like other aromatic herbs and spices, has been used not only as a seasoning but also as an anti-microbial agent to prevent food spoilage. Silva and colleagues (2011) tested the activity of coriander's essential oil against several Gram-positive and Gram-negative bacterial strains, including clinical isolates. These authors registered a wide spectrum of bactericidal activity, with reductions of cell viability near 100 % for nearly all tested strains, using just 0.1 % of oil (except for *Pseudomonas aeruginosa*—ATCC27853, which needed 1.6 % oil for 92 % viability reduction, or 3.2 % oil to kill 99 % of cells). The role of individual compounds in human health and potential uses of coriander's essential oil in the nutraceutical industry have been recently reviewed (Sahib et al. 2013).

5.4.4 Basil (*Ocimum basilicum*)

Basil is originally native to India and most prominently used in Italian cuisine. The leaf contains some protein (3.15 %), lipids (0.64 %: 0.041 % SFA [16:0, 18:0], 0.088 % MUFA [18:1] and 0.389 % PUFA [18:2 and 18:3]), 0.3 % sugars (dextrose, fructose and galactose) and 1.6 % dietary fibre (USDA 2015).

Basil is primarily consumed fresh, sometimes in generous amounts in salads, or added at the end of cooking to preserve its characteristic strong flavour.

[16] IUPAC name: 3,7-dimethylocta-1,6-dien-3-ol.

Basil. *(Ocimum basilicum) is used fresh, sometimes in generous amounts, to season salads and pasta. It is rich in tocopherols and in carotenoids that have been reported to have a direct effect on vision. Photo reprinted with kind permission from T. N. Wassermann*

As other aromatic herbs, basil is a source of minerals, among which Ca (177 mg/100 g) and P (56 mg/100 g), as well as the microelements Mn (1.15 mg/100 g) and Se (0.3 μg/100 g). Basil contains mainly water-soluble vitamins, such as ascorbic acid (18 mg/100 g) and B vitamins (except B_{12}) as well as phylloquinone (414.8 μg/100 g). Basil is rich in carotenoids: β-carotene (3.14 mg/100 g), β-cryptoxanthin[17] (46 μg/100 g) (Fig. 4.5) and lutein + zeaxanthin (6.65 mg/100 g). It also contains α-tocopherol and γ-tocopherol (0.80 and 0.16 mg/100 g, respectively) (USDA 2015).

Xanthophylls and carotenoids are known for their direct influence on vision. Lutein and zeaxanthin have been reported to significantly lower the risk of age-related macular degeneration and cataracts, which can lead to blindness (Hartmann et al. 2004; Mares et al. 2006; Mares-Perlman et al. 2002; Trumbo and Ellwood 2006).

5.4.5 Cumin (*Cuminum cyminum*)

Cumin is native to the eastern Mediterranean and India and belongs to the *Apiaceae* (formerly *Umbelliferae*) family, thus sharing some characteristics that influence biochemical composition with members of this family (e.g. coriander, fennel, parsley). Cumin's seeds (each one contained within a fruit, which is dried) are used both, in whole and ground forms. The seed's water content is about 8 %.

According to the USDA food composition database, cumin seeds contain 22.3 % lipids and 44.2 % carbohydrates, 10 % of which are dietary fibres. Cumin seeds also contain 17.8 % protein and relevant amounts of B vitamins, particularly niacin (4.58 mg/100 g). Other vitamins present in relevant amounts are the carotenoids β-carotene (762 μg/100 g) and lutein + zeaxanthin (448 μg/100 g), as well as some α-tocopherol (3.33 mg/100 g) and phylloquinone (5.4 μg/100 g). The composition of seeds, including the fatty acid profile and secondary metabolites of cumin, varies widely with geographical origin (Bettaieb et al. 2011).

The lipid fraction is dominated by C18:1, which includes oleic acid (Bettaieb et al. 2011), accounting for most of the 14.0 % MUFA (18:1 and traces of 20:1). It also contains 1.54 % SFA (16:0 and 18:0) and PUFA (mainly C18:2), which corresponds to petroselinic acid (C18:1, n-12), a distinctive compound of the cumin's lipid fraction, independent of the seed's origin (Betaied et al. 2011). In addition, 68 mg/100 g of non-specified phytosterols are associated with this lipid fraction (USDA 2015).

Moreover, cumin's essential oil contains other commercially relevant volatile compounds such as cuminaldehyde (4-propan-2-ylbenzaldehyde), which is representative of the cumin aroma and besides its health benefits, it is used commercially in perfumes and other cosmetics. Cumin's aroma fraction also includes volatile terpenoids such as terpinolene[18] and α-bergamotene[19] (Sowbhagyaa 2013). Sowbhagya refers to many nutraceutical properties associated with cumin, including anti-allergenic, antioxidant, anti-platelet aggregation and hypoglycaemic effects.

---

[17] IUPAC name: (3,5,5-trimethyl-4-[(1*E*,3*E*,5*E*,7*E*,9*E*,11*E*,13*E*,15*E*,17*E*)-3,7,12,16-tetramethyl-18-(2,6,6-trimethylcyclohex-1-en-1-yl)octadeca-1,3,5,7,9,11,13,15,17-nonaen-1-yl]cyclohex-3-en-1-ol.

[18] IUPAC name: 1-methyl-4-(propan-2-ylidene)cyclohex-1-ene.

[19] IUPAC name: (1*S*,5*S*,6*R*)-4,6-dimethyl-6-(4-methylpent-3-enyl)bicyclo[3.1.1]hept-3-ene.

Cumin is relevant to food safety, since anti-bacterial and anti-fungal activities have been reported, namely against *Escherichia coli*, *Staphylococcus aureus* and *Streptococcus faecalis* (Allahghadri et al. 2010).

5.4.6 Saffron (*Crocus sativus*)

Saffron is a spice derived from the flower of *Crocus sativus*, of the family *Iridaceae*, from which the pollen (yellow) and the stamen (orange-reddish) are extracted. It is used in various cuisines as a seasoning and colouring agent. In the Mediterranean cuisine the pollen is more often used and 'Paella', a well-known Spanish rice-based dish, is an example of such an utilization, mainly for coloring purposes. Saffron is native to Greece or southwest Asia, and was first cultivated in Greece and later transported to North Africa; it has been traded and used for over four millennia. Dry saffron is highly sensitive to fluctuating pH levels and rapidly breaks down chemically in the presence of light and oxidising agents. It must, therefore, be stored in airtight containers to minimise contact with atmospheric oxygen. Saffron is somewhat more resistant to heat. It contains about 11.9 % water, supplying 310 kcal/100 g: 11.4 % proteins, 5.8 % total lipids (of which 2 g/100 g are PUFA C18:2 and C18:3), 65.4 % carbohydrates (of which 3.9 % are dietary fibres). Saffron contains relevant amounts of minerals, such as Ca (111 mg/100 g), Mg (264 mg/100 g) and Mn (28.4 mg/100 g). It is very rich in kaempferol (205.5 mg) (USDA 2015), an important chemoprotective agent. No carotenoids or tocopherols are registered. On the other hand, relevant concentrations of ascorbic acid (80.8 mg/100 g), vitamin B_6 (1.01 mg/100 g) and niacin (1.46 mg/100 g/100 g) were quantified (USDA 2015). Saffron powder is characterized by its golden yellow-orange colour, bitter taste, and fragrance. All three attributes of this expensive spice are due in large part to products of the degradation of the carotenoid, α-crocin[20] (Ochiai et al. 2007). Saffron's colour is mainly due to crocin's glycosides, one of which—picrocrocin[21]—is mainly responsible for its bitter taste. Safranal[22] is responsible for saffron's aroma. Crocin and its carotenoid derivatives have been reported to have protective effects on neuronal injury (Ochiai et al. 2007). Picrocrocin has been reported to have anti-cancer effects (Samarghandian and Borji 2014) and safranal is noted as protecting agent against retinal degeneration (Fernández-Sánchez et al. 2012).

[20] IUPAC name: bis[(2*S*,3*R*,4*S*,5*S*,6*R*)-3,4,5-trihydroxy-6-[[(2*R*,3*R*,4*S*,5*S*,6*R*)-3,4,5-trihydroxy-6-(hydroxymethyl)oxan-2-yl]oxymethyl]oxan-2-yl] (2*E*,4*E*,6*E*,8*E*,10*E*,12*E*,14*E*)-2,6,11,15-tetramethylhexadeca-2,4,6,8,10,12,14-heptaenedioate.

[21] IUPAC name: (4*R*)-2,6,6-trimethyl-4-[(2*R*,3*R*,4*S*,5*S*,6*R*)-3,4,5-trihydroxy-6-(hydroxymethyl)oxan-2-yl]oxycyclohexene-1-carbaldehyde.

[22] IUPAC name: 2,6,6-trimethylcyclohexa-1,3-diene-1-carbaldehyde.

***Saffron**. Saffron is extracted from the flower of Crocus sativus and has a yellow (pollen) to reddish (stamen) colour. It has been noted as having protective effects on neuronal injury. The high level of carotenoids helps protect against retinal degeneration, as established in epidemiological, clinical and interventional studies. Photo reprinted with kind permission from T. N. Wassermann*

5.4.7 Rosemary (*Rosmarinus officinalis*)

Rosemary is a woody, perennial herb with fragrant, evergreen, needle-like leaves and white, pink, purple or blue flowers, native to the Mediterranean region. Rosemary is reasonably hardy in cool climates and can withstand droughts for lengthy periods. It is a member of the mint family *Lamiaceae*, and is particularly used in roast meats.

Rosemary. *Rosemary is a perennial bush of Rosmarinus officinalis the leaves of which are used as a spice, providing a variety of relevant flavonoids, such as apigenin, luteolin and naringenin, with several documented health benefits. Photo reprinted with kind permission from T. N. Wassermann*

Fresh rosemary contains water (68 %), protein (3.31 %), lipids (5.85 %; mainly C16:0 and C18:1) and about 20 % carbohydrates (of which 14.1 g/100 g are dietary fibres). Rosemary is poor in carotenoids but contains many other bioactive compounds such as apigenin (0.6 mg/100 g) and luteolin (2 mg/100 g), as well as naringin (24.9 mg/100 g) (USDA 2015) which, in humans, is readily metabolized to naringenin[23] (USDA 2015). Naringenin and its metabolites exert a variety of pharmacological effects, such as antioxidant activity, lowering blood lipid's level,

[23] IUPAC name: 5,7-dihydroxy-2-(4-hydroxyphenyl)-2,3-dihydrochromen-4-one.

anti-carcinogenic activity and positive effects on the treatment of metabolic syndrome (Alam et al. 2014; Orhan et al. 2015). Naringin is further discussed in Sect. 5.7.2, as it is typical of citrus fruits, although also found in many other vegetables in variable amounts.

5.4.8 Fennel (*Foeniculum vulgare*)

Fennel, a member of the family *Apiaceae*, is a hardy, perennial, umbelliferous herb with yellow flowers and feathery leaves. It is indigenous to the shores of the Mediterranean but has become widely naturalised in many parts of the world, especially in dry soils near the seacoast and on riverbanks. It is a highly aromatic and flavourful herb with culinary and medicinal uses (Garland 1979). It is one of the primary ingredients of absinthe. Florence fennel or 'finocchio' has a swollen, bulb-like stem base that is used as a vegetable food. Seeds are more commonly used as a spice and provide protein, lipids and carbohydrates as well as some minerals, namely Ca (1.20 g/100 g) and Mn (6.5 mg/100 g). Fennel seeds contain a lipid fraction of 0.480 g/100 g: SFA (16:0), 9.91 g/100 g MUFA (18:1) and 1.69 g/100 g PUFA (18:2). On average, 66 mg/100 g of phytosterols are associated with this lipid fraction (USDA 2015). The presence of abscisic acid[24] (ABA), *p*-anisic acid and phellandrenes has also been registered (Kubo et al. 2003; Li et al. 2011; Piccaglia and Marotti 2001). ABA is an apocarotenoid (isoprenoid) that functions as a plant hormone, regulating many developmental processes, including bud dormancy. In addition, ABA is a universal signalling molecule that stimulates stress responses (to heat and light) in animal cells, immune responses in leukocytes, insulin release from pancreatic β-cells and the expansion of mesenchymal and colon stem cells. ABA also inhibits the growth and induces the differentiation of cancer cells. Therefore, ABA is a promising compound for the treatment of several human diseases (Li et al. 2011). On the other hand, *p*-anisic acid (4-methoxybenzoic acid) is found to have tyrosinase inhibitory activity (Kubo et al. 2003) as well as anti-inflammatory action (Singh et al. 2006). On the other hand, α- and β-phellandrenes[25] are cyclic monoterpenes and are double-bond isomers. Phellandrenes are often used in fragrances because of their pleasant aromas.

5.4.9 Garlic (*Allium sativum*)

Garlic is a bulbous plant of the family *Amaryllidaceae*, subfamily *Alliaceae* (EOL 2015a), and is used worldwide. The flavour varies in intensity and aroma depending

[24] IUPAC name: (2*E*,4*E*)-5-(1-hydroxy-2,6,6-trimethyl-4-oxocyclohex-2-en-1-yl)-3-methylpenta-2,4-dienoic acid.

[25] IUPAC name: 3-methylidene-6-(propan-2-yl)cyclohex-1-ene.

on many factors, including cooking methods. Allicin, a sulphur-containing compound with many different biological properties (described below), is responsible for the typical smell and taste of freshly cut or crushed garlic.

The report corresponding to the average garlic (raw) available in the USDA database in 2015[26] differs considerably from that of the Portuguese database[27] (INSA 2015), suggesting two different varieties/cultivars. Nevertheless, the composition is discussed, for guidance purposes, highlighting the major differences. Raw garlic may contain 58.6–79.8 % of water, a high concentration of carbohydrates (11.3–33 %, mostly starch), some protein (3.8–6.36 %) and a low lipid content (0.5–0.6 %), of which 0.1 % SFA, no MUFA and 0.3 % PUFA, mostly ALA, according to INSA (2015). It also contains Ca (181 mg/100 g) and P (153 mg/100 g), as well as some micronutrients such as Zn (1.16 mg/100 g), Mn (1.7 mg/100 g) and Se (14.2 μg/100 g). In the average composition from the Portuguese database registry, the absence of carotenes is noteworthy (INSA 2015). Conversely, the USDA database mentions traces of β-carotene (5 μg/100 g) and the carotenoids lutein + zeaxanthin (16 μg/100 g).

According to both databases, garlic contains α-tocopherol (0.010–0.080 mg/100 g), most B vitamins (except B_{12}) as well as vitamin C (17–31.2 mg/100 g) and phylloquinone (1.7 μg/100 g). Moreover, garlic contains the flavonols kaempferol (0.3 mg/100 mg), myricetin (1.6 mg/100 g) and quercetin (1.7 mg/100 g) (USDA 2015), which, with allicin, account for garlic's ROS-scavenging activities, resulting in the widely noted anti-ageing and chemoprotective effects. Some mechanisms have been described recently (Chan et al. 2013; Lynett et al. 2011; Shouk et al. 2014).

Allicin[28] has proven pharmacological actions as an anti-infective agent, ROS scavenger, also acting as hypoglycaemic and hypolipidaemic agent (Borlinghaus et al. 2014; Chan et al. 2013; Lynett et al. 2011; Shouk et al. 2014). This nutraceutical compound is also present in onions and other species of the subfamily *Alliaceae*.

Allicin also possesses anti-bacterial and anti-fungal properties that contribute to food safety. According to Borlinghaus and colleagues (2014), the action of allicin concentrates against some human pathogens is comparable to that of synthetic antibiotics, using the plate diffusion assay. Moreover, allicin is reported to act synergistically with some unidentified minor components, as garlic extracts show a wider spectrum and higher potency than that observed for the synthetic allicin. In Mediterranean cuisine, garlic is often paired with onion and tomato, all fried in olive oil, which is likely to facilitate the extraction of many hydrophobic bioactive molecules (such as allicin) to the hot lipid phase.

[26] Report 11215.

[27] Identified as 'alho crú'.

[28] IUPAC name: 3-[(prop-2-ene-1-sulfinyl)sulfanyl]prop-1-ene.

***Garlic & onion**. Garlic (Allium sativa) is a relevant source of the flavonols kaempferol, myricetin and quercetin. Along with allicin (a sulphur-containing compound responsible for garlic's strong aroma) these account for the reported ROS (reactive oxygen species, or free radicals)-scavenger activity, most probably related to the widely noted and proven anti-ageing and chemoprotective effects of garlic. Onions (Allium cepa) exist in different varieties, from white to reddish. The pigmentation is due to flavonoids, and thus red onions are generally higher in flavonoids (mainly anthocyanins) than white or yellow varieties. The pungent effect, observed when onions are chopped, is due to several sulphur-containing compounds, which includes allicin. Onions contain some simple sugars, accounting for its sweet taste, and oligosaccharides, which positively impact the microbiota of the gut. Oleic acid (from olive oil) together with allicin and other thiosulfinates (from garlic and onion) form a potent anti-microbial mixture. Photo reprinted with kind permission from T. N. Wassermann*

5.4.10 Onion (*Allium cepa*)

Onion is a bulb of the same family as garlic (*Amaryllidaceae*, subfamily *Alliaceae*). Although dry and frozen preparations are available in the market (as with garlic), Mediterranean consumers prefer fresh products mainly because of its strong flavour. Like garlic, onions are also cultivated and used around the world, usually being served cooked and often raw in salads. Onions are pungent when chopped, as they contain various sulphur-containing compounds (such as cysteine sulfoxide) that, together with their breakdown products, produce their distinctive odour, flavour and lachrymatory (tear-stimulating) properties (Brewster 1994). Unlike garlic, many varieties of onion are available in the market, such as white and red onions, the composition of which varies, mainly in the presence and concentration of minor components, generally with impact on human health.

According to recent figures (INSA 2015; USDA 2015), which once more are divergent, the average composition of onion (raw) includes water (89.1–93.8 %)

and carbohydrates (3.2–4.2 %, of which dextrose, glucose and fructose appear in decreasing order of concentration); 0.9 % of carbohydrates are oligosaccharides (INSA 2015). Simple sugars account for the sweet taste of onions. They also contain a fair amount of dietary fibre (1.3 %). Fibres and oligosaccharides are important for the survival and prevalence of probiotic bacteria and play a key role in the maintenance of healthy gut microbiota.

Onions are low in lipids (0.1–0.2 %). Records from INSA note the absence of SFA and MUFA; at the same time, they report 0.2 % PUFA, of which 0.1 % is linolenic acid (most probably ALA). On the other hand, report 11282 from the USDA - National Nutrient Database for Standard Reference (USDA 2015) refers to the presence of 0.042 g/100 g SFA (mainly 16:0), 0.013 g MUFA (only 18:1) and 0.017 % PUFA (mainly 18:2 and traces of 18:3). 15 mg/100 g of non-discriminated phytosterols are noted in the above-mentioned USDA report, associated with the lipid fraction.

Onion also provides micronutrients such as Zn (0.17 mg/100 g), Mn (0.129 mg/100 g), Se (0.5 μg/100 g) and F (1.1 μg/100 g) (USDA 2015). Like garlic, no carotenes are present, with the exception of lutein + zeaxanthin (4 μg/100 g). Onion contains α-tocopherol (0.02–0.30 mg/100 g), B vitamins (except B_{12}), as well as vitamin C (8.0 mg/100 g) and phylloquinone (0.4 μg/100 g) (INSA 2015; USDA 2015). Onion bulbs are among the richest sources of dietary flavonoids, particularly quercetin (20.3 mg/100 g), isorhamnetin (5 mg/100 g) and kaempferol (0.6 mg/100 g) (USDA 2015).

As mentioned above, quercetin has been associated with several health benefits, such as chemoprotective and anti-hypertensive effects (Edwards et al. 2007; Graf et al. 2005). The relative bioavailability of quercetin from a single dietary-relevant dose of onions (250 g of lightly fried onions consumed by a 70-kg human subject, i.e. 1–10 mg quercetin/kg body weight) ranges from 1 to 7 % (Graf et al. 2005).

Flavonoids are the predominant pigments in onions; consequently, yellow onions contain fewer flavonoids than red onions (Slimestad et al. 2007). Red onions contain anthocyanins, which may reach 10 % of the total flavonoid content in some cultivars. These are mainly cyanidin glucosides acylated with malonic acid, or non acylated (Slimestad et al. 2007).

Like garlic, onions contain allicin and some poorly described compounds, such as the dihydroflavonol taxifolin[29] and its 3-, 7- and 4′-glucosides (Slimestad et al. 2007).

The above-mentioned herbs and spices are just examples of the wide variety of species that are still used in the preparation of many Mediterranean regional dishes.

Widely spread (and still consumed) weeds in the Mediterranean basin include *Portulaca oleracea*, *Foeniculum vulgare*, *Sonchus oleraceus*, *Silene vulgaris*, *Cichorium intybus*, *Rorippa nasturtiumaquaticum*, *Asparagus acutifolius*, *Malva sylvestris*, *Papaver rhoeas*, *Rubus ulmifolius*, *Allium ampeloprasum*, *Arbutus unedo*, *Crataegus monogyna*, *Scolymus hispanicus* and *Chondrilla juncea* (Rivera

[29] IUPAC name: 2*R*,3*R*)-2-(3,4-dihydroxyphenyl)-3,5,7-trihydroxy-2,3-dihydrochromen-4-one.

et al. 2006). Gathered edible plants should be studied more deeply to understand their dietary impact on health. Their contribution to the traditional MD is qualitatively important, is as yet almost unknown and offers a unique opportunity for discovering and developing potential new products and new crops.

5.5 Starchy Foods

Major sources of starch are grains (such as wheat, corn, rye and rice) and tubers such as potatoes and yams. In Mediterranean countries, wheat, corn and rye are mainly used to bake bread, while potatoes and rice in addition to pasta or couscous (both made from wheat) form the starchy base of most meals in the area.

Bread is a staple food in Europe and in European-derived cultures, as opposed to East Asia, where the equivalent staple food is rice.

5.5.1 Wheat (*Triticum aestivum; Triticum durum*)

Wheat originated from the Middle East and is now the dominant crop in temperate countries. Its success depends partly on the crop adaptability and high yield potential but also on the gluten protein fraction, which confers the viscoelastic properties that allow dough to be processed into bread, pasta, noodles and other food products. Wheat contributes with essential amino acids, minerals and vitamins as well as with phytochemicals and dietary fibre to the human diet.

Bread is usually made from a wheat-flour dough that is fermented with yeast (*Saccharomyces cerevisiae*), allowed to rise, and finally baked in an oven that when heated with wood gives bread a characteristic flavour and consistency. Owing to its high levels of gluten (which give the dough sponginess and elasticity), common wheat (*Triticum aestivum*) is the most common grain used for the preparation of bread. However, bread is also made from the flour of other wheat species (including *Triticum durum*), often in combination with rye (*Secale cereale*) and maize (*Zea mays*). The use of less refined flours (often a blend from different cereals), selected yeast strains, and cooking in a wood-heated oven results in a wide variety of highly appreciated regional breads. It is noteworthy that traditional 'pitta' bread (Greece) includes olive oil, as also does Italian 'focaccia'.

Thirteen types of fresh bread are reported in Portuguese markets, which reflect these combinations of cereals and the use of different types of flour (INSA 2015). As an example, current Portuguese wheat bread provides about 289 kcal/100 g, mainly from carbohydrates (57.3 %), mostly starch (55.2 g/100 g) and including mono- and disaccharides (2.1 g/100 g) and dietary fibre (3.8 g/100 g). It also contains 2.2 % fat and 8.4 % proteins (INSA 2015).

Different milled fractions of wheat have different profiles of both hydrophilic and lipophilic phytochemicals with different expected impacts on health. The total phenolic content of bran/germ fractions was reported to be 15–18-fold higher

($P < 0.01$) than that of respective endosperm fractions, showing the highest dietary value of less refined flours (Adom et al. 2005).

Wheat whole grain flour contains relevant amounts of B vitamins (except vitamin B_{12}) as well as phylloquinone (1.9 μg/100 g), carotenoids lutein + zeaxanthin (220 μg/100 g), traces of β-carotene and the tocopherols α-tocopherol (0.71 mg/100 g), β-tocopherol (0.23 mg/100 g) and γ-tocopherol (1.91 mg/100 g). Wheat is also a source of minerals such as Fe (3.60 mg/100 g), Mn (4.07 mg/100 g) and Se (61.8 μg/100 g) (USDA 2015). Important phytochemicals, present in higher amounts in whole wheat flour, are ferulic acid,[30] catechins and the carotenoid β-cryptoxanthin (Adom et al. 2005).

Wheat bran has been observed to provide protection against colorectal cancer and to improve gut's health, an effect that is attributed to lignans and their metabolites: enterolactone and enterodiol (Qu et al. 2005). These authors observed the inhibition of cancer cell growth with lignan's metabolites, mediated by cytostatic and apoptotic mechanisms.

The content of wheat proteins should not be ignored in the diet, since it includes a wide variety of amino acids, thus complementing other vegetable source of proteins, in the diet. Gluten designation corresponds to a group of proteins accounting for about 80 % of the total grain protein in European wheats (Shewry 2009) and is the cause of a medical condition known as coeliac disease, afflicting about 1 % of the world's population. Coeliac disease is an autoimmune condition triggered by an environmental precipitant that affects genetically predisposed individuals worldwide, although the highest prevalence is registered in North Africa (Mahadov and Green 2011).

On the other hand, mixed wheat and rye bread, which is fairly common in the area (at least in Portugal), provides a little less energy than white wheat bread (270 kcal/100 g). It contains less fat (1.4 %) but higher levels of dietary fibre (4.3 g/100 g) as well as a larger variety and higher levels of vitamins such as α-tocopherol and vitamin B_6 (INSA 2015).

Cornbread is also commonly consumed in the Mediterranean area and may include olive oil in its composition. Simple corn bread has a high average moisture content (51 %), providing only 185 kcal/100 g and containing 37.2 % carbohydrates, of which 3.7 g/100 g are dietary fibre and the remaining portion corresponds to starch. No mono- or disaccharides are commonly found. Protein and fat contents are, respectively, 5.3 and 1.2 % (INSA 2015).

[30] IUPAC name: (*E*)-3-(4-hydroxy-3-methoxyphenyl)prop-2-enoic acid.

***Pasta and couscous**. Wheat (Triticum aestivum, Triticum durum) is a staple food of the Mediterranean region, mainly consumed as bread, pasta or couscous, and constitutes a main energy source. Gluten's protein fraction confers the viscoelastic properties that allow dough to be processed into bread, pasta, noodles and other food products. This protein fraction also contributes with some essential amino acids (those the human body cannot synthesize) thus complementing other vegetable protein sources in the diet. Bread and pasta may be important sources of dietary fibre, particularly if less refined flours are used. Photo reprinted with kind permission from T. N. Wassermann*

Pasta and couscous are two well-known wheat-based foods, respectively, from Italy and Morocco. Both are made from semolina, a course middling of *T. durum*. Pizza was traditionally a healthy food made of wheat dough, olive oil, tomato, fresh mozzarella and herbs, such as oregano and basil. Conversely, its fast food counterpart contains free sugars, less dietary fibre and higher levels of fats, of which the major portion (5 g/100 g) is SFA, besides a non-negligible amount of cholesterol (24 mg/100 g) (USDA 2015).

5.5.2 Rice (*Oryza sativa*)

Rice is consumed in larger amounts in Portugal, followed by Spain and Italy, where it is often the base of complete meals cooked with vegetables, fishes, meats and/or pulses. Olive oil, onion, garlic, tomato and herbs are frequent ingredients, as is wine. Examples of these rice-based meals are 'arroz de polvo' (made from tomato cooked with olive oil, onion and garlic, octopus and fresh coriander) and 'risotto' (cooked with a meat, fish or vegetable-based broth), both types of dishes generally resulting in a creamy consistency. Rice varieties with thick and short grains are preferred in the Mediterranean region (e.g. risotto and the Portuguese-type "arroz carolino") to Asian rice types, with long and thin grains. The Portuguese 'arroz carolino' supplies large amounts of energy (363 kcal/100 g, average value), mainly from starch (79.6 %), also providing proteins (7.4 %), dietary fibre (2.2 %) and relevant amounts of α-tocopherol (0.1 mg/100 g), niacin equivalents (3.6 mg/100 g), vitamin B_6 (0.3 mg/100 g) and folates (20 μg/100 g) (INSA 2015). Different rice varieties are expected to have different compositions. Rice may include relevant amounts of methionine and lysine, thus complementing their absence from pulses. In fact, rice and beans is a nutritionally balanced native combination.

5.5.3 Potato (*Solanum tuberosum*)

Potato, the tubercle of *Solanum tuberosum*, was introduced to Europe some centuries ago by Spanish and Portuguese navigators who brought it from South America, soon becaming an essential crop and a staple food. Although not rich in vitamin C, potato intake by Portuguese and Spanish sailors from the sixteenth century was sufficient to avoid scurvy casualties among the crew. As other complex sources of starch, potatoes supply a moderate amount of energy: 70 kcal (USDA 2015) or 89 kcal (INSA 2015), mainly from starch (18 g/100 g). Potatoes also contain 1.2 g/100 g mono- and disaccharides (sucrose, glucose and fructose), 1.6 g/100g dietary fibre and 2.5 g/100 g protein. All essential amino acids can be found, although in low and variable quantities (USDA 2015); relevant vitamins are α-tocopherol (0.060 mg/100 g), thiamine (0.21 mg/100 g), vitamin B_6 (0.44 mg/100 g), vitamin C (14 mg/100 g), niacin equivalents (2.0 mg/100 g), riboflavin (0.02 mg/100 g) and folates (35 μg/100 g) (INSA 2015), as well as lutein + zeoxanthin (21 μg/100 g) and phylloquinone (2.9 μg/100 g). Relevant phytochemicals are quercetin (0.6 mg/100 g) (USDA 2015) and chlorogenic acid,[31] a hydroxycinnamic acid of a family of naturally occurring esters of caffeic acid. Chlorogenic acid (also present in coffee and other vegetable foods) has been referred to as a carcinogenic inhibitor. It has also been shown to prevent paraquat-induced oxidative stress in rats (Yang et al. 2012).

Potatoes contain a toxic glycoalkaloid, solanine, existing as α-, β- and γ-forms, the concentration of which is higher in the skin (8 mg/100 g) and much lower in the interior of the tuberculum (FDA 2014a). Still according to the US Food and Drug Administration (FDA), the toxic dose (to an average adult) is 20–25 mg, which is rarely reached. Increased amounts of solanine in potatoes have been associated with unusual features, such as a pink colour developing on the cut surface, or a brownish line near the surface, or when sprouting. Exposure of potatoes to light favours the production of solanine, in opposition to ordinary storage in the dark, at room temperature and low moisture. (FDA 2014a).

[31] IUPAC name: (1*S*,3*R*,4*R*,5*R*)-3-{[(2*E*)-3-(3,4-dihydroxyphenyl)prop-2-enoyl]oxy}-1,4,5-trihydroxycyclohexane-1-carboxylic acid.

Potato. *Potato (Solanum tuberosum) is consumed as an energy source (from starch), although it supplies a moderate amount of energy, except when fried. Besides resistant starch, which is important for gut's health, potatoes also contain vitamins and chlorogenic acid, a potent free radical scavenger. Photo reprinted with kind permission from T. N. Wassermann*

Traditionally in the Mediterranean countries, potatoes are peeled and very often boiled in water, strongly decreasing the already small risk of food poisoning. Solanine is soluble in water and diffused by boiling but not eliminated when potatoes are baked in their skin (FDA 2014a).

In addition to the high levels of solanine, raw potato is barely digested due to the predominance of resistant starch.

As with other foods, e.g. tomato, the cooking method can significantly affect nutrient availability by changing the food matrix, promoting chemical reactions and causing alterations in many molecular structures. Water boiling and mashing potatoes increases their content of digestible starch when compared with other cooking methods (García-Alonso and Goñi 2000). Although many nutrients may be destroyed or altered by cooking temperatures, cooking can be beneficial, in this case not only affecting solanine elimination but also improving starch bioavailability.

French fries—the fast food version of potato—are much richer in energy (323 kcal/100 g), as lipids are absorbed during frying (15.47 g/100 g). The composition of the lipid fraction will depend on the frying oil, and it is expected to contain mainly PUFA, but a relatively high proportion of saturated fat may be found (2.27 g/100 g) and total trans-fatty acids reach 0.064 g/100 g of fried potatoes (USDA 2015). Acrylamide (prop-2-enamide) is formed mainly from the reaction between free asparagine and reducing sugars during high-temperature cooking, principally through Maillard reactions. The main source of human dietary exposure are fried potatos (Pedreschi et al. 2014). According to the European Food Safety Authority (EFSA), the risk of accumulation of acrylamide in fried potatoes is a public health concern, and children are the most exposed age group, on a body weight basis (EFSA 2015). Acrylamide and the resulting metabolites in the body are believed to cause mutagenesis and cancer, as well as reproductive and neurotoxic damage, based on animal studies (EFSA 2015; Pedreschi et al. 2014).

5.6 Pulses

A pulse (from the Latin 'puls'), sometimes called a 'grain legume', is an annual leguminous crop yielding from 1 to 12 seeds of variable size, shape and colour within a pod, and used for its grain. Therefore, peanuts and soybeans are not included in this group because both are mainly used as sources of vegetable oils. This food group includes many types of beans, peas, lentils and others but not their fresh pods, which are included in the 'greens' category. Like many leguminous crops, pulses play a key role in crop rotation due to their ability to fix soil nitrogen. To support awareness of this, the United Nations (UN) declared 2016 the UN International Year of Pulses (UN 2013).

Pulses have a unique nutritional profile consistent with several dietary composition factors thought to assist with weight control (McCrory et al. 2010). Pulses are relatively low in energy density and a good source of digestible protein (McCrory et al. 2010) and dietary fibres and are moderate to poor in fats, which are mainly MUFA and PUFA. Pulse carbohydrates are slowly digested (see below), which attributes them some of the lowest glycaemic index scores of carbohydrate-containing foods. On the other hand, pulse proteins fall primarily into the albumin (water-soluble) and globulin (salt-soluble) classes. Storage proteins legumin and vicilin are globulins; albumins comprise the heterogeneous group of enzymes, amylase inhibitors and lectins. Lectins may, in certain cases, cause allergic reactions and food poisoning symptoms in susceptible individuals.

Proteins can be more satiating than carbohydrates or lipids, and pulses have been implicated in providing satiety while keeping weight under control (Pai et al. 2005). Pulses are a good source of several vitamins and provide valuable phytochemicals. They contain several anti-nutrients (such as enzyme inhibitors) that have been suggested to play a role in energy regulation. Phytate, or inositol hexaphosphate, is the major storage form of phosphate in plant cells, and pulses are one of its major sources in the diet. Phytic acid may also contribute to satiety and delay the return of hunger (McCrory et al. 2010). It is noteworthy that phytic acid is an anti-nutrient with the potential to bind mineral micronutrients in food and reduce their bioavailability. Phytic acid content depends on post-harvest processing and boiling of pulses (Thavarajah et al. 2009). Finally, yet importantly, it is noteworthy that in some cases genetic features of consumers (e.g. presence of detoxifying enzymes) are important for safe and easy consumption of the pulse (as may also occur with some other plants).

5.6.1 Common Bean (*Phaseolus vulgaris*)

The common bean is an herbaceous annual plant grown worldwide for its edible fruit; both the dry seed and the unripe fruit are referred to as 'beans'. They are classified botanically into the legume family (*Fabaceae*), most of the members of which acquire nitrogen through an association with *Rhizobium* sp., a nitrogen-fixing bacterium, and its rotation in agriculture improves the sustainability of the systems. Although many varieties of different sizes and colours are available, all share a common nutritional profile, the average contents of which are detailed below. Dry beans have good preservation characteristics due to a low moisture content and strong skin. The compositional data presented below correspond to this category.

Beans are particularly relevant for their high dietary fibre (22.9 %) and oligosaccharide content (3.8 %) (INSA 2015). Dietary fibre has an insoluble fraction, which is associated with faecal bulking through its water-holding capacity, whereas soluble fibre ferments, a feature that is sometimes considered unpleasant (McCrory et al. 2010). On the other hand, and according to the same authors, oligosaccharides present in beans and other pulses are galactosides, derived from sucrose with galactosyl units attached. They are commonly known as raffinose (1 galactosyl unit),[32] stachyose (2 galactosyl units)[33] and verbascose (3 galactosyl units).[34] Oligosaccharides are regarded as a negative attribute of beans (and of other pulses), due to their high fermentability and associated gas production and discomfort. However, oligosaccharides act as prebiotics in promoting a healthy gut microbiota (McCrory et al. 2010).

Common beans contain 21–24 % protein, 1.4 % total fats, 37.5 % starch and 2.6 % mono- and disaccharides; providing 277 kcal/100 g (INSA 2015). Beans also provide a wide range of minerals, including relevant amounts of Ca (143 mg/100 g), Mn (1.02 mg/100 g) and Se (3.2 μg/100 g). The primary vitamins are ascorbic acid (4.5 mg/100 g), α-tocopherol (0.22 mg/100 g) and phylloquinone (19.0 μg/100 g). Beans contain about 127 mg/100 g total phytosterols (USDA 2015); tannins, phenolic acids and flavonoids have also been reported as beneficial compounds of beans (McCrory et al. 2010).

Cowpea or black-eyed pea (*Vigna unguiculata*), also from the *Fabaceae* family, are small beans widely consumed in the Mediterranean region. Compared with common beans, black-eyed peas provide more energy (329 kcal/100 g) and less fibre (9.4 g/100 g) (INSA 2015). Proanthocyanidin oligomers are found in cowpeas and peas (see below) but not in beans (USDA 2015).

---

[32] IUPAC name: (3*R*,4*S*,5*R*,6*R*)-2-[[(2*R*,3*S*,4*S*,5*R*)-6-[(2*S*,3*S*,4*S*,5*R*)-3,4-dihydroxy-2,5-bis(hydroxymethyl)oxolan-2-yl]oxy-3,4,5-trihydroxyoxan-2-yl]methoxy]-6-(hydroxymethyl)oxane-3,4,5-triol.

[33] IUPAC name: (2*S*,3*R*,4*S*,5*R*,6*R*)-2-[[(2*R*,3*R*,4*S*,5*R*,6*S*)-6-[[(2*R*,3*S*,4*S*,5*R*,6*R*)-6-[(2*S*,3*S*,4*S*,5*R*)-3,4-dihydroxy-2,5-bis(hydroxymethyl)oxolan-2-yl]oxy-3,4,5-trihydroxyoxan-2-yl]methoxy]-3,4,5-trihydroxyoxan-2-yl]methoxy]-6-(hydroxymethyl)oxane-3,4,5-triol.

[34] IUPAC name: (2*S*,3*R*,4*S*,5*R*,6*R*)-2-[[(2*R*,3*R*,4*S*,5*R*,6*S*)-6-[[(2*R*,3*R*,4*S*,5*R*,6*S*)-6-[[(2*R*,3*S*,4*S*,5*R*,6*R*)-6-[(2*S*,3*S*,4*S*,5*R*)-3,4-dihydroxy-2,5-bis(hydroxymethyl)oxolan-2-yl]oxy-3,4,5-trihydroxyoxan-2-yl]methoxy]-3,4,5-trihydroxyoxan-2-yl]methoxy]-3,4,5-trihydroxyoxan-2-yl]methoxy]-6-(hydroxymethyl)oxane-3,4,5-triol.

Dry beans require soaking for several hours before cooking, and should be cooked at boiling temperature for over 30 min. There are reasons to ensure that beans are well cooked before being eaten. The main reason is to destroy the toxic lectin phytohaemagglutinin that may cause nausea, vomiting and abdominal pain (Al-Khaldi 2012). Second, protease and amylase inhibitors can negatively affect the digestibility of dietary proteins. These issues are not as important when using pre-cooked canned beans, demonstrating a case where food processing can be advantageous.

5.6.2 Broad Bean (*Vicia faba*)

Broad beans are among the most ancient plants in cultivation and are also among the easiest to grow. Along with lentils, peas and chickpeas, they are believed to have become part of the diet in the area around 6000 BC or earlier. Broad beans, also known as favas or fabas, are eaten while still young and tender. Preparing favas involves first removing the beans from their pods and then peeling the beans of their exterior hard coating before cooking. Broad beans are mostly consumed in Dalmatia (Croatia) as a part of a traditional dish where artichokes are stuffed with fava beans and peas. In Greece, broad beans are eaten in a stew combined with artichokes, while they are still fresh in their pods.

The compositional data presented below correspond to fresh raw fava beans, the edible part of which represents only 40 % of the seed. Fresh fava supplies 69 kcal/100 g from: 7.4 % proteins, 0.5 % total fat (mainly PUFA, particularly linoleic acid) and 8.5 % available carbohydrates, of which 6.4 g/100 g corresponds to starch and 1.6 g/100 g corresponds to mono- and disaccharides. In the dry form, protein percentage increases to the range of 25–26 % as well as the caloric content and other nutrients (in proportion to the decrease in moisture). Like other pulses, fava contains high amounts of oligosaccharides and dietary fibre, exhibiting the health benefits and inconveniences as described above for beans. Fava contains 64 mg/100 g of carotenes, of which 11 μg/100 g have pro-vitamin A activity. It is also a source of water-soluble vitamins. α-tocopherol and phylloquinone are present in trace amounts (INSA 2015; USDA 2015).

Fava beans contain a water-soluble toxic glucoside, vicine[35] (USDA 2015). The genetic basis for the selective toxicity of fava beans ('favism') and the relationship to the anti-malarial drug primaquine was first reported in 1956 (FDA 2014b). Resistance to malaria in some subjects was attributed to a deficiency in the metabolic enzyme glucose-6-phosphate dehydrogenase (G6PD). G6PD deficiency is prevalent throughout tropical and subtropical regions of the world. Vicine and convicine can induce haemolytic anaemia in patients with that hereditary condition, thus revealing protection against malaria. The vicine and convicine content is higher in fresh green cotyledons (moisture content about 80 %) and gradually declines to a constant level (Burbano et al. 1995).

[35] IUPAC name: 2-[(2,4-diamino-6-hydroxypyrimidin-5-yl)oxy]-6-(hydroxymethyl)oxane-3,4,5-triol.

Broad beans are rich in L-dopa,[36] a precursor of dopamine, which is a naturally-occurring psychoactive drug. L-dopa is also found in certain herbs and *Mucuna pruriens*, or velvet bean, and it is used medically in the treatment of Parkinson's disease. L-dopa is also a natural diuretic agent, which might help in controlling hypertension (Apaydin et al. 2000; Ramya and Thaakur 2007).

5.6.3 Lentil (*Lens culinaris*)

The lentil is a bushy annual plant of the legume family grown for its lens-shaped seeds. Lentils vary in size and colour, ranging from yellow and red-orange to green, brown and black. Lentils have been part of the human diet since Neolithic times, being one of the first domesticated crops in the Near East.

The seeds require a cooking time of 10–40 min depending on the variety—shorter for small varieties with the husk removed, such as the common red lentil—and have a distinctive, earthy flavour. It supplies 303 kcal/100 g: 25.2 % proteins, 0.7 % lipids and 52.2 % digestible carbohydrates (of which 43 g/100 g starch). Common red lentils contain 11.8 % dietary fibre and 3.1 % oligosaccharides, with the above-described beneficial effects for the gastrointestinal tract and gut microbiota.

Lentils also contain relevant amounts of water-soluble vitamins such as folates (110 μg/100 g), thiamine (0.43 mg/100 g) and other B vitamins, phylloquinone (5 μg/100 g) and γ-tocopherol (4.23 mg/100 g), as well as minerals, such as Mg (114 mg/100 g) and P (360 mg/100 g). Green lentils contain a higher concentration of fibre than red lentils (about 31 % instead of 11 %) (INSA 2015; USDA 2015). Like other pulses, lentils contain a specific lectin, *Lens culinaris* agglutinin. This lectin has been noted as having a similar structure to that of concanavalin A (Con A), a mannose/glucose-binding lectin found in several pulses (Schwarz et al. 1993). While Con A monomers consist of a continuous polypeptide chain of 237 amino acid residues, the monomers of pea and lentil lectin are each composed of two different subunits of 6 and 18 kDa (Schwarz et al. 1993). Its specific binding properties have been exploited as a diagnostic tool for certain tumours (Romeo et al. 2010; Tateno et al. 2009).

Lentils also contain tannins and flavonols, such as catechin[37] and gallocatechin,[38] with relevant ROS activity and extensively discussed health beneficial effects. The isoflavone genistein is included in this group, showing anti-carcinogenic and anti-inflammatory activities by mimicking and modulating estrogens and inhibiting protein kinases (Xu and Chang 2010; Zou et al. 2011).

[36] IUPAC name: dihydroxiphenylalanine ((2*S*)-2-amino-3-(3,4-dihydroxyphenyl)propanoic acid.

[37] IUPAC name: (2*S*,3*R*)-2-(3,4-dihydroxyphenyl)-3,4-dihydro-2H-chromene-3,5,7-triol.

[38] IUPAC name: (2*R*,3*S*)-2-(3,4,5-trihydroxyphenyl)-3,4-dihydro-2H-chromene-3,5,7-triol.

Lentils also contain proanthocyanidin[39] monomers (0.5 mg/100 g), dimers (1.2 mg/100 g) and trimers (0.1 mg/100 g), which are colourless oligomeric compounds also found in wines, teas, apples and cinnamon (USDA 2015). These flavonoids have been proved to have cardioprotective effects (McCullough et al. 2012), namely against reperfusion-induced injury via their ability to reduce or remove, directly or indirectly, free radicals in myocardium that is reperfused after ischemia (Pataki et al. 2002). Flavonoid intake is also inversely related to the incidence of type II diabetes (Zamora-Ros et al. 2013, 2014) as well as prostate cancer (Wang et al. 2014b), breast cancer (Hui et al. 2013) and stomach and colorectal cancer (Woo and Kim 2013).

Lentils also contain phytic acid, which may interfere with P and Mg bioavailability (Porres et al. 2004), although microbial fermentation of phytates and accumulated end products introduces new variables that may lead to different hypotheses. Regardless, like beans and other pulses, the amount of phytic acid decreases with boiling.

5.6.4 Pea (*Pisum sativum*)

The pea is an herbaceous annual plant in the *Fabaceae* family, originally from the Mediterranean basin and Near East. The pea is most commonly a green, occasionally a purple or golden yellow, pod-shaped vegetable, widely grown as a cool season vegetable crop. There are many varieties of peas, but just a few are commercially processed and suitable for freezing, therefore linking biodiversity to market laws, as happens with other pulses.

Peas are high in fibre, protein, vitamins (folate and vitamin C), minerals (Fe, Mg, P and Zn) and lutein (a yellow carotenoid pigment that benefits vision). Dry weight is about one-quarter protein and one-quarter carbohydrates, mostly sugars (Shereen 2007).

[39] IUPAC name: (3*R*)-2-(3,5-dihydroxy-4-methoxyphenyl)-8-[(2*R*,3*R*,4*R*)-3,5,7-trihydroxy-2-(4-hydroxyphenyl)-3,4-dihydro-2H-chromen-4-yl]-3,4-dihydro-2H-chromene-3,5,7-triol.

Green pea. *Pea (Pisum sativum) and other pulses has a unique nutritional profile of protein, carbohydrates (including a high dietary fibre content), vitamins, minerals and flavonoids. The digestibility of pulses is greatly improved with adequate cooking methods. The intake of pulses may help replace meat and simultaneously contribute to compliance with fibre intake recommendations. Photo reprinted with kind permission from T. N. Wassermann*

Green fresh peas are starchy (5.3 %), high in fibre (4.7 %), oligosaccharides (1.5 %), proteins (6.4 %) and water-soluble vitamins such as folates (33 μg/100 g), B vitamins, the carotenoids α and β-carotene (262 mg/100 g) and lutein + zeaxanthin (2.48 mg/100 g), whereas ascorbic acid is absent from PortFIR compositional data but 40 mg/100 g are reported in the USDA database (INSA 2015; USDA 2015). When in dry form, the protein content increases to about 24 % (the remaining nutrients increase in proportion to the decrease in moisture). Peas contain relevant amounts of α and γ-tocopherols (0.13 and 0.95 mg/100 g, respectively) and phylloquinone (24.8 μg/100 g). Some flavonoids have been reported in peas, such as catechin and epicatechin (about 0.01 mg/100 g each) and quercetin (14.7 mg/100 g) (USDA 2015).

Like other pulses, peas have a lectin, which is believed to be responsible for intolerance to peas. In detail, the *Pisum sativum* lectin is glucose/mannose specific; like lectins from other pulses, it has the ability to preferentially recognise tumour cells and probably intervene in adjuvant therapies (Kavitha and Swamy 2006).

5.6.5 Chickpea (*Cicer arietinum*)

The chickpea, also known as garbanzo bean, is a legume of the family *Fabaceae* and one of the world's three most consumed pulses (the other two are *Phaseolus vulgaris* and *Pisum sativum*). *Cicer arietinum* is one of the earliest cultivated legumes believed to have been domesticated in Neolithic times in the Mediterranean basin and the Middle East area (Abbo et al. 2009, 2011; Vaughan and Geissler 1997). It is popular in the Iberian Peninsula, where chickpeas are eaten in soups, salads, tapas, accompanying 'bacalhau/bacalao' and as part of traditional dishes such as 'rancho' and 'cocido madrileño', respectively, in Portugal and Spain. Like other pulses, dried chickpeas have excellent preservation characteristics, resulting in a long shelf-life but also requiring a long cooking time (1–2 h). When soaked for 12–24 h before use, the cooking time can be shortened by around 30 min. Pre-boiled canned chickpeas are available in the local markets.

When compared with other pulses, chickpeas (herein referring to the dry seed) are lower in protein (19.3–20.5 %) but higher in fat (5–6 %) mostly MUFA 18:1 (1–1.4 %) and PUFA 18:2, including linolenic acid (2.5–2.7 %), and digestible carbohydrates (51.4–62.3 %, of which 45.2 g/100 g is starch) all supplying 332 kcal/100 g. Chickpeas are an important source of dietary fibre (12.2–13.5 %) and oligosaccharides (3.4 %) (INSA 2015), and an excellent source of essential minerals, such as Ca (57–105 mg/100 g), Mn (2.2–21.3 mg/100 g) and Se (8.2 μg/100 g) (INSA 2015; USDA 2015). They contain relevant amounts of several vitamins, providing 3–10 μg/100 g RAE, 40–60 mg/100 g β-carotene and the B vitamins thiamin (0.41–0.48 mg/100 g), riboflavin (0.15–0.21 mg/100 g), niacin (1.5–1.9 mg/100 g), pantothenic acid (1.6 mg/100 g) and vitamin B_6 (0.5 mg/100 g). They also supply 180–557 μg/100 g folates, 3.0–4.0 mg/100 g vitamin C, 0.82–2.7 mg/100 g α-tocopherol and 9.0 μg/100 g phylloquinone (INSA 2015; USDA 2015). Chickpeas contain the isoflavones daidzein[40] (0.2 mg/100 g) and genistein (unusually high at 0.1 mg/100 mg), glycitein[41] (0.2 mg/100 g), biochanin A[42] (1.5 mg/100 g) and formononetin[43] (0.1 mg/100 g) (USDA 2015).

As mentioned above, these isoflavones are reported to have estrogenic and anticancer activities. Genistein is thought to inhibit protein-tyrosine kinase and topoisomerase-II and is therefore used as an anti-neoplastic and anti-tumour agent (Pittaway et al. 2008).

Moreover, preliminary research has shown that chickpea consumption may lower blood cholesterol through small improvements in serum lipid profile and glycaemic control (Pittaway et al. 2008).

In short, pulses can be easily preserved as dry seeds and are a valuable and sustainable source of protein, as it constitutes about 20–25 % of the edible portion.

[40] IUPAC name: 7-hydroxy-3-(4-hydroxyphenyl)chromen-4-one.

[41] IUPAC name: 7-hydroxy-3-(4-hydroxyphenyl)-6-methoxychromen-4-one.

[42] IUPAC name: 5,7-dihydroxy-3-(4-methoxyphenyl)chromen-4-one.

[43] IUPAC name: 7-hydroxy-3-(4-methoxyphenyl)chromen-4-one.

Some essential amino acids are missing or are present in negligible amounts, such as tryptophan, methionine and cysteine, and also—in some cases—threonine, tyrosine and histidine. To correctly address nutritional needs, a diet supplementation with protein from animal sources and/or cereals and nuts is required. It has been argued that phytic acid may act as an anti-nutrient by sequestering minerals (such as Ca, Fe, Mg and Zn ions) and decreasing their bioavailability. However, much is still to be disclosed about mineral absorption during the microbial fermentation processes in the gut, impacting the significance of overall mineral status and of physiological endpoints, such as bone health.

Allergens (such as lectins) and anti-nutrients (such as protease inhibitors that may affect protein bioavailability) are inactivated by heat. Thus, canned pulses can be more nutritious, safer and convenient to use.

5.7 Fresh Fruits

The MD is traditionally rich in seasonal fresh fruits as well as in nuts and dried fruits. While there is ample evidence to indicate the health benefits of diets rich in fruits and nuts, acknowledgment by regulatory authorities of specific health benefits is a slow but rigorous process relying on proven cause–effect relationships. Consequently, only a few food items may exhibit qualified or specific health claims. Health effects have been attributed to specific nutrients and phytochemicals and to the interaction between them as a consequence of food processing, cooking methods and also of meal composition and intake.

The fact that fruits are healthy because of their vitamin content is well-known. Increasing rates of obesity, cancer and coronary and heart disease and their proven relationship with diet has drawn public awareness on the benefits of minor components of fruits, such as phenolics and dietary fibres.

Although many tropical fruits are undoubtedly nutritious and beneficial to health, the current review is focused on traditional fruits of the Mediterranean basin. It is noteworthy that bananas and pineapples are cultivated in Portuguese and Spanish Atlantic islands, where specific cultivars have existed for centuries. Although banana and pineapple have been consumed in the area for decades, these fruits are generally imported commodities and as such are not discussed herein. Thus, most representative fruits of the Mediterranean region are considered to be grapes, oranges, apples, pears, peaches, cherries, plums, figs, melons, watermelon and dates. Wild fruits are also discussed given their contribution to biodiversity and to the scientific knowledge on bioactivities.

5.7.1 Grape (*Vitis vinifera*)

Grapes are the fructifications of some cultivars of *Vitis vinifera*, a plant species native to the Mediterranean region that expands north to France and south towards Germany and East until northern Iran (Nazimuddin and Qaiser 1982). It is cultivated on every continent except for Antarctica, and good-quality wines are now also produced in Chile, California (USA), South Africa and Australia.

The fruit of *Vitis vinifera* is a berry, known as a grape, varying according to the cultivar, climate, location and agronomic techniques. Two major types of vineyards exist: those devoted to wine production, the fruits are smaller and sweeter; and vineyards from which the fruits are intended to be consumed in fresh (table grapes). A third type is for the production of raisins; each type relies on different fruit cultivars and agronomic techniques. Nearly three-quarters of the world's commercial grape production is devoted to wine grapes, around a quarter to table grapes and much smaller amounts to dried grapes (raisins) and non-alcoholic grape juice (EOL 2015b).

Grapes. *Grapes, the fructification from Vitis vinifera, are consumed fresh (table grapes) or dry (raisins) and are a good source of ascorbic acid, carotenoids and other bioactive compounds such as tocopherols and flavonoids, to which anti-carcinogenic activity has been attributed. Photo reprinted with kind permission from T. N. Wassermann*

Although vineyards have been introduced worldwide, quality wines result from fruits of just a few varieties that are cultivated in selected regions using specific agronomic techniques. These fruits have features that are distinct from table grapes. They have thicker skin and larger seeds, providing colour and organoleptic characteristics to the wine, including the balanced astringency from tannins (as proanthocyanidins) and are the major source for resveratrol, a defensin that is produced as a response to injury, such as mechanical trauma, ultraviolet light and infection (Castellarin et al. 2011; Lekli et al. 2010).

5.7.1.1 Table Grapes

Table grapes are large fruits with thin skin and small seeds (or can be seedless) and a watery pulp. The composition of white and red/pink table grapes differs, although it is not possible to include herein a full description of each type, due to the limitations of food composition databases. The USDA database is more extensive; however, table grape varieties available in the USA differ considerably from those available in Europe.

The accessed record in the USDA database was 'Report: 09132, Grapes, red or green (European type, such as Thompson seedless), raw' and was compared with PortFIR data resulting from the analysis of a mixture of several raw grape types, available at Portuguese markets and mainly locally produced (INSA 2015).

Thus, according to the available 2015 edition of food composition's INSA database, 100 g of the average table 'white' (green) grapes supply 72 kcal, mainly from mono- and disaccharides (sucrose, glucose and fructose, 17.3 %), some protein (0.3 %), fats (0.5 %) and dietary fibre (0.8 %). 'White' grapes, on average, also contain low levels of organic acids and relevant amounts of some water-soluble vitamins: ascorbic acid (1 g/100 g), α-tocopherol (0.4 mg/100 g), niacin equivalents (0.3 mg/100 g) as well as folates (2 μg/100 g), vitamin B_6 (0.09 mg/100 g), thiamin (0.02 mg/100 g) and riboflavin (0.02 mg/100 g). According to the same database, red table grapes provide approximately the same amount of energy (77 kcal/100 g) from 0.3 % of, proteins, 0.5 % of fats, di- and mono-saccharides (18.6 %), and dietary fibre (0.9 %), and a noteworthy fraction of organic acids (0.41 %). In addition to identical concentrations of ascorbic acid (1 g/100 g) and α-tocopherol (0.3 mg/100 g), some carotenes (60 mg/100 g, corresponding to 15 μg/100 g RAE) can also be found. Both types of grapes are important sources of minerals and phytochemicals.

Average values for red, pink and green grapes available for US consumers (USDA 2015) differ somewhat from the average sample of table grapes available for Portuguese consumers. The amount of energy provided is smaller (69 kcal/100 g) than that of Portuguese grapes, resulting from sugars (15.5 %; mainly 7.2 % glucose and 8.13 % fructose) and dietary fibre (0.9 %). Apparently, the vitamin concentration is also lower: ascorbic acid (3.2 mg/100 g), RAE (3 μg/100 g) and α-tocopherol (0.19 mg/100 g). The presence of phylloquinone is registered although it is not confirmed in the INSA's database, most probably because this compound is not analysed by the Portuguese Institution. On the other hand, carotenoids are included in the USDA database, such as α- and β-carotenes, lutein and zeaxanthin.

5.7.1.2 Raisins

Raisins are traditionally obtained from drying grapes under the sun, although mechanical dryers have become more popular, resulting in a more rapid process and levelling product quality. Some raisins have large seeds and thick skin and are obtained from excess production of wine's grape, while special seedless cultivars (sultana and corinthian) are grown mainly to supply food industries (breakfast cereals, breads, pastries and snacks). The latter type (mainly corinthian raisins) are dried from about 80 to 15 % moisture, thus substantially reducing water activity

(a_w) and contributing to its preservation. Consequently, the sugar concentrations (mainly glucose and fructose) increases to 59.19 %, also including 3.7 % dietary fibre and providing 299 kcal/100 g (USDA 2015).

The concentration of proteins is more relevant than in fresh raisins (3.07 %), as are minerals, such as Ca (50 mg/100 g), P (101 mg/100 g), Mn (0.3 mg/100 g), Se (0.6 μg/100 g) and F (233.9 μg/100 g). Most vitamins are preserved and concentrated, such as α-tocopherol (0.12 mg/100 g), γ-tocopherol (0.04 mg/100 g) and phylloquinone (3.5 μg/100 g). Flavonoid concentration is higher than in fresh grapes, with relevant amounts of catechin (0.4 mg/100 g) and epicatechin (0.1 mg/100 g) as well as quercetin (0.2 mg/100 g) (USDA 2015).

Catechin[44] and epicatechin (its *cis*-isomer) are antioxidant flavonoids with 37 bioactivity outcomes (NCBI 2015d) mostly for anti-carcinogenic activity (Forester and Lambert 2014; Martinotti et al. 2014).

5.7.2 Citrus (*Citrus* spp.)

The designation 'Citrus' includes a large variety of similar fruits, with a long history of cultivation. Citrus species hybridise readily and most recognised species are probably of hybrid origin, and only known from cultivation, namely *Citrus limon, Citrus reticulata, C. clementina, C. sinensis* and *C. maxima*. Lemon and orange trees are among the most cultivated fruit trees in the world, and they can grow in a wide range of climates. Orange is a non-climacteric fruit, meaning that it needs to be harvested at its full maturity and immediately consumed or processed, mostly to orange juice. Numerous seedless hybrids have been introduced worldwide. Plants of this genus are known for their fragrance and their high citric acid content, which gives them a tangy flavour. In addition to fruit consumption, all species are used in folk medicine (Morton 1987).

Generally, a single orange has ten segments inside the pericarp, and only about 70 % is edible (INSA 2015). Mandarins, clementines and tangerines (*C. reticulata* and others) are small oranges that come from a smaller tree and are generally sweeter and less acidic than larger oranges. Tangerines supply 40 kcal/100 g, mainly from mono- and disaccharides (8.7 %) and organic acids (0.82 %). They contain 1.7 % dietary fibre. In addition, they are rich in carotenes (200 mg/100 g or 33 μg RAE/100 g), α-tocopherol (0.24 mg/100 g) and ascorbic acid (32 mg/100 g) (INSA 2015; USDA 2015).

The USDA database also includes values for other compounds of interest, such as the carotenoids β-cryptoxanthin (407 μg/100 g) and lutein & zeaxanthin (138 μg/100 g) and the flavonoids hesperetin (7.9 mg/100 g) and naringenin (10.0 mg/100 g), the properties of which are discussed below.

When compared with tangerines (on the same basis of 100 g of edible fruit), the common orange (*C. sinensis*) contains more organic acids (0.68 g) and higher

[44] IUPAC name: (2*R*,3*S*)-2-(3,4-dihydroxyphenyl)-3,4-dihydro-2H-chromene-3,5,7-triol.

quantities of vitamin C and hesperetin (57 and 21.9 mg, respectively) (INSA 2015; USDA 2015), while it is poorer in carotenes and β-cryptoxanthin (120 mg and 116 μg/100 g, respectively).

A large fraction of cultivated oranges is commercially used for the production of orange juice and derivatives. Pasteurised raw orange juice contains fewer vitamins, particularly ascorbic acid (40 mg/100 g), carotenes (42 mg/100 g) and α-tocopherol (0.1 mg/100 g), as well as reduced concentrations of dietary fibre (0.4 %) (INSA 2015; USDA 2015). Moreover, most commercially available orange-based drinks often contain 50 % or less of orange juice and generally high levels of added sugars or sweeteners and food preservatives, resulting in a nutritionally low-value production. Pomelo or pummelo (*C. maxima*) is a good source of ascorbic acid (61 mg/100 g) and naringenin (24.7 mg/100 g), while no carotenes are registered. Pummelo contains much lower amounts of β-cryptoxanthin (10 μg/100 g) and hesperetin (8.4 mg/100 g) than other citrus fruits (USDA 2015).

Citrus. *Lemon (Citrus limon) is appreciated for its juice and peel. The peel is scraped off for use as a seasoning or stripped to make infusions. The flavour is mainly due to aromatic oils rich in terpenes such as limonene, which has been considered as anti-carcinogenic. Other citric fruits, such as oranges, pomelo and mandarins are major sources of vitamin C. Photo reprinted with kind permission from T. N. Wassermann*

Generally, citrus fruits contain the flavonoids hesperetin[45] and naringenin and their glycosides. Both flavonoids are considered to have bioactive effects on human health as ROS scavengers, anti-inflammatory agents, carbohydrate metabolism promoters and immune system modulators (Li et al. 2014; Noda et al. 2013; Priscilla et al. 2014).

[45] IUPAC name: (2*S*)-5,7-dihydroxy-2-(3-hydroxy-4-methoxyphenyl)-2,3-dihydrochromen-4-one.

Hesperetin has 31 registered bioactivity outcomes, including confirmatory trials mainly of anti-tumoral activity (NCBI 2015e).

Naringenin has been reported to possess anti-diabetes (Priscilla et al. 2014) and anti-carcinogenic activities (Li et al. 2014). A naringenin chemical derivative, 7-*O*-butylnaringenin exhibits anti-microbial activity against several *Helicobacter pylori* clinical strains. The compound was synthesised after this property was detected in the natural compounds naringenin and hesperetin (Moon et al. 2013). Naringenin also seems to improve intestinal barrier function (Noda et al. 2013). Despite some common nutritional features, differences are found between citrus fruit types and regions of origin.

5.7.3 Apple (*Malus domestica*)

Apple[46] is a small tree in the *Rosaceae* family that originated in western Asia and is now one of the most widely cultivated fruit trees. Apples have been present in the mythology and religions of many cultures, including Norse, Greek and Christian traditions. Through history, several symbolisms have been associated with apples, from the biblical 'forbidden fruit' to the proverb 'an apple a day keeps the doctor away'.

More than 7500 cultivars of applies are currently consumed in some part of the world. Apples have been cultivated in Asia and Europe for many centuries (Courteau 2015a). China was the major world producer in 2011 (FAO 2015), with about half of the global produce, followed by USA, India and Turkey.

In the Mediterranean basin, apples are mostly eaten raw, but they can also be used to make cider, a soft fermented alcoholic beverage popular in Asturias (Spain). Like oranges, apples can be industrially processed, mainly for the production of juice (with lower nutritional value than the raw fruit, as happens with oranges).

Apples are available all year long because some cultivars are harvested in summer (Gala, Golden) and others are harvested in the fall/autumn (Fuji, Red Delicious) or winter (Granny Smith). Moreover, apples are climacteric fruits; before commercialisation, they are usually stored in cold and controlled-atmosphere chambers to delay ethylene-induced ripening.

[46] The figures presented below are a combination of average values of an unknown number of varieties available in Portugal (INSA 2015) and of average values for five varieties commonly found in the USA (USDA 2015).

Apples*. Apples (Malus domestica) along with oranges and grapes are one of the most representative fruits of the Mediterranean basin. Apples are available all year long, depending on the variety. Besides vitamins and dietary fibre, apples also provide many health promoting compounds. Apples can and should be eaten with the skin on, because of its higher content in flavonols, such as catechin and derivatives. Photo reprinted with kind permission from T. N. Wassermann*

Apples can and should be eaten with the skin on, because of their flavonol content, particularly anthocyanidins. The nutritional value of apples and their documented health benefits are vast. According to the INSA and USDA databases (accessed in 2015), 'raw apples with skin' provide about 57 kcal/100 g from: proteins (0.2 %), total fat (0.2–0.5 %; mostly PUFA, of which 0.1 % is linoleic acid), carbohydrates (about 13.6 %, mostly fructose, some glucose, sucrose and seldom starch), organic acids (0.2 %, mostly malic acid) and dietary fibre (2–2.5 %). Apples are a good source of minerals such as Ca, Fe, Mg and the micronutrients Zn, Cu, Mn, Se and F. Apples also provide the vitamins C (5–7 mg/100 g), folates (3–5 μg/100 g), β-carotene (27 μg/100 g), other pro-vitamin A carotenoids (3–4 μg/100 g RAE), α-tocopherol (2–5 mg/100 g), thiamine (0.02 mg/100 g), riboflavin (0.03 mg/100 g), niacin (0.091 mg/100 g), pantothenic acid (0.061 mg/100 g), vitamin B_6 (0.04 mg/100 g) and phylloquinone (2.2 μg/100 g).

In addition, apples supply a wide range of phytochemicals (USDA 2015), mostly known for their antioxidant properties and subsequent health-protective effects. These include β-cryptoxanthin (11 μg/100 g), lutein & zeaxanthin (29 μg/100 g) and the flavonoids catechin (1.3 mg/100 g), as well as their derivatives epigallocatechin (0.3 mg/100 g), epicatechin (7.5 mg/100 g) and epigallocatechin-3-gallate (0.2 mg/100 g), also found in other plant-origin foods, as is the case of teas and wines. A recent research describes a decrease of about 40 % in injury to rat gastric mucosa and a fourfold increase in intracellular antioxidant activity, both outcomes associated with apples, which the authors attributed to catechin and its derivatives

(Graziani et al. 2005). These flavonols are reported to show anti-inflammatory properties (Mackenzie et al. 2004), which could lead to diminished vascular and platelet reactivity. Mackenzie and colleagues also demonstrated that flavonols can regulate the immune response.

Serra and co-workers (2012) registered a significant in vivo decrease of levels of serum triglycerides, total, and LDL cholesterol concentrations in blood. These authors registered reductions of 27.2 %, 21.0 % and 20.4 %, respectively, in relation to the cholesterol-enriched diet group, $P < 0.05$, in rats. The same authors noted that results depended on the variety of apple (in this case, 'Bravo de Esmolfe') and were possibly related to the balance between phenolic and fibre concentrations. In the same study, the bioactive response was correlated with chemical composition, leading the authors to point catechin, epicatechin, procyanidin B1 and carotenes as the major contributors for the reported cholesterol-lowering properties of apples.

The presence of two important health-promoting flavonols glycosides in apple flesh was recently registered (Graziani et al. 2005). One of these is rutin,[47] which, besides its ROS scavenging activity, is the object of a US patent for the treatment of depression and other emotional disorders (Chatterjee et al. 2006). The other is phloridzin,[48] which was observed to increase lipolysis in adipocytes and to suppress in vivo inflammatory response (Huang et al. 2013).

The ingestion of apple seeds should be avoided, as they contain the cyanogenic glycoside amygdalin, although the risk is low, as the fatal dose of cyanide salts is estimated at 200–300 mg for an adult, while 1 g of apple seeds contains about 0.6 mg of hydrogen cyanide (NLM 2015). Hence, inadvertent ingestion of whole seeds is unlikely to result in acute cyanide toxicity.

Apples are industrially processed, mainly into juices and jams, and their polyphenols undergo biochemical and chemical changes, since they are chemically unstable species. Enzymatic oxidation is the first alteration process, which starts as soon as the integrity of the cell is broken. The resulting browning colour is usually detrimental to quality, particularly in juice.

5.7.4 Cherry (*Prunus avium*)

Cherry[49] is a small tree in the *Rosaceae* family that produces a fruit botanically classified as a drupe (a stone fruit) that has been consumed since pre-historical times. This species is indigenous to southern Europe and western Asia, and was first disseminated by the Roman Empire. In the Mediterranean region, fruits are mainly consumed fresh during their very short harvesting season.

[47] IUPAC name: 2-(3,4-dihydroxyphenyl)-5,7-dihydroxy-3-[(2*S*,3*R*,4*S*,5*S*,6*R*)-3,4,5-trihydroxy-6-[[(2*R*,3*R*,4*R*,5*R*,6*S*)-3,4,5-trihydroxy-6-methyloxan-2-yl]oxymethyl]oxan-2-yl]oxychromen-4-one.

[48] IUPAC name: (1-[2,4-dihydroxy-6-[(2*S*,3*R*,4*S*,5*S*,6*R*)-3,4,5-trihydroxy-6-(hydroxymethyl) oxan-2-yl]oxyphenyl]-3-(4-hydroxyphenyl)propan-1-one.

[49] The figures presented below are a combination of average values of four different varieties available in Portugal (PortFIR database, from INSA) and of average values for an unreported number of varieties found in the USA (USDA database).

Sweet cherries provide about 60 kcal/100 g from: proteins (1.06 %), lipids (0.2–0.7 %), mono- and disaccharides (about 13 %; mainly glucose and fructose, but also including sucrose, maltose and galactose), organic acids (0.4 %) and dietary fibre (1.6–2.1 %) (INSA 2015; USDA 2015). Cherries contain 141 mg/100 g of total carotenes (INSA 2015), mainly β-carotene, lutein and zeaxanthin, the properties and bioactivities of which are described above. Cherries also provide phylloquinone (2.1 μg/100 g), α-, β- and γ-tocopherols (about 0.13 mg/100 g, total amount), as well as other vitamins in moderate to small amounts, such as vitamin B_6 (0.04 mg/100 g), and folates (4–5 μg/100 g) as well as the flavonols kaempferol (0.2 mg/100 g) and quercetin (2.3 mg/100 g) (USDA 2015).

Jakobeka and co-workers (2009) reported that the major phenolic acids of sweet cherries are neochlorogenic acid, also known as 5′-O-caffeoylquinic acid,[50] and *p*-coumaric acid derivatives. These compounds have been referred to as ROS scavengers and as having anti-inflammatory and anti-carcinogenic properties; (Seymour and Ou 2011).

Sweet cherries are a relevant source of F (2 μg/100 g). Cherries are interesting from a nutraceutical viewpoint because of the variety and content of flavonoids, particularly anthocyanidins and procyanidins (USDA 2015). Cyanidins are water-soluble plant pigments, the stability of which depends on pH. Under acidic conditions (to which phenolic acids may contribute), coloured cyanidin derivatives are present and range from red to blue, while at a higher pH, the colourless chalcone forms accumulate. Anthocyanidins are characterised by the presence of a 2-phenylbenzopyrylium unit. They are derived along the flavonoid modification pathways and further separated into three types—pelargonidin,[51] cyanidin[52] and delphinidin[53]—because of the different number of hydroxyl groups in the phenyl moiety.

Pelargonidin is present in cherries at an average concentration of 0.3 mg/100 g (USDA 2015), whereas the average value of peonidin[54] is reported to be 1.5 mg/100 g. The related compounds—petunidin[55] and malvidin[56]—were found to be present in trace amounts, and most probably as glucosides. Given their electrophilic character, they are expected to behave as strong ROS scavengers. Proanthocyanidins are present in cherries as monomers (5.1 mg/100 g), dimers (3.2 mg/100 g), trimers (2.4 mg/100 g), 4–6mers (6.5 mg/100 g) and 7–10mers (1.9 mg/100 g) (USDA 2015).

Still according to the 2015 edition of the USDA database, sweet cherries contain relevant amounts of other flavonoids, such as catechin (4.4 mg/100 g), and their

[50] IUPAC name: (1R,3R,4S,5R)-3-[(E)-3-(3,4-dihydroxyphenyl)prop-2-enoyl]oxy-1,4,5-trihydroxycyclohexane-1-carboxylic acid), chlorogenic acid ((1S,3R,4R,5R)-3-[(E)-3-(3,4-dihydroxyphenyl)prop-2-enoyl]oxy-1,4,5-trihydroxycyclohexane-1-carboxylic acid.

[51] IUPAC name: 2-(4-hydroxyphenyl)chromenylium-3,5,7-triol.

[52] IUPAC name: 2-(3,4-dihydroxyphenyl)chromenylium-3,5,7-triol.

[53] IUPAC name: as 2-(3,4,5-trihydroxyphenyl)chromenylium-3,5,7-triol.

[54] IUPAC name: 2-(4-hydroxy-3-methoxyphenyl)chromenylium-3,5,7-triol.

[55] IUPAC name: 2-(3,4-dihydroxy-5-methoxyphenyl)chromenylium-3,5,7-triol.

[56] IUPAC name: 2-(4-hydroxy-3,5-dimethoxyphenyl)chromenylium-3,5,7-triol.

derivatives epigallocatechin (0.3 mg/100 g) and epicatechin (5 mg/100 g), which are similar to members of the same structural class (Fig. 5.2), characterised by the presence of a 2-phenylbenzopyrylium unit. As mentioned above, this class of compounds is also found in other plant foods such as apples and green and black teas (hot water extracts of *Camellia sinensis*). Despite that anthocyanin metabolism is still not fully understood, the bioavailability of each compound is believed to depend on synergistic interactions between certain combinations of compounds. It has been reported that the interaction among catechins may affect intestinal absorption of anthocyanidins, thus affecting their bioavailability (Tagashira et al. 2012).

Oligomeric compounds were reported to be more effective in inducing in vivo tumour cell apoptosis than their individual counterparts (Miura et al. 2008). On the other hand, Fernandes and colleagues (2012) assessed the absorption of cyanidin dimers and monomers through 'Caco-2' cells by performing transepithelial transport assays. These authors found that the monomers seem to be more easily absorbed than dimers, for which the efficiency in crossing the intestinal barrier may be related to the presence of the glucose moiety in its structure. In the ambit of a study with healthy human volunteers, it was observed that, upon ingestion, anthocyanins were glucuro- and sulfo-conjugated and that their main metabolite in human urine was a monoglucuronide of pelargonidin (Felgines et al. 2003).

In short, flavonols are ROS scavengers with anti-inflammatory and anti-metastatic activities, both in vitro and in vivo (Espley et al. 2014; Mackenzie et al. 2004; Serra et al. 2012; Weng and Yen 2012). Flavonoids are attracting a growing attention because of their ability to capture electrons (free radicals).

Due to the high pectin and tannin content, sour cherry (*Prunus cerasus*) is usually used to prepare alcoholic beverages and jams. Sour cherry has about the same sugar and organic acid contents as sweet cherry but has higher levels of vitamin C and folates, as well as a different balance of flavonols.

5.7.5 Fig (*Ficus carica*)

Fig[57] is a tree in the *Moraceae* (mulberry) family that originated in the Mediterranean. It was domesticated over 6000 years ago, making it one of the most ancient fruit crops, and has been widely cultivated since. The fig tree has many symbolisms associated with Hebrew and Christian holy texts and is one of the two sacred trees of Islam. Figs are derived from inflorescences forming a nearly close receptacle: the syconium (fruiting structure), which may range in colour from yellow to green purple or bronze on the outside, with 30–1600 tiny flowers (later seeds) inside. The ripe fruits are very fragile and perishable when fresh, and thus are only available in local markets. They are often eaten as bushfood.

[57] Figures for fresh figs are a combination of the average values for five different varieties available in Portugal (INSA 2015) and of average values for an unreported number of varieties found in the USA (USDA 2015).

Fig trees are pollinated by fig wasps, with which a mutualism relationship is established. The cultivation of different varieties of fig trees thus contributes to increasing biodiversity of both, the plant and the insect (Jandér and Herre 2010).

5.7.5.1 Fresh Figs

Despite fruit skin colours are diverse, i.e. light green, light purple, purple, dark purple and black, the interior is yellowish to reddish. Figs provide about 70–74 kcal/100 g: proteins (0.8–0.9 %), lipids (0.3–0.6 %, of which 0.2 g/100 g correspond to linoleic acid) about 16 % mono- and disaccharides (mainly glucose and fructose), organic acids (0.2 %) and dietary fibre (1.3–2.9 %) (INSA 2015; USDA 2015). Fresh figs contain also carotenes (50 mg/100 g) and α-tocopherol (0.77 mg/100 g) (INSA 2015). Most importantly, a titrable acidity of 24–237 mg/ 100 g of gallic acid has recently been determined in figs from Turkey (Ercisli et al. 2012). Gallic acid (3,4,5-trihydroxybenzoic acid) is readily absorbed by the gastrointestinal tract. Its bioavailability seems to be related to anthocyanins, as gallic acid is most probably obtained via the metabolism of anthocyanins to phenolic acids and aldehydes by gut microbiota (Forester et al. 2014). According to the same authors, the beneficial effects of polyphenols on colon cancer may depend on the local accumulation of gallic acid. It possibly occurs because of the inhibition of the transcription factors that promote tumour cell proliferation, and the induction of apoptosis can in part underlie the observed effects.

5.7.5.2 Dried Figs

Figs have been dehydrated for centuries to preserve them, by reducing a_w. Industrial processes were developed during World War I, when the demand for long-lasting foods increased. Drying processes enabled the export of figs and allowed them to be consumed all year long. The traditional process involved the sun drying of figs carefully laid down on the 'almanxar', a sieve-like apparatus made of wood or open wicker of fennel stems. The air was allowed to circulate, and the racks were covered and preserved in a warehouse from evening to dawn to avoid insect attacks and moisture condensation. The last drying step occurred in the shade, followed by an inspection to eliminate contaminated figs (with undesirable moulds or other defects). The whole process lasted 4–6 days. Modern methods involve the use of industrial air dryers, wherein the figs are exposed in single-layer racks. Dry figs are often stuffed with peeled toasted almonds or nuts. Dry figs only contain about 25–30 % water, conversely to fresh figs carrying 79 % water, thus concentrating their components, including sugars (47–58 %), dietary fibre (10–11 %) and minerals such as Mn (0.51 mg/100 g) and Se (0.6 μg/100 g) (INSA 2015; USDA 2015). Water-soluble vitamins and carotenoids are present in approximately the same range as in fresh figs, despite water evaporation that may have compensated for probable degradation by oxidation reactions. Thus, dry figs also provide the carotenoids lutein + zeaxanthin (32 μg/100 g) and phylloquinone (15.6 μg/100 g) (USDA 2015). Cyanidin glucosides with ROS scavenging activity have been reported in figs (Solomon et al. 2010a, b).

Dried dates and dried figs*. The image represents a dried fig and dried dates. Raisins, plums and apricots also undergo the same preservation process, which mainly consists of decreasing the activity of water (a_w) and generally increasing the content of natural sugars and other relevant compounds such as dietary fibres. Photo reprinted with kind permission from T. N. Wassermann*

5.7.6 Dates (*Phoenix dactylifera*)

The date palm is a monocot plant in the *Arecaceae* family, cultivated in dry tropical regions worldwide for its edible sweet fruit. Its native range is difficult to ascertain, but it is generally considered as originating in the region around the Persian Gulf—northern Africa, the Arabian Peninsula and northwest India (Courteau 2015b).

Dates are oval cylindrical fruits with a single long and thick stone. When ripe, they range from bright red to bright yellow in colour, depending on the variety. Three main cultivar groups of date exist: soft (e.g. Medjool), semi-dry (e.g. Deglet Nour) and dry (e.g. Thoory), with numerous varieties in each group, as well as hybrids with other *Phoenix* species (Courteau 2015b).

Dates can be eaten as is, or stuffed with almonds, nuts, lemon peel, etc. or even cooked. In addition to the edible fruits, seeds can be ground into an edible flour, traditionally used to make bread in times of scarcity. Dates, are very popular among Muslims. In the evening meal during Ramadan, it is traditional to eat a date first. In Morocco and Tunisia, dates are chopped and used in a wide variety of deserts, while in south Spain a popular appetiser is prepared by just wrapping dates in bacon and cooking them in an oven for a few minutes. Dates of Medjool var. supply 277 kcal/100 g from: protein (1.81 %), fats (0.15 %), sugars (66.5 %; mainly glucose and fructose, and small amounts of sucrose and maltose) and dietary fibre (6.7 %) (USDA 2015). Medjool dates are an important source of minerals such as K (696 mg/100 g), Zn (0.44 mg/100 g), copper (0.362 mg/100 g) and Mn (0.296 mg/100 g), simultaneously as supplying moderate amounts of water-soluble vitamins. As expected, variations in composition between varieties are registered. For example, Se (3.0 μg/100 g) can be found in Deglet Nour dates but not in Medjool varieties. The flavonoids quercetin (0.92 mg/100 g) and cyanidin (1.7 mg/100 g) are reported for Deglet Nour but not for other varieties (USDA 2015).

5.7.7 Blackberry (*Rubus fruticosus*)

Blackberries are the fruits of the blackberry bush with thorny stems, of the *Rosaceae* family. Rubus species are considered a weed, despite their ecological significance. Blackberries have been eaten by humans for thousands of years, probably since the Iron Age. *Rubus fruticosus* is widespread through Europe, north-western Africa and certain regions of Asia and America. The fruit is a bramble, not a berry, and its ripening season ranges from early to late summer, according to the local climate. Some cultivars are produced in the European Union and the USA. The gathering of wild fruits is popular as a hobby in certain regions (such as northern Europe) but is a less and less common practice in the Mediterranean basin. The fruits are mainly eaten raw or used to make jams. Nowadays, the consumption of wild blackberries in the Mediterranean area is believed to be marginal.

Blackberry. *Blackberry is the fruit of* Rubus fruticosus*, a bush with thorny stems that often grows as a weed in the Mediterranean region. The blackberry shares some properties with other red fruits, as cherries and raspberries, such as the simultaneous presence of anthocyanins, condensed tannins, as well as catechin and derivatives. These compounds are believed to act synergistically to benefit gut's health, and indirectly on overall wellness. The blackberry and other red fruits also contain relevant levels of different flavonoids, with antioxidant properties. Photo reprinted with kind permission from T. N. Wassermann*

Blackberries supply 43 kcal/100 g from: proteins (1.39 %), fats (0.49 %), sugars (4.88 %; mainly glucose and fructose, and small amounts of sucrose, maltose and galactose) and dietary fibre (5.3 %) (USDA 2015). Blackberries are notable for their vitamin C content (21 mg/100 g) in addition to a wide range of water-soluble vitamins, β-carotene (128 μg/100 g) and lutein + zeaxanthin (118 μg/100 g). Blackberries also provide α-tocopherol (1.17 mg/100 g) as well as the less common β-, γ- and δ-tocopherols (2.28 mg/100 g), and relevant amounts of phylloquinone (19.8 μg/100 g) and the microelements Mn (0.646 mg/100 g) and Se (0.4 μg/100 g) (USDA 2015).

The presence of a wide range of relevant flavonoids is noteworthy. These include anthocyanins, which are responsible for the typical colour of blackberries. Thus, blackberries contain, on average, pelargonidin (0.4 mg/100 g) and peonidin (0.2 mg/100 g), along with proanthocyanidin monomers (3.7 mg/100 mg), dimers (4.4 mg/100 g), trimers (2.1 mg/100 g), 4–6mers (7.3 mg/100 g), 7–10mers (4.2 mg/100 g) and polymers of more than 10 units (1.5 mg/100 g). The flavonols catechin (37 mg/100 g), epigallocatechin (0.1 mg/100 g), epicatechin (4.7 mg/100 g) and epigallocatechin-3-gallate (0.7 mg/100 g) (USDA 2015) have been previously noted to synergistically interact with tannins. Other quantified flavonols are kaempferol (0.3 mg/100 g), myricetin (0.7 mg/100 g) and quercetin (3.6 mg/100 g) (Fig. 5.2).

The health benefits of condensed tannins have raised consumer awareness of the health properties of red and dark berries; these features are mentioned above and further discussed in Chap. 8. Despite several studies presenting evidence on the effects of dark berries (such as cherries and blackberries) on coronary disease, cancer and brain functions, no health claims have been approved to date.

5.8 Nuts

Nuts are fruits comprising a hard shell and a seed and can be found in a wide variety of genera. Herein, the term 'nut' should be regarded in the culinary sense, as it includes both, fruits and seeds, when considering the botanical classification.

Native nuts from the Mediterranean region include walnuts (from *Juglans regia*), hazelnuts (from *Corylus avellana*), chestnuts (from *Castanea sativa*), almonds (from *Prunus dulcis*), pistachio (from *Pistacia vera*) and pine nuts (from *Pinus pinea*). The kernel of the seed usually constitutes the edible portion. Traditionally, in the Mediterranean basin, nuts have been mostly consumed raw or used to stuff dry figs or dates, as appetizers. Nuts are also used as ingredients in a few desserts and pastries. During the evening meal of Ramadan, Muslims consume typical nut-based high caloric desserts.

Storage conditions are critical for the quality and safety of nuts. Moulds can grow rapidly in warm environments with about 75 % air humidity. These moulds may include the aflatoxin-producing *Aspergillus flavus* and some strains of *A. niger*. Aflatoxin exhibits acute toxicity, as well as long-term carcinogenic effects that may be attenuated by terpenes (Eaton and Schaupp 2014). Examination tests for aflatoxins (in nuts, grains and feedstocks) are included in food safety plans.

In some individuals, nuts can trigger life-threatening allergic reactions. In general, allergens are minor constituents of tree nuts, generally lipid-transfer proteins or major seed-storage proteins. Many nut allergens have been identified and characterized; immunoglobulin (Ig)-E-reactive epitopes have been described for some of these allergens (Roux et al. 2003).

Regarding health benefits, in the USA, the FDA approved the following statement as a qualified health claim for labelling purposes: 'Scientific evidence suggests but does not prove that eating 1.5 ounces per day of most nuts [such as name of specific nut] as part of a diet low in saturated fat and cholesterol may reduce the risk of heart disease' (FDA 2003). Nuts are a good source of ALA (C18:3, n-3)[58] (Fig. 5.5). As mentioned in Chap. 4, (n-3) PUFA are beneficial with respect to the maintenance, prevention and treatment of a wide range of health conditions, including mental and visual health, cardiovascular disease and inflammatory disorders, and also play a key role in foetal and infant development (Madden et al. 2009).

5.8.1 Walnut (*Juglans regia*)

The Latin name for the walnut was '*nux gallica*', suggesting that *Juglans regia* was originally from Turkey (ancient Galatia was in the highlands of Turkey), where wild species may still be found. The Roman Empire and Alexander the Great are thought to have strongly contributed to dissemination of the species (Kitsteiner 2011). In 2012, China was the major world producer followed by Iran, the USA and Turkey (FAO 2015).

The whole fruit, including the husk, falls in autumn; the seed is large, with a relatively thin shell, and is edible, with a rich flavour. Harvested walnuts need to evaporate some water before storage and consumption. Freshly harvested raw walnuts, with a water content of 2–8 %, offer the best colour, flavour and nutrient density.

[58] IUPAC name: (9Z,12Z,15Z)-octadeca-9,12,15-trienoic acid.

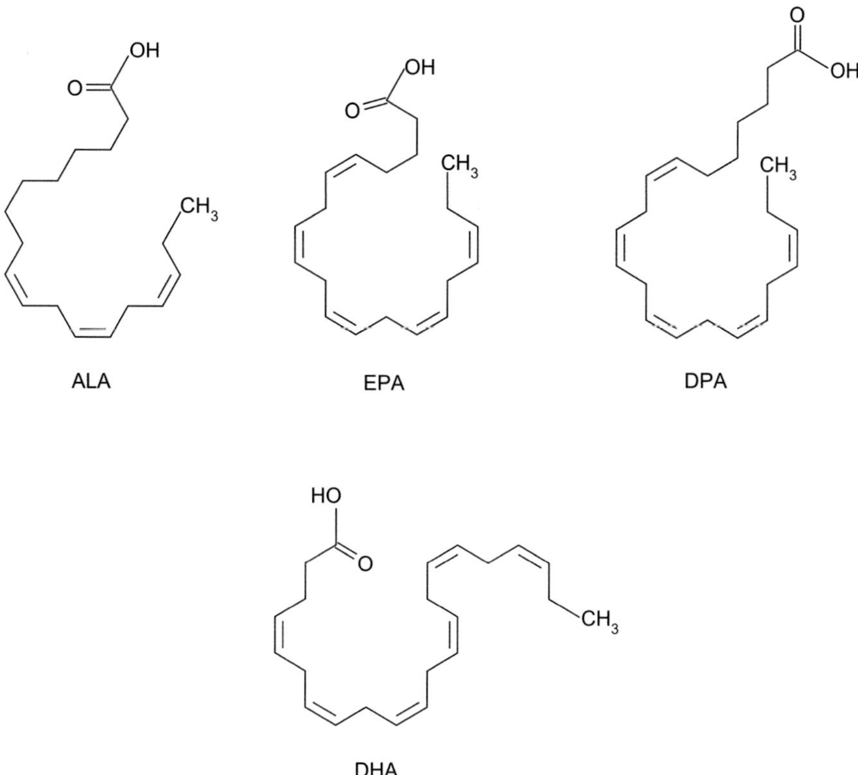

Fig. 5.5 Long-chain n-3 polyunsaturated fatty acids that play important biological roles: α-linolenic acid (ALA), eicosapentaenoic acid (EPA), docosapentaenoic acid (DPA) and docosahexaenoic acid (DHA). ALA is of plant origin, while seafood, dairy products and some meat are dietary sources of EPA and DHA. BKchem version 0.13.0, 2009 (http://bkchem.zirael.org/index.html) was used to draw these structures

Walnut. *Nuts from Juglans regia are mainly consumed raw as appetizers or included in some desserts. Walnuts have a unique lipid profile and are a source of essential fatty acids, flavonoids, phytosterols and squalene. A health claim was approved by the European Food Safety Authority (EFSA) for walnuts, based on their contribution to decreasing the risk of coronary heart disease. Photo reprinted with kind permission from T. N. Wassermann*

On average, walnuts supply 654–698 kcal/100 g from: proteins (15.2–16.7 %), fats (65.2–67.5 %; SFA [5.4 %, palmitic and stearic acids], MUFA [15 %, oleic acid] and PUFA [47 %, 38 % of which are linoleic and linolenic acids, including ALA]), starch (0.06–1 %), sugars (2.6–3.6 %, mainly sucrose, fructose and glucose) and dietary fibre (5.2–6.7 %) (INSA 2015; Ozkan and Koyuncu 2005; USDA 2015). Walnuts supply Ca (90–98 mg/100 g), P (288–346 mg/100 g), Mg (158–160 mg/100 g), Mn (3.4 mg/100 g) and Se (4.9 μg/100 g) (INSA 2015; USDA 2015). Walnuts are low in carotenes but provide high levels of tocopherols: α-tocopherol (0.7 mg/100 g), β-tocopherol (0.15 mg/100 g), γ-tocopherol (20.83 mg/100 g) and δ-tocopherol (1.89 mg/100 g) (USDA 2015). Walnuts contain also the B vitamins: thiamine (0.33–0.34 mg/100 g), niacin (0.9–1.1 mg/100 g), riboflavin (0.14–0.15 mg/100 g), pantothenic acid (0.57 mg/100 g), folates (66–98 μg/100 g) and vitamin B_6 (0.54–0.67 mg/100 g), as well as phylloquinone (2.7 μg/100 g) (INSA 2015; USDA 2015). Walnuts, like other nuts, are abundant in proanthocyanidins, mostly highly polymerised: monomers (3.6 mg/100 g), dimers (5.7 mg/100 g), trimers (7.2 mg/100 g), 4–6mers (22.0 mg/100 g), 7–10mers (5.4 mg/100 g) and polymers with more than 10 units (20.0 mg/100 g). In short, long polymers (of more than 10 units) and oligomers of 4–6 units are predominant.

Proanthocyanidins, also known as condensed tannins, are thought to have relevant roles in the organism, most probably involving gut microbiota and in relation to phytosterols (Sects. 5.7.1, 5.7.4, 5.7.7 and Chap. 8).

Walnuts also possess a wide range of phytosterols associated with the lipid fraction (USDA 2015) (Fig. 5.1):

- Campesterol[59] with an average concentration of 5 mg/100 g
- β-sitosterol[60] with a concentration of 87 mg/100 g
- δ-5-avenasterol[61] (7.3 mg/100 g)
- Campestanol[62] (2.3 mg/100 g)
- and other non-identified phytosterols (8.6 mg/100 g)

As mentioned previously, phytosterols contribute to the lowering of LDL cholesterol levels in the bloodstream. On the other hand, the main MUFA in walnuts is oleic acid, whereas the main PUFA are linoleic acid and ALA. This specific fatty acid profile, combined with the antioxidant properties of γ-tocopherol, may (at least partly) explain the beneficial effects of nuts in relation to coronary heart disease (Maguire et al. 2004; Tapsell et al. 2004).

Recent research has reported that diets supplemented with walnuts (10–24 % of total calories) resulted in a significant decrease in total cholesterol and in LDL cholesterol concentrations, with no adverse effects on body weight (Banel and Hu 2009). The same paper also mentions a positive impact on inflammatory markers. These observations were also validated for patients with type II diabetes, as the dietary fatty acid profile is believed to be associated with the risk of diabetes (Tapsell et al. 2004). Ingestion of SFA increases the risk of type II diabetes, whereas the intake of linoleic acid may decrease that risk (Hodge et al. 2007).

In 2011, the EFSA approved the health claim 'Walnuts contribute to the improvement of endothelium-dependent vasodilation' based on the evidence of a cause-and-effect relationship between walnut consumption and improvement of endothelium-dependent vasodilation. An important contribution came from intervention studies that showed walnut consumption had a sustained effect on endothelium-dependent vasodilation and a positive effect on endothelium-mediated vasodilation. The EFSA considers that, to obtain the claimed effect, 30 g of walnuts should be consumed daily (EFSA 2011).

5.8.2 Hazelnuts (*Corylus avellana*)

Hazelnuts are roughly spherical, about 10–15 mm in diameter and have a fibrous outer husk surrounding a smooth shell. As other edible botanical species, *Corylus avellana* was disseminated by the Roman empire through Europe and the Middle

[59] IUPAC name: (3*S*,8*S*,9*S*,10*R*,13*R*,14*S*,17*R*)-17-[(2*R*,5*R*)-5,6-dimethylheptan-2-yl]-10,13-dimethyl-2,3,4,7,8,9,11,12,14,15,16,17-dodecahydro-1H-cyclopenta[a]phenanthren-3-ol.

[60] IUPAC name: (3*S*,8*S*,9*S*,10*R*,13*R*,14*S*,17*R*)-17-[(2*R*,5*R*)-5-ethyl-6-methylheptan-2-yl]-10,13-dimethyl-2,3,4,7,8,9,11,12,14,15,16,17-dodecahydro-1H-cyclopenta[a]phenanthren-3-ol.

[61] IUPAC name: (3*S*,8*S*,9*S*,10*R*,13*R*,14*S*,17*R*)-10,13-dimethyl-17-[(*Z*,2*R*)-5-propan-2-ylhept-5-en-2-yl]-2,3,4,7,8,9,11,12,14,15,16,17-dodecahydro-1H-cyclopenta[a]phenanthren-3-ol.

[62] IUPAC name: (3*S*,5*S*,10*S*,13*R*)-17-[(2*R*,5*R*)-5,6-dimethylheptan-2-yl]-10,13-dimethyl-2,3,4,5,6,7,8,9,11,12,14,15,16,17-tetradecahydro-1H-cyclopenta[a]phenanthren-3-ol.

East. Hazelnuts are harvested in autumn. Spain, Italy, Greece and Cyprus are among the primary producers of hazelnuts (FAO 2015).

Hazelnuts are generally eaten raw or ground into a paste that is used in the preparation of some dishes, including desserts. Some very popular commercial food products are available, such as praline, chocolate and hazelnut pastes and liqueurs. Hazelnuts are used abundantly in Turkish and Italian cuisine, mainly because hazelnut oil contains an aromatic ketone, filbertone,[63] also used in perfumery (NCBI 2015f). Natural filbertone is an unbalanced mixture of enantiomers, and the composition can vary widely (particularly with the cultivar, agronomic techniques and harvesting conditions).

On average, hazelnuts supply 628–677 kcal/100 g from: proteins (14–15 %), fats (60.8–66.3 %: SFA 4.9 %, MUFA 52.2 % and PUFA 6.2 %, mainly oleic and linoleic acids), starch (0.48–2.1 %), sugars (3.9–4.3 %, mainly sucrose, fructose and glucose) and dietary fibre (6.1–9.7 %). Besides the high levels of protein (that includes some essential amino acids) and unsaturated fats, hazelnuts are good sources of oil-soluble vitamins such as α-tocopherol (15–25 mg/100 g), β-tocopherol (0.33 mg/100 g) and phylloquinone (14.2 μg/100 g) (INSA 2015; USDA 2015). Hazelnuts also contain the water-soluble B vitamins thiamine (0.30–0.64 mg/100 g), niacin (1.8 mg/100 g), riboflavin (0.11–0.16 mg/100 g), folates (73–113 μg/100 g) and vitamin B_6 (0.56–0.59 mg/100 g) (INSA 2015; USDA 2015) and supply relevant amounts of the micronutrients Mn (6.18 mg/100 g) and Se (2.4 μg/100 g).

Some phytosterols detected in hazelnuts can also be found in walnuts, including δ-5-avenasterol (2.6 mg/100 g), campestanol (3.0 mg/100 g) and sitostanol[64] (3.9 mg/100 g), among other non-discriminated phytosterols (2.5 mg/100 g) (USDA 2015).

Flavonoids commonly found in hazelnuts include catechin (1.2 mg/100 g), epigallocatechin (2.8 mg/100 g), epicatechin (0.2 mg/100 g), and epigallocatechin-3-gallate (1.1 mg/100 g). Like with other nuts and some of the fruits described above, the intake of catechins from hazelnuts may elicit interactions between these compounds and condensed tannins with an impact on anthocyanin bioavailability and metabolism (Chap. 8). The reported direct beneficial effects of flavonoids are also noteworthy (Sects. 5.7.1, 5.7.4 and 5.7.7). Hazelnuts are rich in proanthocyanidin monomers (9.8 mg/100 g), dimers (12.5 mg/100 g), trimers (13.6 mg/100 g), 4–6mers (67.7 mg/100 g), 7–10mers (74.6 mg/100 g) and polymers with more than 10 units (322.4 mg/100 g). The dominance of large oligomers and polymers is noteworthy. These structures (with molecular weights >500 Da) are generally designated as tannins, and proanthocyanidins are normally referred to as condensed tannins (Cheynier 2005), all having the ability to interact with proteins. According to the same author, proanthocyanidins differ in the nature of their constitutive units, their sequences, the positions of interflavanic linkages, their chain lengths and the presence of substituents (e.g. galloyl or glucosyl groups). Given the many chances for

[63] IUPAC name: (*E*)-5-methylhept-2-en-4-one.

[64] IUPAC name: 17-(5-ethyl-6-methylheptan-2-yl)-10,13-dimethyl-2,3,4,5,6,7,8,9,11,12,14,15,16,17-tetradecahydro-1H-cyclopenta[*a*]phenanthren-3-ol.

substitution, the number of possible isomers increases exponentially with the chain length. The ability of tannins to interact with proteins is responsible for astringency perception (resulting from interactions of tannins with salivary proteins), for the formation of haze and precipitates in beverages and for the inhibition of enzymes and reduced digestibility of some dietary proteins (Cheynier 2005). These properties are primarily related to the number and accessibility of *o*-diphenol moieties. The oxygen radical-scavenging capacity of procyanidin dimers and trimers was shown to increase with galloylation, and to a lesser extent with (longer) chain length. Nevertheless, it is also influenced by the position of galloyl substituents. Conversely, in the lipid phase, radical-scavenging activity decreased as the molecular weight increased (Cheynier 2005). Properties of polyphenols are greatly affected by their interactions with other constituents of the food matrix. Colour intensification, resulting from interactions of anthocyanins with other compounds (i.e. co-pigmentation), is well documented and easily observed in roasted hazelnuts. The astringency of tannins may be altered by the release of polysaccharides and proteins during the heating process.

Hazelnuts & Chestnuts*. The image represents two different types of nuts: the oil-rich hazelnuts (Corylus avellana) and the starch-rich chestnuts (Castanea sativa). Hazelnuts share the same general nutritional profile as the other oily nuts (e.g. walnuts, pine nuts and almonds), such as the presence of phytosterols, catechins and condensed tannins. Hazelnuts show some peculiarities typical of the species, as expected in vegetable foods. Hazelnuts are rich in the aromatic ketone filbertone, which is also used in perfumery and is responsible for the aroma. Photo reprinted with kind permission from T. N. Wassermann*

5.8.3 Chestnut (*Castanea sativa*)

Chestnut was a staple food in Europe, particularly in the Mediterranean basin, before the introduction of the potato in the sixteenth century. The Romans planted chestnut trees across Europe while on their various campaigns, to ensure their armies would have a source of carbohydrates. Chestnut trees, despite requiring a mild climate and adequate moisture for good growth and a good nut harvest, were very abundant, particularly in the highlands of the north of Portugal, Spain, Italy and Turkey (Anonymous 2011).

With the introduction of potatoes, chestnuts started to decrease in popularity, passing from a secondary source of carbohydrates to a 'famine food'. A so-called ink disease of chestnut trees spread from Italy, reaching Portugal and France and discouraging replanting for many years.

However, chestnuts never disappeared from the Mediterranean basin because of their link to old Christian and Celtic traditions. The harvesting season for chestnuts overlaps annual wine production. Thus, traditionally in Italy, chestnuts are soaked in wine and eaten on Saint Simon's day; in Portugal, chestnuts are roasted on an open fire and eaten to accompany the tasting of new wine during the celebrations of St. Martin's day.

Chestnuts can be dried and further ground to produce flour, which could almost exclusively feed isolated populations for half the year as a temporary but complete substitution for cereals. In some parts of Italy, a cake made of chestnuts was used as a substitute for potatoes. Roasted chestnuts are traditionally sold in streets, markets and fairs by street vendors with mobile braziers and they are much appreciated during fall/autumn and winter, e.g. in Portugal, where they are consumed mainly as appetisers because of their pleasant sweet taste and starchy texture.

***Roasted chestnuts**. Before the arrival of the potato in Europe, chestnuts (Castanea sativa) were widely distributed in southern Europe and used as an energy source. Chestnuts are poor in oils but rich in starch (including resistant types), dietary fibres and water-soluble vitamins such as B-complex and ascorbic acid. Chestnuts are consumed boiled or roasted (often seasoned with fennel) and also accompanying roasted meats. Chestnuts can be dried and ground to a flour from which traditional cakes are made in some regions of the Mediterranean basin. Photo reprinted with kind permission from T. N. Wassermann*

Chestnuts followed the way of other foods of the past. Once popular and cheap foods that fed many, they have now become expensive delicacies. Nowadays, chestnut orchards are profitable for the highly valued fruit and the wood; Turkey, Italy, Greece, Portugal and Spain are among the main chestnuts producers in the world (FAO 2015).

On average, raw unpeeled chestnuts supply 185 kcal/100 g from: protein (3.1 %), fats (1.1 %: SFA 0.2 %, MUFA 0.4 % and PUFA 0.4 %, mainly oleic and linoleic acids), starch (30 %), mono- and disaccharides (9.8 %) and dietary fibre (6.1 %). Chestnuts supply vitamin C (51 mg/100 g), vitamin B_6 (0.35 mg/100 g), thiamine (0.22 mg/100 g), folates (61 μg/100 g), carotenes (64 mg/100 g), α-tocopherol (1.2 mg/100 g) and some minerals, namely Mg (33 mg/100 g) (INSA 2015).

Food nutrient databases lack information about the composition of chestnuts. Despite some published works on the antioxidant properties of chestnuts (Barreira et al. 2008; Barros et al. 2011; Vasconcelos et al. 2010), well-established information about minor components of chestnuts and their possible health effects is scarce. Moreover, as chestnuts are rarely eaten raw, its dietary value is affected by cooking temperature, with some thermal degradation of vitamins B and C, as well as the concentration of some components and the occurrence of chemical reactions.

5.8.4 Almond (*Prunus dulcis*)

Prunus dulcis is a small tree in the *Rosaceae* family, native to the Arabian peninsula and western Asia, but later naturalised throughout the Mediterranean regions and temperate Asia. Referred to in some classifications as *Amygdalis communis*, *Amygdalus communis*, *Prunus amygdalis* (*amygdalus*) and *Prunus communis*, it is intensively cultivated in California, which is the leading almond's producer in the world. There are two main varieties: the sweet and the bitter almond. The first has been consumed since ancient times and is mentioned in the Bible, in Hebrew sacred texts and in the Alcoran. As with other nuts, improper storage conditions may favour the development of aflatoxin-producing moulds. They may also cause allergic reactions in sensitive individuals.

The fruit is but a drupe with an outer hull and a hard shell protecting the seed kernel, which is the edible part (Ladizinsky 1999). Almonds are eaten raw, roasted or as a component of various dishes. As mentioned above, almonds are often eaten with dates, figs and table olives, as stuffing. Almond cookies, with regional variations, are found all over the Mediterranean basin.

On average, raw almonds (with skin) supply 579–619 kcal/100 g from: proteins (21 %), fats (50–56 %: SFA 3.8–4.7 %, MUFA 31.6–34.5 % and PUFA 12.3–14.3 %, of which 13.9 % are conjugated linoleic and linolenic acid forms), starch (0.72–2.6 %), mono- and disaccharides (4.4–4.6 %, mainly sucrose, small amounts of glucose and fructose as well as traces of maltose and galactose) and dietary fibre (about 12 %) (INSA 2015; USDA 2015).

Almonds are low in water-soluble vitamins, and the absence of vitamin C or carotenes must be noted. However, they are rich in some oil-soluble vitamins such as α-tocopherol (24–25.6 mg/100 g), β-tocopherol (0.23 mg/100 g), γ-tocopherol (0.64 mg/100 g), and δ-tocopherol (0.07 mg/100 g) (USDA 2015). Despite this, almonds provide the B vitamins thiamine (0.21 mg/100 g), riboflavin (0.75–1.14 mg/100 g, niacin (2.2–3.6 mg/100 g), pantothenic acid or vitamin B_5 (0.47 mg/100 g), folates (44–49 μg/100 g) and vitamin B_6 (0.14–0.15 mg/100 g).

Almonds are rich in minerals, supplying relevant amounts of Ca (266–269 mg/100 g), Mg (259–270 mg/100 g), P (405–481 mg/100 g), Mn (2.2 mg/100 g) and Se (4.1 μg/100 g) (INSA 2015; USDA 2015).

Many flavonoids can be found in almonds, including catechin (1.3 mg/100 g), epigallocatechin (2.6 mg/100 g) and epicatechin (0.6 mg/100 g), as well as eriodictyol[65] (0.2 mg/100 g) and naringenin (0.4 mg/100 g), the molecular structures of which are shown in Fig. 5.2. Isorhamnetin (2.6 mg/100 g), kaempferol (0.4 mg/100 g) and quercetin (0.4 mg/100 g) are also present in the noted average concentrations (USDA 2015). These important flavonoids are found in other edible plants (particularly of the *Brassicaceae* family), but are uncommon in nuts. Almond phytosterols (also found in other nuts) are δ-5-avenasterol (21 mg/100 g), campestanol (2.0 mg/100 g), sitostanol (4.0 mg/100 g) (Fig. 5.1) in addition to non-identified phytosterols (32 mg/100 g).

Roasting almonds causes alterations in their composition, altering the flavonoid profile and increasing the rate of chemical reactions, including those between tannins and proteins.

Almonds contain no phylloquinone nor anthocyanidins containing tannins instead, in the form of proanthocyanidin monomers (7.8 mg/100 g), dimers (9.5 mg/100 g), trimers (8.8 mg/100 g), 4–6mers (40 mg/100 g), 7–10 mers (37.7 mg/100 g) and polymers of more than 10 units (80.3 mg/100 g) (USDA 2015). Large oligomers dominate, as is also the case with hazelnuts.

Almonds are nutritionally very rich, providing an extensive list of beneficial compounds, the roles of which are described above and include anti-inflammatory, anti-cholesterol, anti-cancer and antioxidant properties. In addition, eriodictyol was found to increase insulin-stimulated glucose uptake, suggesting anti-diabetes properties (Zhang et al. 2012). It may also attenuate associated pathologies, such as diabetic retinopathy, where eriodictyol was demonstrated to reduce the degree of retinal inflammation and plasma lipid peroxidation preserving the blood–retinal barrier in early diabetic rats (Bucolo et al. 2012).

[65] IUPAC name: (2*S*)-2-(3,4-dihydroxyphenyl)-5,7-dihydroxy-2,3-dihydrochromen-4-one.

Pistachios & Almonds. *Nuts, in general, can be an important source of the amino acids tryptophan, methionine and cysteine in a diet poor in or without meat and dairy products. Pistachios (from Pistacia vera) and almonds (Prunus dulcis) are botanically classified as drupes, the seeds of which are the edible portion. They are good dietary sources of carotenoids and tocopherols. Both, pistachios and almonds contain catechin and condensed tannins, which positively impact gut's health. Pistachios contain the flavonoids daidzein and genistein, which are commonly found in pulses but are not so common in nuts. On the other hand, almonds contain Isorhamnetin, kaempferol and quercetin, which are typical of other botanical families (such as* Brassicaceae*) but uncommon in nuts. Almonds are rich in phytosterols, such as δ-5-avenasterol, campestanol and sitostanol, with positive impacts on cardiovascular health. Photo reprinted with kind permission from T. N. Wassermann*

5.8.5 Pistachio (*Pistacia vera*)

The pistachio is a small tree originally from Central Asia and the Middle East, a member of the cashew family (*Anacardiaceae*) and native to eastern Mediterranean (Cyprus and Turkey to Israel and Syria). The ancient Romans introduced the seed into Italy and Spain at about the same time.

Like the almond, the pistachio is a drupe, the seed of which is the edible part. When the fruit ripens, the shell changes from green to beige and abruptly splits partway open. Commercial cultivars vary in how consistently they split open. Greece and Italy are among the larger world producers in a list led by Iran and the USA (FAO 2015).

Pistachios are eaten raw as an appetiser (generally roasted with salt) or included in desserts and often in ice-cream.

On average, raw pistachios (with skin) supply 562 kcal/100 g from: proteins (20.3 %), fats (45.4 %: SFA 5.6 %, MUFA 23.8 % and PUFA 13.7 %), starch (1.7 %), mono- and disaccharides (7.7 %, mainly sucrose, glucose and fructose and traces of maltose) and dietary fibre (about 10.3 %) (USDA 2015).

Like other nuts, pistachios are a good source of B vitamins, supplying thiamine (0.87 mg/100 g), niacin (1.3 mg/100 g), riboflavin (0.16 mg/100 g), pantothenic acid (0.52 mg/100 g), folates (51 μg/100 g) and vitamin B_6 (1.7 mg/100 g). Other vitamins present in relevant amounts are the tocopherols α-tocopherol (2.3 mg/100 g), γ-tocopherol (22.6 mg/100 g) and δ-tocopherol (0.8 mg/100 g). Pistachios also contain the fat-soluble carotenes β-carotene (249 μg/100 g) and lutein + zeaxanthin (1.4 mg/100 g). Pistachios are a good source of P (490 mg/100 g), Mn (1.2 mg/100 g), Se (7.0 μg/100 g) and F (3.4 μg/100 g) (USDA 2015).

Flavonoids found in pistachios are catechin (3.6 mg/100 g), epigallocatechin (2 mg/100 g), epicatechin (0.8 mg/100 g) and epigallocatechin-3-gallate (0.4 mg/100 g), which, besides their antioxidant properties, may affect the absorption and bioavailability of anthocyanidins, as noted above.

In addition, pistachios contain quercetin (1.5 mg/100 g), referred to as a powerful anti-cancer compound (Fig. 5.2) and 3.6 mg/100 g of total isoflavones, including daidzein (1.9 mg/100 g) and genistein (1.8 mg/100 g), which are commonly found in pulses but are not so common in nuts (USDA 2015).

In addition, pistachios contain proanthocyanidins in the form of monomers (10.9 mg/100 g), dimers (13.3 mg/100 g), trimers (10.5 mg/100 g), 4–6mers (42.2 mg/100 g), 7–10mers (37.9 mg/100 g) and polymers with more than 10 units (122.5 mg/100 g). As with other nuts, large oligomers dominate; pistachios contain larger amounts of tannins than other nuts.

Like other members of the *Anacardiaceae* family, pistachios contain urushiol,[66] a catechol that is a natural toxin and a known irritant and allergen (NCBI 2015g).

5.8.6 Pine Nut (*Pinus pinea*)

Pine nuts are the edible seeds of pines (family *Pinaceae*, genus *Pinus*). Pine trees are distributed worldwide, and about 20 species of pine produce seeds large enough to be worth harvesting. Pine nuts have been eaten in Europe and Asia since the Palaeolithic period. They are eaten raw but are more frequently added to meat, fish, salads and vegetable dishes, baked into bread, included in desserts and even added to green tea preparations.

Throughout Europe and the Middle East, the pine nuts that are used are traditionally from *Pinus pinea* (stone pine). Asian pine nuts (from *Pinus koraiensis* and *Pinus armandii*) are often used in replacement, due to their lower price. Recently, a metallic taste disturbance was reported and attributed to the consumption of pine nuts. This syndrome is known as metallogeusia. It is typically observed 1–3 days after ingestion, being worse on the second day and lasting typically up to 2 weeks. Investigations hypothesised that nuts from a particular species of pine occurring mostly in China, *Pinus armandii*, were the cause of the problem. A scientific study found results consistent with this hypothesis and suggested that the cause might be linked to the

[66] IUPAC name: 3-[(8*E*,11*E*)-pentadeca-8,11-dienyl]benzene-1,2-diol.

chemicals used in the shelling process (Destaillats et al. 2011). European pine nuts are easily distinguished from the Asian pine nuts by their more slender shape and more homogeneous flesh.

Pine nuts. *Like other nuts, pine nuts (from Pinus pinea) have an important lipid fraction with a peculiar carbohydrate profile including several soluble cyclitols such as chiro and meso-inositols and fagopyritol, and pinitol, which have been suggested to interfere with insulin sensitivity and thus may help in controlling type II diabetes. The wide variety of phytosterols found in pine nuts also contribute to wellness, by lowering the levels of LDL cholesterol in the blood. Photo reprinted with kind permission from T. N. Wassermann*

European pine nuts (kernel) supply, on average, 618 kcal/100 g of edible portion from: proteins (about 33.2 %), fats (51.7 %: SFA 3.5 %, MUFA 15.0 % and PUFA 31.0 %, mostly linoleic acid), starch (2.6 %), mono- and disaccharides (2.4 %) and dietary fibre (about 1.9 %) (INSA 2015). Ruiz-Aceituno and colleagues (2012) reported low-molecular-weight carbohydrates to be mainly sugars (glucose, fructose, sucrose, raffinose, galactose, maltose and planteose) but also cyclitols, such as chiro-inositol, or meso-inositol (an important glucose isomer), which are present at average concentrations of 126.7–222.1 mg/100 g. According to the same authors, other detected unusual saccharides are fagopyritol B1[67] (94.2–177.1 mg/100 g) and pinitol[68] (51.2–282.8 mg/100 g).

Pine nuts are good dietary sources of the B vitamins thiamine (0.39 mg/100 g), niacin (2.7 mg/100 g), riboflavin (0.22 mg/100 g), folates (57 μg/100 g) and vitamin B_6 (0.11 mg/100 g), as well as fat-soluble vitamins such as α-tocopherol (10 mg/100 g) and phylloquinone. Pine nuts are important sources of minerals: Mg (270 mg/100 g), P (350 mg/100 g) and Zn (6.5 mg/100 g) (INSA 2015), as well as Mn. Data from the 2015 edition of the USDA food database refer to sample mixtures of nuts from *P. pinea* and *P. koraiensis*. This database is more extensive than PortFIR (INSA 2015), and the additional information is relevant, despite the differences in the food item. The USDA food database also mentions the presence of γ-tocopherol and the phytosterols campesterol, β-sitosterol, 5-avenasterol, sitostanol and campestanol, among other minor phytosterols, associated with pine nut lipid fraction.

Like other nuts, pine nuts exhibit nutraceutical properties. Thus, phospholipid-containing meso-inositol plays important roles in signal transduction (NCBI 2015h). D-pinitol has been suggested to play a role in the control of type II diabetes by influencing whole-body glucose tolerance and insulin sensitivity indexes (Hernández-Mijares et al. 2013; Kim et al. 2012) and the translocation of glucose in skeletal muscle (Dang et al. 2010).

Based on in vivo studies, Lee and co-workers (2014) suggest that the consumption of pine nut's oil prevents oxidative stress and atherosclerosis via the improvement of serum lipoprotein (lowering LDL cholesterol, in relation to high-density lipoprotein cholesterol) and the inhibition of lipid oxidation, certainly contributing to the anti-cholesterol and antioxidant activities of pine nuts and other nuts as discussed above.

In short, products of vegetable origin should account for the majority of the ingested food. These food items include cereals, pulses, green vegetables, aromatic herbs, fresh fruits, nuts, olive oil and table olives, and others. It is noteworthy that excessive industrial processing may affect their nutritional value e.g. their mineral and fibre contents. A growing tendency to use highly refined flours and to add simple sugars to industrial bakery products has been observed worldwide as well as in the area.

A range of different-coloured vegetables should be present on the plate: some are better raw (e.g. lettuce, carrots) and others release beneficial compounds upon being cooked (e.g. tomato and the liberation of lycopene, particularly when fried in olive oil). The variety of colours and textures is an indication of the diversity of vitamins, flavonoids, phytosterols and other health-beneficial compounds.

Aromatic herbs and spices provide flavour, anti-microbial protection for foods and important phytochemicals.

[67] IUPAC name: (1*S*,2*R*,4*R*,5*S*)-6-[(2*R*,3*R*,4*S*,5*R*,6*R*)-3,4,5-trihydroxy-6-(hydroxymethyl)oxan-2-yl]oxycyclohexane-1,2,3,4,5-pentol.

[68] IUPAC name: (1*S*,2*S*,4*S*,5*R*)-6-methoxycyclohexane-1,2,3,4,5 pentol.

Pulses are a sustainable and healthy alternative to animal protein sources, contributing to the reduction of meat consumption. They provide satiety, supplying much less energy than is commonly believed, because of their high fibre content. The digestibility of pulses is greatly improved with adequate cooking methods.

Fruits should always be provided at the end of each main meal instead of desserts. The Mediterranean basin provides a large variety of fruits each season, each one with their specificities. Finally, olive oil and walnuts have been granted health claims regarding their contribution to the control of coronary heart disease.

References

Abbo S, Saranga Y, Peleg Z, Kerem Z, Lev-Yadun S, Gopher A (2009) Reconsidering domestication of legumes versus cereals in the ancient Near East. Q Rev Biol 84(1):29–50. doi:10.1086/596462

Abbo S, Rachamim E, Zehavi Y, Zezak I, Lev-Yadun S, Gopher A (2011) Experimental growing of wild pea in Israel and its bearing on Near Eastern plant domestication. Ann Bot 107(8): 1399–1404. doi:10.1093/aob/mcr081

Abu-Yousif AO, Smith KA, Getsios S, Green KJ, Van Dross RT, Pelling JC (2008) Enhancement of UVB-induced apoptosis by apigenin in human keratinocytes and organotypic keratinocyte cultures. Cancer Res 68(8):3057–3065. doi:10.1158/0008-5472.CAN-07-2763

Adom KK, Sorrells ME, Liu RH (2005) Phytochemicals and antioxidant activity of milled fractions of different wheat varieties. J Agric Food Chem 53(6):2297–2306. doi:10.1021/jf048456d

Alam MA, Subhan N, Rahman MM, Uddin SJ, Reza HM, Sarker SD (2014) Effect of citrus flavonoids, naringin and naringenin, on metabolic syndrome and their mechanisms of action. Adv Nutr 5(4):404–417. doi:10.3945/an.113.005603

Al-Khaldi S (2012) Phytohaemagglutinin (kidney bean lectin). In: Lampel KA, Al-Khaldi S, Cahill SM (eds) Bug book, foodborne pathogenic microorganisms and natural toxins, 2nd edn. Food and Drug Administration, Silver Spring

Allahghadri T, Rasooli I, Owlia P, Nadooshan MJ, Ghazanfari T, Taghizadeh M, Astaneh SDA (2010) Antimicrobial property, antioxidant capacity, and cytotoxicity of essential oil from cumin produced in Iran. J Food Sci 75(2):H54–H61. doi:10.1111/j.1750-3841.2009.01467.x

Anonymous (2011) Castanea sativa Mill. Flora of Pakistan, http://www.tropicos.org/Name/13100366?projectid=32. Accessed 5 Nov 2015

Apaydin H, Ertan S, Ozekmekçi S (2000) Broad bean (Vicia faba)-a natural source of L-dopa—prolongs 'on' periods in patients with Parkinson's disease who have 'on-off' fluctuations. Mov Disord 15(1):164–166. doi:10.1002/1531-8257(200001)15:1<164::AID-MDS1028>3.0.CO;2-E

Banel DK, Hu FB (2009) Effects of walnut consumption on blood lipids and other cardiovascular risk factors: a meta-analysis and systematic review. Am J Clin Nutr 90(1):56–63. doi:10.3945/ajcn.2009.27457

Barreira JCM, Ferreira ICFR, Oliveira MBPP, Pereira JA (2008) Antioxidant activities of the extracts from chestnut flower, leaf, skins and fruit. Food Chem 107(3):1106–1113. doi:10.1016/j.foodchem.2007.09.030

Barros AIRNA, Nunes FM, Gonçalves B, Bennett RN, Silva AP (2011) Effect of cooking on total vitamin C contents and antioxidant activity of sweet chestnuts (Castanea sativa Mill.). Food Chem 128(1):165–172. doi:10.1016/j.foodchem.2011.03.013

Bentley AR, Kritchevsky SB, Harris TB, Holvoet P, Jensen RL, Newman AB, Lee JS, Yende S, Bauer D, Cassano PA (2012) Dietary antioxidants and forced expiratory volume in 1 s decline: the Health, Aging and Body Composition study. Eur Respir J 39(4):979–984. doi:10.1183/09031936.00190010

Bettaieb I, Bourgou S, Sriti J, Msaada K, Limam F, Marzouk B (2011) Essential oils and fatty acids composition of Tunisian and Indian cumin (*Cuminum cyminum* L.) seeds: a comparative study. J Sci Food Agric 91:2100–2107

Bjarnsholt T, Jensen PO, Rasmussen TB, Christophersen L, Calum H, Hentzer M, Hougen HP, Rygaard J, Moser C, Eberl L, Høiby N, Givskov M (2005) Garlic blocks quorum sensing and promotes rapid clearing of pulmonary Pseudomonas aeruginosa infections. Microbiology 151 (12):3873–3880. doi:10.1099/mic.0.27955-0

Borlinghaus J, Albrecht F, Gruhlke MCH, Nwachukwu ID, Slusarenko AJ (2014) Allicin: chemistry and biological properties. Molecules 19(8):12591–12618. doi:10.3390/molecules190812591

Brewster JL (1994) Onions and other vegetable alliums. CAB, Wallingford

Bucolo C, Leggio GM, Drago F, Salomone S (2012) Eriodictyol prevents early retinal and plasma abnormalities in streptozotocin-induced diabetic rats. Biochem Pharmacol 84(1):88–92. doi:10.1016/j.bcp.2012.03.019

Burbano C, Cuadrado C, Muzquiz M, Cubero JI (1995) Variation of favism-inducing factors (vicine, convicine and L-DOPA) during pod development in Vicia faba L. Plant Foods Hum Nutr 47(3):265–275. doi:10.1007/BF01088335

Burdock GA, Carabin IG (2009) Safety assessment of coriander (Coriandrum sativum L.) essential oil as a food ingredient. Food Chem Toxicol 47(1):22–34. doi:10.1016/j.fct.2008.11.006

Byun S, Lee KW, Jung SK, Lee EJ, Hwang MK, Lim SH, Bode AM, Lee HJ, Dong Z (2010) Luteolin inhibits protein kinase cε and c-src activities and UVB-induced skin cancer. Cancer Res 70(6):2415–2423. doi:10.1158/0008-5472

Carvalho AM, Morales R (2005) Persistence of wild food and wild medicinal plant knowledge in a northeastern region of Portugal. In: Santayana MP, Pieroni A, Puri RK (eds) The New Europe: people, health and wild plant resources. Berghahn, Oxford

Castellarin SD, Gambetta GA, Wada H, Shackel KA, Matthews MA (2011) Fruit ripening in Vitis vinifera: spatiotemporal relationships among turgor, sugar accumulation, and anthocyanin biosynthesis. J Exp Bot 62:4345–4354. doi:10.1093/jxb/err150

Chan JY, Yuen AC, Chan RY, Chan SW (2013) A review of the cardiovascular benefits and antioxidant properties of allicin. Phytother Res 27:637–646

Chatterjee S, Nöldner M, Schötz K (2006) Use of rutin and isorhamnetin for treating depressive states and depression and other emotion disorders. US Patent 20060198914 A14, 7 Sept 2006

Cheynier V (2005) Polyphenols in foods are more complex than often thought. Am J Clin Nutr 81(1 Suppl):223S–2239S

Chiang CT, Way TD, Lin JK (2007) Sensitizing HER2-overexpressing cancer cells to luteolin-induced apoptosis through suppressing p21WAF1/CIP1 expression with rapamycin. Mol Cancer Ther 6(7):2127–2138. doi:10.1158/1535-7163.MCT-07-0107

Colle IJP, Lemmens L, Buggenhout SV, Van Loey AM, Hendrickx ME, Tian D, McIntee EJ, Hecht SS, Chung F-L (2013) Modeling lycopene degradation and isomerization in the presence of lipids. Food Bioprocess Technol 6(4):909–918. doi:10.1007/s11947-011-0714-4

Conaway CC, Wang C-X, Pittman B, Yang Y-M, Schwartz JE (2005) Phenethyl isothiocyanate and sulforaphane and their N-acetylcysteine conjugates inhibit malignant progression of lung adenomas induced by tobacco carcinogens in A/J mice. Cancer Res 65(18):8548–8557. doi:10.1158/0008-5472.CAN-05-0237

Courteau J (2015a) Malus domestica. Encyclopedia of Life. Available at http://eol.org/pages/629094/details. Accessed 5 Nov 2015

Courteau J (2015b) Phoenix dactylifera. Encyclopedia of Life. Available http://eol.org/pages/1135088/details. Accessed 5 Nov 2015

Crocoll C, Asbach J, Novak J, Gershenzon J, Degenhardt J (2010) Terpene synthases of oregano (Origanum vulgare L.) and their roles in the pathway and regulation of terpene biosynthesis. Plant Mol Biol 73(6):587–603. doi:10.1007/s11103-010-9636-1

Dang NT, Mukai R, Yoshida K, Ashida H (2010) D-pinitol and myo-inositol stimulate translocation of glucose transporter 4 in skeletal muscle of C57BL/6 mice. Biosci Biotechnol Biochem 74(5):1062–1067. doi:10.1271/bbb.90963

Destaillats F, Cruz-Hernandez C, Giuffrida F, Dionisi F, Mostin M, Verstegen G (2011) Identification of the botanical origin of commercial pine nuts responsible for dysgeusia by gas-liquid chromatography analysis of fatty acid profile. J Toxicol 2011:316789. doi:10.1155/2011/316789

Eaton DL, Schaupp CM (2014) Of mice, rats, and men: could Nrf2 activation protect against aflatoxin heptocarcinogenesis in Humans? Cancer Prev Res 7(7):653–657. doi:10.1158/1940-6207.CAPR-14-0119

Edwards RL, Lyon T, Litwin SE, Rabovsky A, Symons D, Jalili T (2007) Quercetin reduces blood pressure in hypertensive subjects. J Nutr 137(11):2405–2411

EFSA (2011) Scientific opinion on the substantiation of health claims related to walnuts and maintenance of normal blood LDL-cholesterol concentrations (ID 1156, 1158) and improvement of endothelium-dependent vasodilation (ID 1155, 1157) pursuant to Article 13(1) of Regulation (EC) No 1924/2006. EFSA J 9(4):2074–2093. doi:10.2903/j.efsa.2011.2074

EFSA (2015) Acrylamide. European Food Safety Authority, Parma. Available at http://www.efsa.europa.eu/en/topics/topic/acrylamide.htm. Accessed 15 Feb 2015

EOL (2015a) Allium sativum. Available at http://eol.org/pages/1084926/details. Accessed 5 Nov 2015

EOL (2015b) Vitis vinifera. Available at http://eol.org/pages/582304/details. Accessed 5 Nov 2015

Ercisli S, Tosun M, Karlidag H, Dzubur A, Hadziabulic S, Aliman Y (2012) Color and antioxidant characteristics of some fresh fig (Ficus carica L.) genotypes from northeastern Turkey. Plant Foods Hum Nutr 67(3):271–276. doi:10.1007/s11130-012-0292-2

Espley RV, Butts CA, Laing WA, Martell S, Smith H, McGhie TK, Zhang J, Paturi G, Hedderley D, Bovy A, Schouten HJ, Putterill J, Allan AC, Hellens RP (2014) Dietary flavonoids from modified apple reduce inflammation markers and modulate gut microbiota in mice. J Nutr 144(2):146–154. doi:10.3945/jn.113.182659

FAO (2015) Statistics Division. Trade. http://faostat3.fao.org/browse/T/. Accessed 16 Mar 2015

FDA (2003) Qualified Health Claims: Letter of Enforcement Discretion—Nuts and Coronary Heart Disease (Docket No 02P-0505). Food and Drug Administration, Silver Spring. Available at http://www.fda.gov/Food/IngredientsPackagingLabeling/LabelingNutrition/ucm072926.htm. Accessed 5 Nov 2015

FDA (2014a) FDA Poisonous plant database. http://www.accessdata.fda.gov/scripts/plantox/detail.cfm?id=1364. Accessed 10 Oct 2014

FDA (2014b) FDA Poisonous plant database. http://www.accessdata.fda.gov/scripts/plantox/detail.cfm?id=18235. Accessed 10 Oct 2014

Felgines C, Talavéra S, Gonthier MP, Texier O, Scalbert A, Lamaison JL, Rémésy C (2003) Strawberry anthocyanins are recovered in urine as glucuro- and sulfoconjugates in humans. J Nutr 133(5):1296–1301

Fernandes I, Nave F, Goncalves R, Freitas V, Mateus N (2012) On the bioavailability of flavanols and anthocyanins: flavanol-anthocyanin dimers. Food Chem 135(2):812–818. doi:10.1016/j.foodchem.2012.05.037

Fernández-Sánchez L, Lax P, Esquiva G, Martín-Nieto J, Pinilla I, Cuenca N (2012) Safranal, a saffron constituent, attenuates retinal degeneration in P23H rats. PLoS One 7(8), e43074

Fielding JM, Rowley KG, Cooper P, O'Dea K (2005) Increases in plasma lycopene concentration after consumption of tomatoes cooked with olive oil. Asia Pac J Clin Nutr 14(2):131–136

Firouzi R, Shekarforoush SS, Nazer AH, Borumand Z, Jooyandeh AR (2007) Effects of essential oils of oregano and nutmeg on growth and survival of Yersinia enterocolitica and Listeria monocytogenes in barbecued chicken. J Food Prot 70(11):2626–2630

Fitzpatrick TB, Basset GJC, Borel P, Carrari F, DellaPenna D, Fraser PD, Hellmann H, Osorio S, Rothan C, Valpuesta V, Caris-Veyrat C, Fernie AR (2012) Vitamin deficiencies in humans: can plant science help? Plant Cell 24(2):395–414. doi:10.1105/tpc.111.093120

Forester SC, Lambert JD (2014) Synergistic inhibition of lung cancer cell lines by (-)-epigallocatechin-3-gallate in combination with clinically used nitrocatechol inhibitors of catechol-O-methyltransferase. Carcinogenesis 35(2):365–372. doi:10.1093/carcin/bgt347

Forester SC, Choy YY, Waterhouse AL, Oteiza PI (2014) The anthocyanin metabolites gallic acid, 3-O-methylgallic acid, and 2,4,6-trihydroxybenzaldehyde decrease human colon cancer cell viability by regulating pro-oncogenic signals. Mol Carcinog 53(6):432–439. doi:10.1002/mc.21974

Foster RH, Hardy G, Alany RG (2010) Borage oil in the treatment of atopic dermatitis. Nutrition 26(7–8):708–718. doi:10.1016/j.nut.2009.10.014

Franzen CA, Amargo E, Todorović V, DEsai BV, Huda S, Mirzoeva S, Chiu K, Grzybowski BA, Chew T-L, Green KJ, Pelling JC (2009) The chemopreventive bioflavonoid apigenin inhibits prostate cancer cell motility through the focal adhesion kinase/Src signaling mechanism. Cancer Prev Res 2:830–841

Gajendragadkar PR, Hubsch A, Maki-Petaja KM, Serg M, Wilkinson IB, Cheriyan J (2014) Effects of oral lycopene supplementation on vascular function in patients with cardiovascular disease and healthy volunteers: a randomised controlled trial. PLoS One 9(6):e99070. doi:10.1371/journal.pone.0099070

García-Alonso A, Goñi I (2000) Effect of processing on potato starch: in vitro availability and glycaemic index. Nahrung 44:19–22

Garland S (1979) The complete book of herbs & spices: an illustrated guide to growing and using culinary, aromatic, cosmetic and medicinal plants. Frances Lincoln, London

Gossell-Williams M, Davis A, O'Connor N (2006) Inhibition of testosterone-induced hyperplasia of the prostate of Sprague-Dawley rats by pumpkin seed oil. J Med Food 9:284–286

Graf BA, Mullen W, Caldwell ST, Hartley RC, Duthie GG, Lean MEJ, Crozier A, Edwards CA (2005) Disposition and metabolism of [2-14C]Quercetin-4′-glucoside in rats. Drug Metab Dispos 33(7):1036–1043. doi:10.1124/dmd.104.002691

Graziani G, D'Argenio G, Tuccillo C, Loguercio C, Ritieni C, Morisco AF, Blanco CDV, Fogliano V, Romano M (2005) Apple polyphenol extracts prevent damage to human gastric epithelial cells in vitro and to rat gastric mucosa in vivo. Gut 54(2):193–200. doi:10.1136/gut.2004.046292

Hartmann D, Thürmann PA, Spitzer V, Schalch W, Manner B, Cohn W (2004) Plasma kinetics of zeaxanthin and 3′-dehydro-lutein after multiple oral doses of synthetic zeaxanthin. Am J Clin Nutr 79(3):410–417

Hernández-Mijares A, Bañuls C, Peris JE, Monzó N, Jover A, Bellod L, Victor VM, Rocha M (2013) A single acute dose of pinitol from a naturally-occurring food ingredient decreases hyperglycaemia and circulating insulin levels in healthy subjects. Food Chem 141:1267–1272

Herrmann M, Joppe H, Schmaus G (2002) Thesinine-4′-O-beta-D-glucoside the first glycosylated plant pyrrolizidine alkaloid from Borago officinalis. Phytochemistry 60(4):399–402. doi:10.1016/S0031-9422(02)00069-9

Hodge AM, English DR, O'Dea K, Sinclair AJ, Makrides M, Gibson RA, Giles GG (2007) Plasma phospholipid and dietary fatty acids as predictors of type 2 diabetes: interpreting the role of linoleic acid. Am J Clin Nutr 86(1):189–197

Horinaka M, Yoshida T, Shiraishi T, Nakata S, Wakada M, Sakai T (2006) The dietary flavonoid apigenin sensitizes malignant tumor cells to tumor necrosis factor-related apoptosis-inducing ligand. Mol Cancer Ther 5(4):945–951. doi:10.1158/1535-7163.MCT-05-0431

Huang WC, Chang WT, Wu SJ, Xu PY, Ting NC, Liou CJ (2013) Phloretin and phlorizin promote lipolysis and inhibit inflammation in mouse 3T3-L1 cells and in macrophage-adipocyte co-cultures. Mol Nutr Food Res 57(10):1803–1813. doi:10.1002/mnfr.201300001

Hui C, Qi X, Qianyong Z, Xiaoli P, Jundong Z, Mantian M (2013) Flavonoids, flavonoid subclasses and breast cancer risk: a meta-analysis of epidemiologic studies. PLoS One 8(1):e54318. doi:10.1371/journal.pone.0054318

INSA (2015) Tabela da Composição de Alimentos (TCA). http://www.insa.pt/sites/INSA/Portugues/AreasCientificas/AlimentNutricao/AplicacoesOnline/TabelaAlimentos/PesquisaOnline/Paginas/PorPalavraChave.aspx. Accessed 5 Nov 2015

Jakobeka L, Šerugaa M, Voćab S, Šindrakc Z, Dobričevićb N (2009) Flavonol and phenolic acid composition of sweet cherries (cv. Lapins) produced on six different vegetative rootstocks. Sci Hort 123(1):23–28. doi:10.1016/j.scienta.2009.07.012

Jandér KC, Herre EA (2010) Host sanctions and pollinator cheating in the fig tree–fig wasp mutualism. Proc Biol Sci 277(1687):1481–1488. doi:10.1098/rspb.2009.2157

Jeyabal PV, Syed MB, Venkataraman M, Sambandham JK, Sakthisekaran D (2005) Apigenin inhibits oxidative stress-induced macromolecular damage in N-nitrosodiethylamine (NDEA)-induced hepatocellular carcinogenesis in Wistar albino rats. Mol Carcinog 44(1):11–20. doi:10.1002/mc.20115

Jung SK, Lee KW, Byun S, Lee EJ, Kim J-E, Bode AM, Dong Z, Lee HJ (2010) Myricetin inhibits UVB-induced angiogenesis by regulating PI-3 kinase in vivo. Carcinogenesis 31(5):911–917. doi:10.1093/carcin/bgp221

Jutooru I, Guthrie AS, Chadalapaka G, Pathi S, Kim K, Burghardt R, Jin UH, Safe S (2014) Mechanism of action of phenethylisothiocyanate and other reactive oxygen species-inducing anticancer agents. Mol Cell Biol 34(13):2382–2395. doi:10.1128/MCB.01602-13

Kavitha K, Swamy MJ (2006) Thermodynamic studies on the interaction of water-soluble porphyrins with the glucose/mannose-specific lectin from garden pea (Pisum sativum). IUBMB Life 58(12):720–730. doi:10.1080/15216540601069761

Kim HJ, Park KS, Lee SK, Min KW, Han KA, Kim YK, Ku BJ (2012) Effects of pinitol on glycemic control, insulin resistance and adipocytokine levels in patients with type 2 Diabetes mellitus. Ann Nutr Metab 60(1):1–5. doi:10.1159/000334834

Kitsteiner J (2011) Permaculture plants: walnut trees. temperate climate permaculture, http://tcpermaculture.blogspot.it/2011/10/permaculture-plants-walnut-trees.html. Accessed 5 Nov 2015

Kohler T, Weidenmaier C, Peschel A (2009) Wall teichoic acid protects Staphylococcus aureus against antimicrobial fatty acids from human skin. J Bacteriol 191(13):4482–4484. doi:10.1128/JB.00221-09

Kubo I, Chen QX, Nihei K, Calderon JS, Cespedes CL (2003) Tyrosinase inhibition kinetics of anisic acid. Z Naturforsch C58(9–10):713–718

Labbé D, Provençal M, Lamy S, Boivin D, Gingras D, Béliveau R (2009) The flavonols quercetin, kaempferol, and myricetin inhibit hepatocyte growth factor-induced medulloblastoma cell migration. J Nutr 139(4):646–652. doi:10.3945/jn.108.102616

Ladizinsky G (1999) On the origin of almond. Gen Res Crop Evol 46(2):143–147. doi:10.1023/A:1008690409554

Lee KW, Kang NJ, Rogozin EA, Kim H-G, Cho YY, Bode AM, Lee HJ, Surh Y-J, Bowden GT, Dong Z (2007) Myricetin is a novel natural inhibitor of neoplastic cell transformation and MEK1. Carcinogenesis 28(9):1918–1927. doi:10.1093/carcin/bgm110

Lee JS, Kim DJ, Kang YH, Kim TW, Kim KK, Choe M (2014) Antiatherosclerosis effect of pine nut oil in HCHF diet-fed rats. FASEB J 28(1 suppl):959.12

Lekli I, Ray D, Das DK (2010) Longevity nutrients resveratrol, wines and grapes. Genes Nutr 5(1):55–60. doi:10.1007/s12263-009-0145-2

Li HH, Hao RL, Wu SS, Guo PC, Chen CJ, Pan LP, Ni H (2011) Occurrence, function and potential medicinal applications of the phytohormone abscisic acid in animals and humans. Biochem Pharmacol 82:701–712

Li H, Zhu F, Chen H, Cheng KW, Zykova T, Oi N, Lubet RA, Bode AM, Wang M, Dong Z (2014) 6-C-(E-phenylethenyl)-naringenin suppresses colorectal cancer growth by inhibiting cyclooxygenase-1. Cancer Res 74:43–252

Lin X, Racette SB, Ma L, Wallendorf M, Spearie CA, Ostlund RE Jr (2015) Plasma biomarker of dietary phytosterol intake. PLoS One 10(2):e0116912. doi:10.1371/journal.pone.0116912

Lu J, Papp LV, Fang J, Rodriguez-Nieto S, Zhivotovsky B, Holmgren A (2006) Inhibition of mammalian thioredoxin reductase by some flavonoids: implications for myricetin and quercetin. Anticancer Activity Cancer Res 66:4410–4418

Lynett PT, Butts K, Vaidya V, Garrett GE, Pratt DA (2011) The mechanism of radical-trapping antioxidant activity of plant-derived thiosulfinates. Org Biomol Chem 9(9):3320–3330. doi:10.1039/c1ob05192j

Mackenzie GG, Carrasquedo F, Delfino JM, Keen CL, Fraga CG, Oteiza PI (2004) Epicatechin, catechin, and dimeric procyanidins inhibit PMA-induced NF-kappaB activation at multiple steps in Jurkat T cells. FASEB J 18(1):167–169. Epub 2003 Nov 20

Madden SMM, Garrioch CF, Holub BJ (2009) Direct diet quantification indicates low intakes of (n-3) fatty acids in children 4 to 8 years old. J Nutr 139(3):528–532. doi:10.3945/jn.108.100628

Maguire LS, O'Sullivan SM, Galvin K, O'Connor TP, O'Brien NM (2004) Fatty acid profile, tocopherol, squalene and phytosterol content of walnuts, almonds, peanuts, hazelnuts and the macadamia nut. Int J Food Sci Nutr 55(3):171–178. doi:10.1080/09637480410001725175

Mahadov S, Green PHR (2011) Celiac disease: a challenge for all physicians. Gastroenterol Hepatol 7(8):554–556

Mares JA, LaRowe TL, Snodderly DM, Moeller SM, Gruber MJ, Klein ML (2006) Predictors of optical density of lutein and zeaxanthin in retinas of older women in the carotenoids in age-related eye disease study, an ancillary study of the Women's Health Initiative. Am J Clin Nutr 84(5):1107–1122

Mares-Perlman JA, Millen AE, Ficek TL, Hankinson SE (2002) The body of evidence to support a protective role for lutein and zeaxanthin in delaying chronic disease. Overview. J Nutr 132(3): 518S–524S

Martinotti S, Ranzato E, Parodi M, Vitale M, Burlando B (2014) Combination of ascorbate/epigallocatechin-3-gallate/gemcitabine synergistically induces cell cycle deregulation and apoptosis in mesothelioma cells. Toxicol Appl Pharmacol 274(1):35–41. doi:10.1016/j.taap.2013.10.025

McCrory MA, Hamaker BR, Lovejoy JC, Eichelsdoerfer PE (2010) Pulse consumption, satiety, and weight management. Adv Nutr 1:17–30. doi:10.3945/an.110.1006

McCullough ML, Peterson JJ, Patel R, Jacques PF, Shah R, Dwyer JT (2012) Flavonoid intake and cardiovascular disease mortality in a prospective cohort of US adults. Am J Clin Nutr 95(2): 454–464. doi:10.3945/ajcn.111.016634

Miura T, Chiba M, Kasai K, Nozaka H, Nakamura T, Shoji T, Kanda T, Ohtake Y, Sato T (2008) Apple procyanidins induce tumor cell apoptosis through mitochondrial pathway activation of caspase-3. Carcinogenesis 29(3):585–593

Moon SH, Lee JH, Kim KT, Park YS, Nah SY, Ahn DU, Paik HD (2013) Antimicrobial effect of 7-O-butylnaringenin, a novel flavonoid, and various natural flavonoids against Helicobacter pylori strains. Int J Environ Res Public Health 10(11):5459–5469. doi:10.3390/ijerph10115459

Morton JF (1987) Orange. In: Morton JF (ed) Fruits of warm climates. Julia F. Morton, Miami

Nazimuddin S, Qaiser M (1982) Vitaceae. Flora Pak 147:1–20

NCBI (2015a) PubChem compound Database. CID 445354. National Center for Biotechnology Information. http://pubchem.ncbi.nlm.nih.gov/compound/445354. Accessed 5 Nov 2015

NCBI (2015b) PubChem compound Database. CID 3035211. National Center for Biotechnology Information. http://pubchem.ncbi.nlm.nih.gov/compound/3035211. Accessed 5 Nov 2015

NCBI (2015c) PubChem compound Database. CID 446925. National Center for Biotechnology Information. http://pubchem.ncbi.nlm.nih.gov/compound/446925. Accessed 5 Nov 2015

NCBI (2015d) PubChem compound Database. CID 1203. National Center for Biotechnology Information. http://pubchem.ncbi.nlm.nih.gov/compound/1203. Accessed 5 Nov 2015

NCBI (2015e) PubChem compound Database. CID 72281. National Center for Biotechnology Information. http://pubchem.ncbi.nlm.nih.gov/compound/72281. Accessed 5 Nov 2015

NCBI (2015f) PubChem compound Database. CID 5362588. National Center for Biotechnology Information. http://pubchem.ncbi.nlm.nih.gov/compound/5362588. Accessed 5 Nov 2015

NCBI (2015g) PubChem compound Database. CID 5281862. National Center for Biotechnology Information. http://pubchem.ncbi.nlm.nih.gov/compound/5281862. Accessed 5 Nov 2015

NCBI (2015h) PubChem compound Database. CID 892. National Center for Biotechnology Information. http://pubchem.ncbi.nlm.nih.gov/compound/892. Accessed 5 Nov 2015

Nishimura M, Ohkawara T, Sato H, Takeda H, Nishihira J (2014) Pumpkin seed oil extracted from Cucurbita maxima improves urinary disorder in human overactive bladder. J Tradit Complement Med 4(1):72–74. doi:10.4103/2225-4110.124355

NLM (2015) TOXNET—network of databases on toxicology, hazardous chemicals and environmental health. U.S. National Library of Medicine. Available at http://toxnet.nlm.nih.gov/. Accessed 5 Nov 2015

Noda S, Tanabe S, Suzuki T (2013) Naringenin enhances intestinal barrier function through the expression and cytoskeletal association of tight junction proteins in Caco-2 cells. Mol Nutr Food Res 57(11):2019–2028. doi:10.1002/mnfr.201300045

O'Byrne SM, Blaner WS (2013) Retinol and retinyl esters: biochemistry and physiology. Thematic review series: fat-soluble vitamins: vitamin A. J Lipid Res 54(7):1731–1743. doi:10.1194/jlr.R037648

Ochiai T, Shimeno H, Mishima K, Iwasaki K, Fujiwara M, Tanaka H, Shoyama Y, Toda A, Eyanagi R, Soeda S (2007) Protective effects of carotenoids from saffron on neuronal injury in vitro and in vivo. Biochim Biophys Acta 1770(4):578–584. doi:10.1016/j.bbagen.2006.11.012

Orhan IE, Nabavi SF, Daglia M, Tenore GC, Mansouri K, Nabavi SM (2015) Naringenin and atherosclerosis: a review of literature. Curr Pharm Biotechnol 16(3):245–251. doi:10.2174/1389201015666141202110216

Ozkan G, Koyuncu MA (2005) Physical and chemical composition of some walnut (Juglans regia L) genotypes grown in Turkey. Grasas Aceites 56(2):141–146. doi:10.3989/gya.2005.v56.i2.122

Pai S, Ghugre PS, Udipi SA (2005) Satiety from rice-based, wheat-based and rice-pulse combination preparations. Appetite 44(3):263–271. doi:10.1016/j.appet.2005.01.004

Palaniswamy UR, McAvoy RJ, Bible BB, Stuart JD (2003) Ontogenic variations of ascorbic acid and phenethyl isothiocyanate concentrations in watercress (Nasturtium officinale R.Br.) leaves. J Agric Food Chem 51(18):5504–5509. doi:10.1021/jf034268w

Pataki T, Bak I, Kovacs P, Bagchi D, Das DK, Tosaki A (2002) Grape seed proanthocyanidins improved cardiac recovery during reperfusion after ischemia in isolated rat hearts. Am J Clin Nutr 75(5):894–899

Pedras MS, Montaut S, Suchy M (2004) Phytoalexins from the crucifer rutabaga: structures, synthesis, biosynthesis, and antifungal activity. J Org Chem 69(13):4471–4476. doi:10.1021/jo049648a

Pedreschi F, Mariotti MS, Granby K (2014) Current issues in dietary acrylamide: formation, mitigation and risk assessment. J Sci Food Agric 94(1):9–20. doi:10.1002/jsfa.6349

Piccaglia R, Marotti M (2001) Characterization of some Italian types of wild fennel (Foeniculum vulgare Mill.). J Agric Food Chem 49(1):239–244. doi:10.1021/jf000636+

Pittaway JK, Robertson IK, Ball MJ (2008) Chickpeas may influence fatty acid and fiber intake in an ad libitum diet, leading to small improvements in serum lipid profile and glycemic control. J Am Diet Assoc 108(6):1009–1013. doi:10.1016/j.jada.2008.03.009

Porres JM, Lopez-Jurado M, Aranda P, Urbano G (2004) Bioavailability of phytic acid-phosphorus and magnesium from lentils (Lens culinaris) in growing rats: influence of thermal treatment and vitamin-mineral supplementation. Nutrition 20(9):794–799. doi:10.1016/j.nut.2004.05.018

Priscilla DH, Roy D, Suresh A, Kumar V, Thirumurugan K (2014) Naringenin inhibits α-glucosidase activity: a promising strategy for the regulation of postprandial hyperglycemia in high fat diet fed streptozotocin induced diabetic rats. Chem Biol Interact 210:77–85. doi:10.1016/j.cbi.2013.12.014

Psaltopoulou T, Naska A, Orfanos P, Trichopoulos D, Mountokalakis T, Trichopoulou A (2004) Olive oil, the Mediterranean diet, and arterial blood pressure: the Greek European prospective investigation into cancer and nutrition (EPIC) study. Am J Clin Nutr 80(4):1012–1018

Qu H, Madl RL, Takemoto DJ, Baybutt RC, Wang W (2005) Lignans are involved in the antitumor activity of wheat bran in colon cancer SW480 cells. J Nutr 135(3):598–602

Ramprasath VR, Jenkins DJ, Lamarche B, Kendall CW, Faulkner D, Cermakova L, Couture P, Ireland C, Abdulnour S, Patel D, Bashyam B, Srichaikul K, de Souza RJ, Vidgen E, Josse RG, Leiter LA, Connelly PW, Frohlich J, Jones PJ (2014) Consumption of a dietary portfolio of cholesterol lowering foods improves blood lipids without affecting concentrations of fat soluble compounds. Nutr J 13:101. doi:10.1186/1475-2891-13-101

Ramya KB, Thaakur S (2007) Herbs containing L-Dopa: an update. Anc Sci Life 27(1):50–55

Rivera D, Obón C, Heinrich M, Inocencio C, Verdea A, Fajardo J (2006) Gathered Mediterranean food plants—ethnobotanical investigations and historical development. In: Heinrich M, Müller WE, Galli C (eds) Local Mediterranean food plants and nutraceuticals. Forum Nutr 59:18–74

Romeo R, Sangiovanni A, Iavarone M, Vavassori S, Della Corte C, Colombo M (2010) Diagnostic value of Lens culinaris agglutinin isoform 3 fraction (AFP-L3 %) and des-gamma-carboxy prothrombin (DCP) for the diagnosis of hepatocellular carcinoma in cirrhotic patients. ASCO Meet Abstr 28:4119

Roux KH, Teuber SS, Sathe SK (2003) Tree nut allergens. Int Arch Allergy Immunol 131(4): 234–244. doi:10.1159/000072135

Ruiz-Aceituno L, Ramos L, Martinez-Castro I, Sanz ML (2012) Low molecular weight carbohydrates in pine nuts from Pinus pinea L. J Agric Food Chem 60(19):4957–4959. doi:10.1021/jf2048959

Sahib NG, Anwar F, Gilani AH, Hamid AA, Saari N, Alkharfy KM (2013) Coriander (Coriandrum sativum L.): a potential source of high-value components for functional foods and nutraceuticals—a review. Phytother Res 27(10):1439–1456. doi:10.1002/ptr.4897

Samarghandian S, Borji A (2014) Anticarcinogenic effect of saffron (*Crocus sativus* L.) and its ingredients. Pharmacognosy Res 6:99–107

Schieber A, Carle R (2005) Occurrence of carotenoid cis-isomers in food: technological, analytical, and nutritional implications. Trends Food Sci Technol 16(9):416–422. doi:10.1016/j.tifs.2005.03.018

Schonewille M, Brufau G, Shiri-Sverdlov R, Groen AK, Plat J (2014) Serum TG-lowering properties of plant sterols and stanols are associated with decreased hepatic VLDL secretion. J Lipid Res 55(12):2554–2561. doi:10.1194/jlr.M052407

Schwarz FP, Puri KD, Bhat RG, Surolia A (1993) Thermodynamics of monosaccharide binding to concanavalin A, pea (Pisum sativum) lectin, and lentil (Lens culinaris) lectin. J Biol Chem 268:7668–7677

Serra AT, Rocha J, Sepodes B, Matias AA, Feliciano RP, Carvalho A, Bronze MR, Duarte CM, Figueira ME (2012) Evaluation of cardiovascular protective effect of different apple varieties—correlation of response with composition. Food Chem 135(4):2378–2386. doi:10.1016/j.foodchem.2012.07.067

Seymour EM, Ou B (2011) Phytochemical and diverse antioxidant capacity profile of whole tart cherries (Prunus cerasus). FASEB J 25(meeting abstract supplement):773.14

Shereen S (2007) Peas, nutrition. About.com, www.about.com

Shewry PR (2009) Wheat. J Exp Bot 60(6):1537–1553. doi:10.1093/jxb/erp058

Shi J, Le Maguer M (2000) Lycopene in tomatoes: chemical and physical properties affected by food processing. Crit Rev Biotechnol 20(4):293–334. doi:10.1080/07388550091144212

Shi R, Huang Q, Zhu X, Ong Y-B, Zhao B, Lu J, Ong C-N, Shen H-M (2007) Luteolin sensitizes the anticancer effect of cisplatin via c-Jun NH2-terminal kinase-mediated p53 phosphorylation and stabilization. Mol Cancer Ther 6(4):1338–1347. doi:10.1158/1535-7163.MCT-06-0638

Shouk R, Abdou A, Shetty K, Sarkar D, Eid AH (2014) Mechanisms underlying the antihypertensive effects of garlic bioactives. Nutr Res 34(2):106–115. doi:10.1016/j.nutres.2013.12.005

Siegelin MD, Reuss DE, Habel A, Herold-Mende C, von Deimling A (2008) The flavonoid kaempferol sensitizes human glioma cells to TRAIL-mediated apoptosis by proteasomal degradation of survivin. Mol Cancer Ther 7(11):3566–3574. doi:10.1158/1535-7163.MCT-08-0236

Silva WS, Harney JW, Kim BW, Li J, Bianco SDC, Crescenzi A, Christoffolete MA, Huang SA, Bianco AC (2007) The small polyphenolic molecule kaempferol increases cellular energy expenditure and thyroid hormone activation. Diabetes 56(3):767–776. doi:10.2337/db06-1488

Silva F, Ferreira S, Queiroz J, Domingues FC (2011) Coriander (Coriandrum sativum L.) essential oil: its antibacterial activity and mode of action evaluated by flow cytometry. J Med Microbiol 60(Pt10):1479–1486. doi:10.1099/jmm.0.034157-0

Silva N, Alves S, Gonçalves A, Amaral JS, Poeta P (2013) Antimicrobial activity of essential oils from Mediterranean aromatic plants against several foodborne and spoilage bacteria. Food Sci Technol Int 19(6):503–510. doi:10.1177/1082013212442198

Singh N, Jabeen T, Pal A, Sharma S, Perbandt M, Betzel C, Singh TP (2006) Crystal structures of the complexes of a group IIA phospholipase A2 with two natural anti-inflammatory agents, anisic acid, and atropine reveal a similar mode of binding. Proteins 64(1):89–100. doi:10.1002/prot.20970

Slimestad R, Fossen T, Vågen IM (2007) Onions: a source of unique dietary flavonoids. J Agric Food Chem 55(25):10067–10080. doi:10.1021/jf0712503

Solomon A, Golubowicz S, Yablowicz Z, Bergman M, Grossman S, Altman A, Kerem Z, Flaishman MA (2010a) EPR studies of O(2)(*-), OH, and (1)O(2) scavenging and prevention of glutathione depletion in fibroblast cells by cyanidin-3-rhamnoglucoside isolated from fig (Ficus carica L.) fruits. J Agric Food Chem 58(12):7158–7165. doi:10.1021/jf100153z

Solomon A, Golubowicz S, Yablowicz Z, Bergman M, Grossman S, Altman A, Kerem Z, Flaishman MA (2010b) Protection of fibroblasts (NIH-3T3) against oxidative damage by cyanidin-3-rhamnoglucoside isolated from fig fruits (Ficus carica L.). J Agric Food Chem 58 (11):6660–6665. doi:10.1021/jf100122a

Sowbhagyaa HB (2013) Chemistry, technology, and nutraceutical functions of cumin (*Cuminum cyminum* L). Crit Rev Food Sci Nutr 53:1–10

Suzuki T, Tanabe S, Hara H (2011) Kaempferol enhances intestinal barrier function through the cytoskeletal association and expression of tight junction proteins in Caco-2 cells. J Nutr 141(1): 87–94. doi:10.3945/jn.110.125633

Tagashira T, Choshi T, Hibino S, Kamishikiryou J, Sugihara N (2012) Influence of gallate and pyrogallol moieties on the intestinal absorption of (−)-epicatechin and (−)-epicatechin gallate. J Food Sci 77(10):H208–H215. doi:10.1111/j.1750-3841.2012.02902.x

Tang L, Jin T, Zeng X, Wang J-S (2005) Lycopene inhibits the growth of human androgen-independent prostate cancer cells in vitro and in BALB/c nude mice. J Nutr 135(2):287–290

Tang N-Y, Huang Y-T, Yu C-S, Ko Y-C, Wu S-H, Ji B-C, Yang J-C, Yang J-L, Hsia T-C, Chen Y-Y, Chung JG (2011) Phenethyl Isothiocyanate (PEITC) promotes G2/M phase arrest via p53 expression and induces apoptosis through caspase- and mitochondria-dependent signaling pathways in human prostate cancer DU 145 cells. Anticancer Res 31:1691–1702

Tapsell LC, Gillen LJ, Patch CS, Batterham M, Owen A, Bar M, Kennedy M (2004) Including walnuts in a low-fat/modified-fat diet improves HDL cholesterol-to-total cholesterol ratios in patients with type 2 Diabetes. Diabetes Care 27(12):2777–2783. doi:10.2337/diacare.27.12.2777

Tasset-Cuevas I, Fernández-Bedmar Z, Lozano-Baena MD, Campos-Sánchez J, de Haro-Bailón A, Muñoz-Serrano A, Alonso-Moraga A (2013) Protective effect of borage seed oil and gamma linolenic acid on DNA: in vivo and in vitro studies. PLoS One 8(2):e56986. doi:10.1371/journal.pone.0056986

Tateno H, Nakamura-Tsuruta S, Hirabayashi J (2009) Comparative analysis of core-fucose-binding lectins from Lens culinaris and Pisum sativum using frontal affinity chromatography. Glycobiology 19(5):527–536. doi:10.1093/glycob/cwp016

Thavarajah P, Thavarajah D, Vandenberg A (2009) Low phytic acid lentils (Lens culinaris L.): a potential solution for increased micronutrient bioavailability. J Agric Food Chem 57(19): 9044–9049. doi:10.1021/jf901636p

Torkin R, Lavoie J-F, Kaplan DR, Yeger H (2005) Induction of caspase-dependent, p53-mediated apoptosis by apigenin in human neuroblastoma. Mol Cancer Ther 4(1):1–11

Trumbo PR, Ellwood KC (2006) Lutein and zeaxanthin intakes and risk of age-related macular degeneration and cataracts: an evaluation using the Food and Drug Administration's evidence-based review system for health claims. Am J Clin Nutr 84(5):971–974

UN (2013) Resolution adopted by the General Assembly on 20 December 2013. United Nations, General Assembly, Sixty-eighth session, Agenda item 25 Distr.: General, 7 February 2014. United Nations, New York. Available at http://www.un.org/en/ga/search/view_doc.asp?symbol=A/RES/68/231. Accessed 18 Nov 2015

USDA (2015) Agricultural Research Service National Nutrient Database for Standard Reference Release 27, Software v.2.2.4, The National Agricultural Library. http://ndb.nal.usda.gov/ndb/foods. Accessed 5 May 2015

Vacillotto G, Favretto D, Seraglia R, Pagiotti R, Traldi P, Mattoli L (2013) A rapid and highly specific method to evaluate the presence of pyrrolizidine alkaloids in Borago officinalis seed oil. Mass Spectrom 48(10):1078–1082. doi:10.1002/jms.3251

Vasconcelos MC, Bennett RN, Rosa EA, Ferreira-Cardoso JV (2010) Composition of European chestnut (Castanea sativa Mill.) and association with health effects: fresh and processed products. J Sci Food Agric 90(10):1578–1589. doi:10.1002/jsfa.4016

Vaughan JG, Geissler CA (1997) The New Oxford book of food plants. Oxford University Press, Oxford

Wang Y, Chung SJ, McCullough ML, Song WO, Fernandez ML, Koo SI, Chun OK (2014a) Dietary carotenoids are associated with cardiovascular disease risk biomarkers mediated by serum carotenoid concentrations. J Nutr 144(7):1067–1074. doi:10.3945/jn.113.184317

Wang Y, Stevens VL, Shah R, Peterson JJ, Dwyer JT, Gapstur SM, McCullough ML (2014b) Dietary flavonoid and proanthocyanidin intakes and prostate cancer risk in a prospective cohort of US men. Am J Epidemiol 179(8):974–986. doi:10.1093/aje/kwu006

Weng CJ, Yen GC (2012) Flavonoids, a ubiquitous dietary phenolic subclass, exert extensive in vitro anti-invasive and in vivo anti-metastatic activities. Cancer Metastasis Rev 31(1–2):323–351. doi:10.1007/s10555-012-9347-y

Williams MT, Hord NG (2005) The role of dietary factors in cancer prevention: beyond fruits and vegetables. Nutr Clin Pract 20(4):451–459. doi:10.1177/0115426505020004451

Woo HD, Kim J (2013) Dietary flavonoid intake and risk of stomach and colorectal cancer. World J Gastroenterol 19(7):1011–1019. doi:10.3748/wjg.v19.i7.1011

Xu B, Chang SK (2010) Phenolic substance characterization and chemical and cell-based antioxidant activities of 11 lentils grown in the northern United States. J Agric Food Chem 58:1509–1517

Yang Y, Kayan B, Bozer N, Pate B, Baker C, Gizir AM (2007) Terpene degradation and extraction from basil and oregano leaves using subcritical water. J Chromatogr A 1152(1–2):262–267. doi:10.1016/j.chroma.2006.11.037

Yang MD, Lai KC, Lai TY, Hsu SC, Kuo CL, Yu CS, Lin ML, Yang JS, Kuo HM, Wu SH, Chung JG (2010) Phenethyl isothiocyanate inhibits migration and invasion of human gastric cancer AGS cells through suppressing MAPK and NF-κB signal pathways. Anticancer Res 30(6):2135–2144

Yang JS, Liu CW, Ma YS, Weng SW, Tang NY, Wu SH, Ji BC, Ma CY, Ko YC, Funayama S, Kuo CL (2012) Chlorogenic acid induces apoptotic cell death in U937 leukemia cells through caspase- and mitochondria-dependent pathways. In Vivo 26(2):971–978

Zamora-Ros R, Knaze V, Luján-Barroso L, Romieu I, Scalbert A, Slimani N, Hjartåker A, Engeset D, Skeie G, Overvad K, Bredsdorff L, Tjønneland A, Halkjær J, Key TJ, Khaw KT, Mulligan AA, Winkvist A, Johansson I, Bueno-de-Mesquita HB, Peeters PH, Wallström P, Ericson U, Pala V, de Magistris MS, Polidoro S, Tumino R, Trichopoulou A, Dilis V, Katsoulis M, Huerta JM, Martínez V, Sánchez MJ, Ardanaz E, Amiano P, Teucher B, Grote V, Bendinelli B, Boeing H, Förster J, Touillaud M, Perquier F, Fagherazzi G, Gallo V, Riboli E, González CA (2013) Differences in dietary intakes, food sources and determinants of total flavonoids between Mediterranean and non-Mediterranean countries participating in the European Prospective Investigation into Cancer and Nutrition (EPIC) study. Br J Nutr 109(8):1498–1507. doi:10.1017/S0007114512003273

Zamora-Ros R, Forouhi NG, Sharp SJ, González CA, Buijsse B, Guevara M, van der Schouw YT, Amiano P, Boeing H, Bredsdorff L, Fagherazzi G, Feskens EJ, Franks PW, Grioni S, Katzke V, Key TJ, Khaw KT, Kühn T, Masala G, Mattiello A, Molina-Montes E, Nilsson PM, Overvad K, Perquier F, Redondo ML, Ricceri F, Rolandsson O, Romieu I, Roswall N, Scalbert A, Schulze M, Slimani N, Spijkerman AM, Tjonneland A, Tormo MJ, Touillaud M, Tumino R, van der A DL, van Woudenbergh GJ, Langenberg C, Riboli E, Wareham NJ (2014) Dietary intakes of individual flavanols and flavonols are inversely associated with incident type 2 diabetes in European populations. J Nutr 144(3):335–343. doi:10.3945/jn.113.184945

Zhang WY, Lee JJ, Kim Y, Kim IS, Han JH, Lee SG, Ahn MJ, Jung SH, Myung CS (2012) Effect of eriodictyol on glucose uptake and insulin resistance in vitro. J Agric Food Chem 60: 7652–7658

Zhou Q, Yan B, Hu X, Li X-B, Zhang J, Fang J (2009) Luteolin inhibits invasion of prostate cancer PC3 cells through E-cadherin. Mol Cancer Ther 8(6):1684–1691. doi:10.1158/1535-7163.MCT-09-0191

Ziesenitz S, Eldridge A, Antoine J-L, Coxam V, Flynn A, Fox K, Gray J, Macdonald I, Maughan R, Samuels F, Sanders T, Tomé D, van Loveren C, Williamson G (2012) Healthy lifestyles diet, physical activity and health. ILSI Europe, International Life Sciences Institute, Belgium. Available at http://www.ilsi.org/Europe/Publications/ILSIcm11-004_Diet08.pdf. Accessed 4 Nov 2015

Zou Y, Chang SK, Gu Y, Qian SY (2011) Antioxidant activity and phenolic compositions of lentil (Lens culinaris var. Morton) extract and its fractions. J Agric Food Chem 59(6):2268–2276. doi:10.1021/jf104640k

6 Milk and Dairy Products

Abstract

Milk is the first food of mammals, and breastfeeding has recently been found to be determinant in building up a healthy microbiome. A healthy adult is thought to carry about 2 kg of bacteria in the intestine, affecting immunity, weight balance and even mood. Small ruminants are well adapted to the mountains of the Mediterranean region and are easily handled by small rural communities. Traditional dairy products of these communities are yoghurt and cheese produced from the milk of these small ruminants. Cheese is an ancient fermented food valued for its portability, its longer shelf life than milk, and also for its nutritional value resulting from high amounts of protein, fat, calcium and phosphorus. Cheese also contains essential fatty acids, short-chain fatty acids (such as butyrate) and bioactive peptides (released from the breakdown of caseins). These compounds accumulate during ripening and some of them result from microbial metabolism. In particular, short chain fatty acids and bioactive peptides have recently been noted as beneficial to health and wellness. Many cheeses hold protected designations of origin (PDO), granted by the European Commission, such as 'Mozzarella di Bufala Campana', 'Feta', 'Queijo Serra da Estrela', or 'Queso de Murcia'. On the other hand, yoghurt is a typical example of a probiotic dairy food. Even if no bifidobacteria or other selected probiotic strains are added, the normal combination of lactic acid bacteria used to ferment milk to yoghurt has the ability to positively affect gut microbiota. Moreover, fermented dairy products are low in lactose, the digestibility of which usually decreases with age. In short, moderate consumption of dairy products, particularly fermented foods, helps provide essential amino acids and some vitamins that may be rare or absent in vegetables. On the other hand, oligosaccharides and dietary fibres (from vegetables) improve the survival of probiotics of dairy origin.

A.M. Delgado et al., *Chemistry of the Mediterranean Diet*,
DOI 10.1007/978-3-319-29370-7_6

6.1 Milk and Dairy Products: An Introduction

Milk, the secretion of the mammary gland, is the primary food source for mammals before they are able to digest other types of food. Early-lactation milk (colostrum) contains the mother's antibodies, which help decrease the risk of disease and aid the newborn in building up his/her gut microbiota. An average healthy adult contains up to 2 kg of thousands of different species of bacteria, some more beneficial than others, that strongly affect individual health. The original source for probiotic bacterial strains is milk.

Milk used directly or indirectly for human consumption is mostly that of cattle. The domestication of other mammals, such as goat, sheep, buffalo and camel, led to the diversification of milk for human consumption.

Before the development of dairy industries, milk was a very unsafe product and was frequently associated with many infectious diseases. The manufacture of cheese and yoghurt aimed to increase the time span for safe consumption by means of natural fermentation processes.

Moderate consumption of dairy products is recommended, in the context of the Mediterranean diet (MD). It is noteworthy that some dairy products, particularly yoghurt and cheese, are valuable sources of nutrients, nutraceuticals and probiotics as discussed below.

6.2 Gut Microbiome

Herein the designation 'microbiota' refers to a combination of the microbial communities of symbionts (commensals and mutualists) that colonise the human gastrointestinal tract (GI). Its collective genomes are known as a 'microbiome'. Microbiota is a complex community that includes members of the domains Bacteria, Archaea and Eukarya (Kelly and Mulder 2012; Rajilic-Stojanovic and De Vos 2014). This complex microbial community is part of us, affecting our immunity, our capacity to metabolise certain molecules, and influencing how we gain body weight and even our mood. For this reason, it has recently been considered as an organ (Chen et al. 2014; De Vos and De Vos 2012; Sandrini et al. 2015; Logan 2015).

Recent discoveries have revealed the importance of the two-way interaction between microbiota/microbiome and their host. These two-way interactions influence the integrity of the host's intestinal epithelium and immune system, bacterial gene regulation and many other functions (Burcelin 2012; Duca et al. 2014; Giorgetti et al. 2015; Kelly and Mulder 2012; Ostaff et al. 2013; Velasquez-Manoff 2015). To protect the gut from excessive bacterial exposure, the intestinal epithelium is protected by a thick and continuous mucin layer that limits and restricts the exposure of gut surfaces to luminal gut microbes.

The microbial composition and diversity of the gut microbiota depend on host-related factors such as the birth mode, genotype, age, geographical location, living environment, diet and health status (Kelly and Mulder 2012; Kieffer et al. 2014;

Waworuntu et al. 2014). The epithelium's N-linked and O-linked glycosylation patterns, as well as its mucin content, change with early feeding methods and during child's growth. Carbohydrate structures of mucin provide important binding sites for gut bacteria, simultaneously representing a rich source of nutrients. Moreover, the mucosal immune system distinguishes between harmful pathogens and beneficial commensal bacteria, delivering completely opposing downstream responses to these distinct bacterial groups (Kelly and Mulder 2012). The factors that enable a commensal bacterium to colonise a host's gut are not yet fully elucidated, but there is a general acceptance that the mutual benefits provide the key to this successful partnership.

The microbiota of an individual comprises more than 1000 species, representing a unique signature for that individual. Healthy twins living in different environments have similar microbiota at the genus level but different microbiota at the finer levels of resolution, such as the bacterial species and strains. These data illustrate the relevance of environmental factors in dictating bacterial colonisation (Kelly and Mulder 2012; Rajilic-Stojanovic and De Vos 2014).

The formation of human GI microbiota starts during the lactation period. Milk, the secretion of the mammary gland, is the primary food source for mammals before they are able to digest other types of food. Early-lactation milk (colostrum) contains mother's antibodies that improve the child's immunity, thus decreasing the risk of diseases, and helping build up the newborn's gut microbiota (Burcelin 2012; Rubio et al. 2012). Rubio and colleagues argue that maternal factors influence the composition of breast milk and consequently the child's microbiota. According to those authors, bacteria found in breast milk are from other sources besides environmental contamination (Rubio et al. 2012). Moreover, microbial composition of milk changes during lactation and in response to maternal factors (Kelly and Mulder 2012; Rubio et al. 2012).

Generally, the composition of human breast milk is dominated by bacilli (>75 %). The most common genera found in colostrum are *Weissella* and *Leuconostoc* (both lactic acid bacteria) followed by *Staphylococcus*, *Streptococcus* and *Lactococcus* (Giorgetti et al. 2015). Later in lactation, the proportions of other bacteria increase in the milk, particularly those from genera *Veillonella*, *Leptotrichia* and *Prevotella* (Rubio et al. 2012), diversifying GI microbial populations. It is noteworthy that breast milk contains oligosaccharides that support the growth of *Bacteroidetes* and *Bifidobacterium* species (Burcelin 2012; Rubio et al. 2012), which are considered vital to the building of a healthy microbial gut community. *Bifidobacterium* spp. are more abundant in lean mothers, influencing the child's microbiota composition, and are highly prevalent throughout life (Burcelin 2012; Kelly and Mulder 2012; Rajilic-Stojanovic and De Vos 2014).

Species from environmental origins that colonise the gut right after infant delivery mainly belong to genera *Bifidobacterium*, *Clostridium*, *Ruminococcus*, *Enterococcus*, *Enterobacter* and *Bacteroides* (Burcelin 2012; Voreades et al. 2014).

Host genetics interact with the specific profile of symbiont microbiota that successfully colonises the gut (Chen et al. 2014; Kelly and Mulder 2012; Ostaff et al. 2013), namely those contributed by the host's innate immunity and the expression of obesity-related genes. Genetic factors interplay with external factors (such as the lactation mode and the use of antibiotics early in life) in selecting a customised GI community. Infancy is a critical period for intestinal colonisation. Inadequate gut microbiota composition in early life seems to account for the deviant programming of later immunity and overall health status (Arumugam et al. 2011; Kelly and Mulder 2012; Velasquez-Manoff 2015).

Recent research has identified that, although gut bacterial composition differs widely between individuals, it is possible to simplify this analysis because all people, including those with chronic intestinal disease, can be classified into the broad 'enterotypes' *Bacteroides* and *Prevotella* (both genera belonging to the phylum Bacteroidetes), or a third enterotype *Ruminococcus*, which is rare and consequently generally ignored (Gophna 2011). Enterotypes seem to consolidate during infancy and to be determined by long-term diet.

The *Bacteroides* enterotype is positively associated with a frequent intake of animal protein and saturated fats and favoured by a high-fat/low-fibre diet, whereas the *Prevotella* enterotype is associated with predominantly plant-based nutrition (including high levels of complex carbohydrates) and low meat and dairy consumption (low-fat/high-fibre diet).

The *Bacteroides* enterotype responds to short-term changes in diet, whereas the *Prevotella* enterotype does not seem to respond. These microbial profiles are stable, and short-term dietary changes are not sufficient to alter them permanently.

Nevertheless, it has been observed that the average Firmicutes-to-Bacteroidetes ratio changes throughout life. Despite some fluctuations, particularly in the first 2 years, the load and diversity of gut microbiota increases until adulthood (Chen et al. 2014). In infancy, Firmicutes-to-Bacteroidetes ratio is around 0.4, increasing to 10.9 during adulthood and decreasing to 0.3 in old age (Chen et al. 2014).

Generally, members of Firmicutes may account for 50–80 % of GI microbiota in healthy adults, and belong to four classes: Bacilli, Clostridia, Erysipelotrichi and Negativicutes. The Firmicutes include Gram-positive bacteria belonging to the genera *Lactobacillus*, *Enterococcus* and *Streptococcus*, as well as spore-formers such as *Clostridium* spp., *Bacillus* spp. and related species. This ability to form spores confers special survival skills within and outside the GI (Rajilic-Stojanovic and De Vos 2014). While the class Bacilli includes many health-promoting and probiotic strains, the second cluster (spore-forming agents) is normally associated with disease and includes pathogens such as *Clostridium difficile*. However, members of the *Clostridium* clusters – namely, IV, XIVa and XVIII – have been found to be remarkably important to gut's health, as they promote the integrity of the intestinal barrier and maintain a balanced immune system (Rajilic-Stojanovic and De Vos 2014). These populations are thought to ensure a healthy flow of mucus and to favour certain bacterial groups thus shaping the gut ecosystem. Conversely, defects in the mucous layer are associated with the depletion of these members of *Clostridium* clusters and to the rise of aberrant communities of microbes that generally cause disease (Arumugam et al. 2011; Ostaff et al. 2013; Velasquez-Manoff 2015).

The adult's GI microbiota is a diversified, complex and quite steady community, including many non-cultivable microbes, the functions of which are yet to be discovered (Rajilic-Stojanovic and De Vos 2014; Ostaff et al. 2013). Normally, the adult's microbiota resists most changes, returning to its equilibrium state. However, if the perturbations overwhelm its capacity to resist, shifts in microbial populations will occur and may result in a range of diseases (Ostaff et al. 2013). In this regard, *Prevotella* enterotype seems more enduring but also generally associated with good health (Burcelin 2012; Chen et al. 2014).

Aging is characterised by a deterioration of energy homeostasis, with a loss of muscle mass. The human gut microbiota also undergoes substantial alterations, as generally does the functionality of the immune system, resulting in a greater susceptibility to infections. The core microbiota of elderly subjects is thus distinct from that previously established for young adults. With elderly subjects, a greater proportion of *Bacteroides* spp. and a distinct abundance pattern of other groups are observed. An increase in the proportion of *Escherichia coli* and other *Enterobacteriaceae* has generally been identified (Burcelin 2012; Rajilic-Stojanovic and De Vos 2014), as well as a decrease in anti-inflammatory symbionts, namely *Bifidobacterium* spp. and *Faecalibacterium prausnitzii*, as well as other members of *Clostridium* clusters (Rajilic-Stojanovic and De Vos 2014; Voreades et al. 2014). GI microbiota within the elderly is generally characterised by decreased butyrate production capacity (and other short-chain fatty acids playing important roles in host–microbe interactions), thus reflecting an increased risk of degenerative diseases (Chen et al. 2014; Rajilic-Stojanovic and de Vos 2014).

Diets rich in certain fats and sugars tend to deplete anti-inflammatory bacteria from the GI, thin the mucous layer, and foster systemic inflammation, while fibre-rich diets tend to favour healthy gut microbial populations (Arumugam et al. 2011; Chen et al. 2014; Voreades et al. 2014).

Obesity has been associated with an increased proportion of Firmicutes (e.g. gen. *Bacillus, Staphylococcus* and *Clostridium*) versus Bacteroidetes (e.g. gen. *Bacteroides* and *Prevotella*) (Duca et al. 2014; Kieffer et al. 2014; Schnorr et al. 2014; Suzuki and Worobey 2014) and a decrease in *Methanobrevibacter smithii* (Chen et al. 2014; Voreades et al. 2014). Moreover, it has been shown that the gut microbiota of lean individuals is more diverse than that of obese individuals (Chen et al. 2014).

Food interacts intimately with the microbiota, with a wide range of impacts. Modern high-processed foods have been noted to alter the balance of microbial groups, causing a decrease in immunity, the augmented virulence of otherwise asymptomatic opportunist pathogens and inflammatory responses (Kelly and Mulder 2012; Selhub et al. 2014; Rajilic-Stojanovic and De Vos 2014). High-fat diets (particularly saturated and trans-fats) trigger alterations in microbiota promoting the development of Gram-negatives, which often produce toxic lipopolysaccharides (LPS). Excessive LPS levels (defined as metabolic endotoxemia) are related to gut, hepatic and adipose tissue inflammation, as well as with diabetes (Alkanani et al. 2014; Burcelin 2012; Chen et al. 2014).

Some commonly used food additives, such as emulsifiers and sweeteners, can alter microbial population dynamics independently of the host's enterotype (Chassaing et al. 2015; Palmnas et al. 2014; Suez et al. 2014). Evidence has been showing that artificial sweeteners in general are responsible for alterations in microbial metabolic pathways, affecting host susceptibility to metabolic disease and causing dysbiosis and glucose intolerance (Palmnas et al. 2014). Aspartame, in particular, has been reported to increase the abundance of *Enterobacteriaceae* and *Clostridium leptum* and possibly also interferes in the Firmicutes-to-Bacteroidetes ratio (Palmnas et al. 2014). In terms of emulsifiers, Chassaing and colleagues reported that the consumption of polysorbate 80 and carboxymethylcellulose, even at moderate doses, changed gut microbiota, stimulating pro-inflammatory bacteria, and facilitating their translocation across epithelial cells. Changes also resulted in increased expression of flagellin and eventually toxic LPS (Chassaing et al. 2015).

Conversely, fermented foods (such as yoghurt, cheese and wine) provide an array of microbial metabolites and other compounds (e.g. short-chain fatty acids, bioactive peptides and flavonoids) that may act positively upon the microbiota profile. The effects of the *Bacteroides* enterotype may be more pronounced.

Many indirect effects of fermented foods on health have been reported, including a positive contribution to mood and mental health. Bacterial metabolites and products of biochemical reactions (including many as-yet unidentified compounds) are thought to be players in a gut–brain–microbiota intersection, with favourable outcomes for the host's mental status (Selhub et al. 2014). Yoghurt is the most classical example of a probiotic food.[1] Even if no other cultures are added, the starter cultures *Streptococcus salivarius* var. *thermophilus* and *Lactobacillus delbrueckii* ssp. *bulgaricus* (independent of the strain) are recognised as enhancing lactose digestion in lactose-intolerant individuals (Kieffer et al. 2014). The use of probiotics has been promoted by many researchers and is known to be more effective if complemented by the ingestion of prebiotics, generally oligosaccharides (such as inulin and oligofructose), which improve probiotic viability and stimulate beneficial gut microbes, namely *F. prausnitzii*, a key symbiont (Kieffer et al. 2014; Sokol et al. 2008; Velasquez-Manoff 2015).

In short, microbiota–host interactions include the breaking down of otherwise non-usable dietary components (e.g. resistant starch and fibres), the creation and modulation of the immune system, the modulation of hormone secretion and the degradation of toxins (Arumugam et al. 2011; Velasquez-Manoff 2015). Gut microbiota also regulates insulin sensitivity, phytochemical metabolism and end-product release, such as gases (Arumugam et al. 2011; Chen et al. 2014; Ostaff et al. 2013; Velasquez-Manoff 2015; Voreades et al. 2014). Further insights into diet–microbiota relationships and the immune system will eventually have an impact on nutritional guidelines for both healthy individuals and patients with chronic intestinal diseases and metabolic diseases, such as obesity and diabetes. It is important, therefore, to establish how far the intestinal microbiota can be thought of as static, or the extent to which it can be subjected to dietary control. The ability to manipulate the microbiota through diet and the regular intake of probiotics should provide a route for delivering health benefits.

[1] Probiotics are 'live microorganisms that, when administered in adequate amounts, confer a health benefit to the host' (FAO/WHO 2002).

6.3 Milk

Milk, in the liquid form, is marketed in a variety of types, including reduced fat and fortified formulas (with added calcium [Ca] and/or vitamins), lactose-free versions and with various added ingredients (from chocolate to fibres). Milk must comply with strict food safety regulations, which include pasteurisation or sterilisation (there are several options for time/temperature combinations). Like any heat processing, pasteurisation may affect minor components in milk; however, as it is a mandatory standard procedure, the data and discussion that follows refer to pasteurised milk. Raw milk can only be used in a few situations, such as cheese making.

Milk is simultaneously an emulsion of fat in water, a colloidal suspension of proteins and an aqueous solution of minerals, vitamins and sugars, also containing typical groups of bacteria. Somatic cells are sometimes found. Milk composition varies according to many factors, including animal feed, breed and climate. These variations affect not only the flavour but also the proteins and fatty acid profile, vitamin content and other minor components.

Milk density ranges from about 1.02 to 1.04 kg/m^3 and one portion corresponds to 244–258 g (1 cup). The values presented below are expressed per 100 g of milk (less than half a portion), and percentage values are presented on a weight-to-weight basis.

Whole milk supplies 62–64 kcal/100 g from: proteins (3.0–3.3 %), total fat (3.5–3.7 %) and carbohydrates (4.7 %, mostly lactose). Milk is a good source of Ca (109–119 mg/100 g), phosphorus (P: 77–93 mg/100 g) and also contains manganese (Mn: 4 μg/100 g) and selenium (Se: 2 μg/100 g).

Milk is a dietary source of the fat-soluble vitamins retinol (31–59 μg/100 g), vitamin D (0.05 μg/100 g) and vitamin E (29 mg/100 g). In addition, milk contains the B complex: vitamins B_6 (0.039–0.042 mg/100 g), B_{12} (0.18–0.36 μg/100 g), thiamine (0.038–0.040 mg/100 g), niacin (0.084–0.20 mg/100 g), riboflavin (0.14–0.16 mg/100 g), folates (1–5 μg/100 g) and pantothenic acid (0.31 mg/100 g) (INSA 2015; USDA 2015).

Milk fat (triacylglycerides, generally of myristic, palmitic and oleic acids) is enclosed in globules surrounded by a membrane of phospholipids and proteins, acting as an emulsifier and protecting fats from the action of enzymes (including those present in milk). Fat globules may vary in size and composition. Fat-soluble vitamins (A, D, E and K), along with essential fatty acids (such as conjugated isomers of linoleic and linolenic acids), cholesterol and minor components, are found in the fat portion. Milk carries an active form of vitamin A (retinol) as well as some vitamin D (cholecalciferol and ergocalciferol), which can be relevant in climates with reduced exposure to sunlight. People in southern countries are usually exposed to sunlight throughout most of the year, a necessary condition for the synthesis of vitamin D_3 (cholecalciferol) from cholesterol, via a photochemical reaction that takes place in the skin (Berg et al. 2002).

The composition of milk fat varies in response to many factors, as mentioned above, with a predominance of SFA (2.0–2.3 %) over MUFA (0.8–1 %) and PUFA (0.10–0.14 %). Predominant SFA are C14:0 to C20:0 as well as C4:0 (butyric acid, which may act as a signalling molecule within the gut, influencing host-microbiota interactions, see De Vos and De Vos 2012; Rajilic-Stojanovic and De Vos 2014; Schnorr et al. 2014). Predominant MUFA are C16:1 and C18:1. Main PUFA is linoleic acid (INSA 2015). Whole milk carries cholesterol (13–14 mg/100 g) and trans-fatty acids (TFA, 0.1 g/100 g) (INSA 2015; USDA 2015), to which negative health effects such as hypercholesterolemia, atherogenicity and increased cardiovascular risk have been attributed (EFSA 2004). The consumption of TFA increases low-density lipoprotein cholesterol (LDL-c) and decreases the proportion of high-density lipoprotein cholesterol (HDL-c), in the bloodstream. It thus contributes to increased rates of coronary heart disease and of mortality and morbidity (USDA and USDHHS 2010).

Milk (and some dairy products) may contribute to the intake of a high level of TFA, originating from the bacterial transformation of fats in the rumen of cows, goats and sheep and accumulating in their milk and meat. The World Health Organization (WHO) and the European Food Safety Authority (EFSA) recommend the intake of all-*trans* fats to be reduced, regardless of origin (EFSA 2004). However, some authors claim that TFA from ruminants are not equivalent to those found in margarines and refined vegetable oils or fast foods, and do not pose health risks (FIL/IDF 2005). This discussion extends to the meat of ruminants.

Conjugated linoleic acid (CLA) consists of a family of linoleic acid positional isomers with conjugated dienes in the *cis*- and *trans*-configurations, and a wide spectrum of isomers has been described, with varying positions (from 6,8- to 12,14-) and geometry—*trans,trans; trans,cis; cis,trans; cis,cis*. In total, 20 different CLA isomers occur naturally in food (Martins et al. 2007). It is noteworthy that CLA can be classified as either *cis*- or *trans*-fats. CLA are present in some foods of vegetable origin but are found mostly in meat and dairy products obtained from ruminants.

The major CLA isomer, *cis*-9, *trans*-11 (c9,t11),[2] as well as the usually second most prevalent isomer (t7,c9), are produced in the rumen during microbial biohydrogenation of dietary 18:2n-6, and in the tissues through *D*9-desaturation of the rumen-derived *trans*-octadecenoate (*trans*-11-18:1). The origin of all other CLA isomers is thought to arise from ruminal biohydrogenation of dietary unsaturated C18-fatty acids; however, the metabolic pathways are yet to be elucidated (Martins et al. 2007).

CLA have been associated with health benefits, some of which are not well proven and that most likely vary according to the isomer (Koba and Yanagita 2014).

Some CLA isomers are incorporated into cholesterol esters and triglycerides and accumulate in adipose tissue (FIL/IDF 2005), whereas others are precursors of eicosanoids. Some CLA mixtures have been associated to health-promoting effects such as in the control of diabetes, carcinogenesis, atherosclerosis and obesity.

[2] Also known as rumenic acid.

Bioactive compounds have been noted to include c9,t11 and t10,c12 CLA (FIL/IDF 2005; Kim et al. 2012). The t10,c12 isomer may play an important role in lipid metabolism, whereas c9,t11 and t10,c12 isomers seem to be equally effective in anti-carcinogenesis. t10,c12 CLA may have adverse effects, such as increasing oxidative stress and inflammation in adipose tissue (Martins et al. 2007).

Animal fats are a richer source of CLA than are vegetable oils. Products from ruminant animals, including milk, dairy products and meat, are the most important sources of CLA in the human diet (Martins et al. 2007); 90 % of CLA in ruminant fat is represented by c9,t11 CLA (FIL/IDF 2005).

TFA from ruminants are conjugated isomers of oleic and linoleic acids, particularly elaidic (Δ9 trans [t9]) and vaccenic (t11) acids, which are noted as having effects on human health that are totally different from those of TFA of industrial origin precursors (Brouwer et al. 2010; EFSA 2004; FIL/IDF 2005; Gayet-Boyer et al. 2014; Tyburczy et al. 2009). The same can be confirmed for rumenic acid and for trans-vaccenic acid: C18:1, t11, n-7 (FIL/IDF 2005). Nevertheless, all molecules are included in the same TFA category, despite a distinction between CLA precursors and industrially produced TFA has been recognised (Brouwer et al. 2010; Doell et al. 2012).

The reduction of fat in milk removes several components in a non-proportional form and concentrates others. Table 6.1 shows the effect of fat removal on the concentration of some milk components. As expected, fat content is strongly reduced in low-fat milk; consequently, TFA are removed and only traces of cholesterol remain. It is noteworthy that retinol is absent in low-fat milk and the concentration of vitamin B_{12} is also strongly reduced, when compared with whole milk. The concentration of other vitamins is also reduced, whereas lactose and protein concentrations increase due to volume changes caused by fat removal. These alterations in milk composition, when fat is removed, affect the nutritional value of milk and may decrease its digestibility, particularly in the elderly, who will not benefit from the removed vitamins and are more likely to suffer from hypolactasia

Table 6.1 Changes in milk composition as a consequence of fat reduction

Nutrient	Whole milk	Reduced fat milk	Low fat milk
Total fat	2.8–3.6 %	1.6 %	0.2 %
Cholesterol per 100 g	3.4–10.2 mg	1.5–8.3 mg	1 mg
Trans-fatty acids	0.1 %	0.1 %	0
Ca per 100 g	114 mg	112 mg	114 mg
Total protein	3 %	3.3 %	3.4 %
Lactose	4.7 %	4.9 %	4.9 %
Retinol per 100 g	56 μg	22 μg	0
Vitamin B_6 per 100 g	0.06 mg	0.05 mg	0.05 mg
Vitamin B_{12} per 100 g	0.39 μg	0.12 μg	0.11 μg

Only some representative nutrients are shown and are expressed in a weight-to-weight basis for 100 g of milk; figures refer to UHT pasteurized or treated milk and were retrieved from different works (INSA 2015; Scherr and Ribeiro 2010). Available ranges, when mentioned, are calculated with a confidence level of 95 %

(see below). Dairy industries have developed several strategies to overcome these issues, such as fortifying milk with vitamins, hydrolysing lactose and offering a wide range of products in the ‘low-fat milk’ category.

Milk proteins are either in solution (albumins and lactoglobulin) or in suspension as large, roughly spherical sub-micelles (caseins) that are bound together by calcium phosphate to form the colloidal casein micelles. Four different types of casein proteins (αs1, αs2, β and κ) account for the majority of milk proteins. κ-casein is located in the outermost layer of micelles and, because of its net negative charge, contributes to micelle stability, keeping them separated via repulsion forces. These features are very important in cheese manufacturing. Soluble protein fractions (also known as whey proteins) account for 20 % of the total protein content. Whey protein fraction is heterogeneous and dominated by β-lactoglobulin, a potential allergen, as listed in Annex IIIa of Directive 2000/13/EC (EC 2000). It also contains α-lactalbumin, which—when present in a multimeric form—strongly binds Ca and zinc (Zn) ions and may possess anti-tumor activity (Brinkmann et al. 2011; Nakamura et al. 2013; Rammer et al. 2010; Yamaguchi et al. 2014). Whey proteins also include bioactive peptides and enzymes.

Lactose is the dominant sugar in milk. It is a disaccharide of glucose and galactose. The enzyme needed to digest lactose, a β-galactosidase called lactase, is at its highest levels in the small intestine after birth and then begins to slowly decline with age. Hypolactasia (the inability to breakdown lactose) causes a painful digestive condition due to reduced levels of the enzyme lactase. This condition may lead the individual to avoid milk and/or milk products and is an example of diet/genome interaction. Rural communities that were more dependent on cattle than agriculture adapted to milk consumption; consequently, the prevalence of lactose intolerance is low, as is the case in north European countries. On the other hand, communities that were more dependent on agriculture traditionally consumed less milk in adulthood and have a higher prevalence of lactose intolerance. The haplotype conferring lactose tolerance has a frequency of about 86 % in the northern European population, but only 36 % in southern European populations (Enattah et al. 2002); it is thought to be even lower in Africa.

Until recently, milk has been recommended by health authorities as a source of protein and Ca and good for bone health. A serving of three glasses of milk (about 500 ml) per day has been recommended for adults (WHO 2012).

However, even in populations able to digest lactose, milk effects on bone health may not be as good as previously thought. D-galactose induces changes that resemble natural aging caused by oxidative stress damage, leading to a mechanism of age-related bone loss and sarcopenia. As a consequence, recommendations to increase milk intake for prevention of fractures became a conceivable contradiction (Michaëlsson et al. 2014). These authors hypothesised that high milk consumption (three or more glasses of milk per day) may increase oxidative stress, which in turn affects the risk of mortality and fracture. The authors observed that milk intake was positively associated with 8-iso-PGF2α (a biomarker of oxidative stress) in both genders and with interleukin 6 (a biomarker for inflammation) in men.

The underlying mechanism was attributed to D-galactose, which is absent—or present in very small amounts—in fermented dairy products and in vegetables.

Michaëlsson and colleagues recently presented results from two Swedish cohorts over a time span of about 20 years. This country has been noted as having few lactose-intolerant individuals and a high average milk intake. These authors hypothesised that, based on meta-data analysis of extensive observational results, milk may increase the risk of hip fractures among women and cardiovascular and overall mortality in both genders. Effects were found to be dose dependent, with different hazard rates in female and male cohorts (Michaëlsson et al. 2014). Thus, for every glass of milk, the adjusted hazard ratio of all-cause mortality was 1.15 (95 % confidence interval [CI] 1.13–1.17) in women and 1.03 (95 % CI 1.01–1.04) in men. Only milk is associated with the observed deleterious effects, which were not been demonstrated for fermented dairy products. The same authors mentioned that—unlike milk—cheese and yoghurt were associated with a decreased risk of cardiovascular diseases because of their low content or complete absence of lactose and galactose, and the presence of probiotics as well as antioxidant and anti-inflammatory effects, which all positively impact gut microbiota.

Whereas, in the past, milk production was related to each population, dairy industries are now global companies that promote per capita milk consumption, encouraged by medical advice. State-of-the-art knowledge now questions this practice. Moreover, ecologists and the United Nations Food and Agriculture Organization (FAO) are highlighting the green-house gas emissions of cattle (Gerber et al. 2013), and economists are raising questions about the sustainability of the current model (Economist 2014).

Despite the high prevalence of lactose intolerance, the consumption of milk in the Mediterranean basin has apparently increased in the past decades. The availability of milk and dairy products for national consumption significantly increased between 1961 and 2011, at least in the countries included in the current work (FAO 2015a). European food policies, namely the Common Agriculture Policy established post World War II, led to an increased production of milk and dairy products, the consumption of which has increased, including in MD countries.

Apart from cow milk, which largely dominates the market, milk from other ruminants is also traded, mostly for the production of cheese.

The milk of small ruminants can be similar to that of cows (e.g. goat) or differ significantly in major components (e.g. sheep), although this depends on many factors. Regardless, cow milk differs significantly in terms of minor constituents and some features relevant to cheese making. It is noteworthy that the intensification of breeding systems, feeding and genetic selection has changed the composition of milk, particularly in Western Europe.

Goat milk has higher levels of total fat and proteins and a generally lower concentration of lactose than cow milk (Raynal-Ljutovac et al. 2008). According to updated data by Raynal-Ljutovac and co-workers, the goat milk with the highest fat content is found in Greece (5.63 %) and is produced by a local breed, also showing the highest level of total solids (14.8 %), when compared with breeds from other countries. That same milk also displays one of the higher values for casein content (3.05 %), corresponding to the highest 'caseins to total protein' ratio. On the other hand, milk from the 'Murciano-Granadina' goat in Spain has the highest total protein content (4.09 %), with 3.21 % of caseins (Raynal-Ljutovac et al. 2008). On average, goat milk has less sodium (380 mg/l) than milk from sheep and cows, and more iron (Fe), copper (Cu), Mn and iodine than cow milk. Caprine milk has about twice the level of niacin and about four times more sialic acid (an oligosaccharide that promotes probiotic growth) than cow milk; moreover, about 25 oligosaccharides have been reported in milk from 'Murciano-Granadina' goats (Raynal-Ljutovac et al. 2008).

On the other hand, sheep milk contains an average of 14.4–20.7 % total solids, 3.6–9.97 % fat, 4.75–7.2 % total proteins (of which 3.72–5.01 % are caseins) and 4.11–5.51 % lactose. When compared with the average composition of raw cow milk (3.5 % proteins, 3.0–4.0 % total fat and 5 % lactose), sheep milk is quite similar in composition, but has higher levels of fat and protein than goat and cow milk (FAO 2015b). Recent studies on milk from high-altitude sheep have indicated a high CLA and omega 3 (n-3) content in the lipid fraction. In terms of mineral composition, sheep milk provides more Ca and a generally higher Ca/P ratio (1.3–1.6) than goat or cow milk (1.3) (Raynal-Ljutovac et al. 2008), meaning that Ca from sheep milk can be more easily absorbed. According to the same authors, on average, sheep milk also contains higher levels of Mg, Fe, Cu, iodine and Se, as well as twice the levels of retinol and vitamin D than goat or cow milk. The same applies to most B vitamins.

Buffala's milk has a very high fat content that is, on average, twice that of cow milk (FAO 2015b). The fat-to-protein ratio in buffala's milk is about 2:1. Buffala's milk also has a higher casein-to-protein ratio than cow milk. The high Ca content of casein facilitates cheese making.

The primary characteristics of small ruminant milk fat are the smaller fat globules and the higher short- and medium-chain fatty acid content, especially in goat milk, which has at least twice as many C6–C10 fatty acids than cow milk fat (Raynal-Ljutovac et al. 2008). The metabolism of these fatty acids differs from that of long-chain fatty acids, as they are rapidly oxidised and constitute a rapid energy supply; consequently, they are less involved in LDL-c blood levels.

It is noteworthy that 2012 statistical data (FAO 2015a) reveal that some of the countries included in the UN Education, Scientific and Cultural Organization (UNESCO) 'Mediterranean diet' list are among the largest world producers of milk from goats (Spain and Greece), sheep (Greece, Spain and Italy) and buffala (Italy). These different types of milk have been successfully used to produce a large variety of valued cheeses. Examples of these traditional cheeses are presented and discussed in Sect. 6.5.

6.4 Butter

As a dairy product, the composition of butter is strongly linked to that of milk, fat being its major component. Butter manufacturing originated in northern Europe and was later disseminated throughout southern Europe in the twentieth century, with the implementation and development of dairy industrial plants. It was not very popular at first because of the rancid taste that easily developed in these mild climates without proper refrigeration.

Butter has been used in the Mediterranean basin mainly as a spread for bread, although olive oil is still preferred for that purpose in many regions. As with milk, the consumption of butter seems to be steadily increasing in the area.

Butter is usually produced from fermented cream, previously extracted from whole milk. As milk is pasteurised, fermentation (or cream maturation) is achieved by adding started cultures of lactic acid bacteria generally belonging to gen. *Lactococcus* and *Leuconostoc*. Selected strains can accumulate secondary metabolites of the fermentation of lactose and citrate, such as diacetyl and certain aldehydes, responsible for the typical aroma of butter.

Butter is produced by agitating cream, which disrupts the phospholipid membranes that enclose triglycerides and allows the milk fat to combine. Churning produces small butter grains that float in the water-based portion of the cream, called buttermilk, which is drained. The fat grains are then pressed and kneaded together. The resulting butter contains fat in three separate forms: free butterfat, butterfat crystals and undamaged fat globules. Different proportions of these forms result in different consistencies of the finished product; when butterfat crystals predominate, butter is hard, while butters with many free fat's globules are soft.

Butter is thus a water-in-oil emulsion, whereas milk is an oil-in-water emulsion. Remaining milk proteins act as emulsifiers. Butter is a hard solid when refrigerated and softens to a spreadable consistency at room temperature, melting at 32–35 °C. Its natural colour varies from pale to deep yellow, depending on cattle feed, but it can be manipulated industrially, usually by the addition of carotenes or annatto. Salt is generally added to increase the shelf-life and to activate flavours.

Other industrial manipulations include increasing the proportion of water in the emulsion to produce the so-called low-fat butters.

Regular butter supplies 717–739 kcal/100 g from: proteins (0.1–0.85 %), lactose (0.06–0.7 %) and total fat (81.8 %) (INSA 2015; USDA 2015). This total fat is composed of SFA (46.3–51.4 %), MUFA (18.9–21.0 %), PUFA (2.4–3.0 %, mainly linoleic acid), TFA (3.3 %, mainly trans-monoenoic fatty acids) and cholesterol (230 mg/100 g). The most common trans-monoenoic fatty acids in ruminants are elaidic and vaccenic acids. The negative implications of the lipid and lipoprotein metabolism of these have been contested by several authors, as discussed in Sect. 6.3.

On the other hand, butter is a rich source of fat-soluble vitamins, particularly vitamins A, D and E. On average, it contains retinol (565–671 μg/100 g), cholecalciferol (0.74–1.5 μg/100 g) and α-tocopherol (2.0–2.32 mg/100 g), in addition to β-carotene (0.16–45 mg/100 g), which transmits the natural yellowish colour to milk fat (INSA 2015; USDA 2015).

Butter provides much less Ca than milk, as most of it is removed with the protein fraction, and contains only traces of lactose. Moderate consumption of butter is not a problem for lactose-intolerant individuals. People with milk allergies may still need to avoid butter, as it may contain enough allergen proteins to cause a reaction.

6.5 Cheese

Whereas cheese is an ancient food that originated either in Europe, Central Asia or the Middle East, cheese making became a sophisticated enterprise during the Roman Empire, which spread these techniques across Europe. Cheese was valued for its portability and longer shelf life than milk as well as for its nutritional value, given the high levels of protein, fat, Ca and P.

There are hundreds of types of cheeses with a wide range of flavours, textures and forms. Rigorous hygienic control measures apply in addition to the features that characterize a certain protected designation of origin (PDO),[3] in the case of most of the cheeses described below.

In Mediterranean countries, cheese fermentation often relies on environmental bacteria, rather than on industrial starters of a defined genotype. Thus, many producers in the area have been striving to comply with the requirements that will confer both PDO designation and the European Union organic food logo.

Raw milk, including that of small ruminants, is often used to best preserve certain organoleptic characteristics. Cheeses made from the milk of small ruminants are still preferred by consumers due to their specific taste, texture, typicity and commercial image of a natural product.

Cheese making involves the coagulation of caseins that aggregate with fat (curd), the draining of the liquid phase (whey) and the maturation process, which usually involves enzymatic and microbiological alterations.

The first step often requires milk acidification, which is usually accomplished via the fermentation of lactose to lactic acid, by lactic acid bacteria (LAB) normally from the genus *Lactococcus*, *Lactobacillus* and/or *Streptococcus*. During fermentation, LAB produce secondary metabolites that contribute to flavour. Recently, some in-depth studies have focused on artisanal cheeses manufactured in Mediterranean countries, although primarily aiming at characterising the microbial groups related to sanitary issues in order to comply with regulations. Because *Enterobacteriaceae* and coliforms are indicator microorganisms for the sanitary quality of foods, their presence in cheeses is sometimes misjudged. Current routine methodologies deliver poor characterisation and discrimination of these microbial groups in

[3] PDO covers agricultural products and foodstuffs that are produced, processed and prepared in a given geographical area using recognized knowledge (http://ec.europa.eu/agriculture/quality/schemes/index_en.htm).

ripened cheese. Some commensal *Enterococcus* strains of proven relevance in cheese ripening (Sects. 6.5.1, 6.5.3, 6.5.4 and 6.5.5) are counted and included within the group generally linked to public health threats. This may represent a handicap for the commercialisation of Mediterranean cheeses, despite scientific evidence demonstrating the opposite (Banwo et al. 2012; Christoffersen et al. 2012; Cuív et al. 2013; Szabó et al. 2009).

The second major step in cheese making is casein coagulation, involving the addition of salt and proteases, to the already acidified milk. These proteases can be obtained from animal gastric juice (rennet and chymosin) or by macerating certain plants (as is the case of cinarase, cardosins or cyprosins that are obtained from thistle flower). Denatured molecules of casein form large aggregates in the presence of Ca ions, entrapping fat globules and forming the so-called curd, from which the whey is drained. The curd mostly contains insoluble and hydrophobic compounds, whereas whey contains soluble proteins and hydrophilic solutes from milk.

Milk of small ruminants shows a typical casein's micelle organization and mineralisation, and both, goat and sheep milk micelles are highly mineralised. Caprine milk-fat micelles are significantly larger than bovine or ovine milk micelles (Raynal-Ljutovac et al. 2008). These factors are crucial in cheese making.

The third major step is maturation, which heavily influences the special characteristics of each cheese, and occurs under widely varying conditions. It encompasses a cascade of chemical and enzymatic reactions, involving residual clotting enzymes and microorganisms, primarily mesophylic bacteria and moulds.

Cheese maturation is not yet fully understood, particularly in many traditional cheeses. Temperature, moisture, pH and curd texture are key factors that may also contribute to the selection of the dominant microbial populations and affect the final result.

Most flavour notes that specifically characterise each type of cheese develop during ripening as a result of the action of free enzymes (from milk, rennet or microbial origin) and microbial activity producing organic acids (e.g. butyrate and propionate, which may act as signalling molecules in the gut; please see Sect. 6.1), free fatty acids, and free amino acids through glycolysis, lipolysis and proteolysis reactions. Some of these compounds are further converted into amines, aldehydes, esters, phenols, indole and alcohols (Tavaria et al. 2002), as well as into diacetyl and butanedione.

During maturation, the primary reactions are lipolysis (resulting in the accumulation of free fatty acids and secondary metabolites), proteolysis (resulting in the accumulation of short bioactive peptides and free amino acids), and the production of primary and secondary microbial fermentation products; such accumulated metabolites are lactic, acetic and propionic acids, as well as diacetyl, butanedione and several esters and aldehydes. Texture and flavour undergo important alterations along the maturation process (or ripening), mainly due to the accumulation of secondary microbial fermentation products (Raynal-Ljutovac et al. 2008). However, it has been reported (in a study of ewe and goat cheeses) that there are no significant differences between the proportion of the different fatty acids present in milk and those present in the cheese made from it (Raynal-Ljutovac et al. 2008). Thus, according to these authors, cheese making has no impact on CLA isomer content. Given this strong relationship between milk and cheese, cheeses made from the milk of small ruminants are expected to contain higher levels of short- and medium-chain fatty acids than those made from cow milk (Raynal-Ljutovac et al. 2008), with the above-mentioned health advantages (Sect. 6.3).

Both, fat-soluble and water-soluble vitamins are present in fresh and ripened cheeses; quantities depend on many factors, including the technological process, type of milk and region (Raynal-Ljutovac et al. 2008). Since these vitamins are heat sensitive, the loss induced by pasteurisation is about 10–20 % for ascorbic acid, around 10 % for thiamine, 5–7 % for pyridoxine, folates and cobalamin, <1 % for riboflavin, and 0 % for niacin, pantothenic acid and biotin. Moreover, water-soluble vitamins are easily lost in the whey. Nevertheless the alterations actually found in cheeses suggest the net consumption of some vitamins and the production of others by microorganisms during ripening.

Proteolysis comprises a primary series of reactions involved in maturation, typically mediated by LAB, and divided into three steps:

1. Reactions such as decarboxylation, deamination, transamination, desulphuration and hydrolysis of amino acid side chains
2. Conversion of the resulting compounds (mainly amines, α-ketoacids and free amino acids) to aldehydes
3. Reduction of aldehydes to alcohols, or their oxidation to carboxylic acids

Sulphur-containing residues may undergo specific chemical reactions that are primarily responsible for the generation of methanethiol and a few other sulphur derivatives. Volatiles evolving from the metabolism of amino acids apparently contribute to the final organoleptic characteristics of ripened cheeses (Tavaria et al. 2002).

Milk protein-derived peptides (particularly casein hydrolysates) have been shown to exert various health-promoting activities that affect the following systems:

(a) Cardiovascular system. Responsible molecules are anti-thrombotic, anti-oxidant, anti-hypertensive and hypocholesterolemic peptides (Korhonen and Pihlanto 2006; Mao et al. 2011; Mills et al. 2011; Pihlanto 2006; Silva et al. 2006)
(b) Immune system. Responsible chemicals are immunomodulatory, cyto-modulatory and anti-microbial peptides (Hayes et al. 2007; Mills et al. 2011; Phelan et al. 2009)
(c) Nervous system. Responsible molecules are opioid peptides (Hayes et al. 2007; Mills et al. 2011)

Some of these bioactive peptides result from LAB activity (Delgado et al. 2005, 2007; Hayes et al. 2007; Turpin et al. 2010).

Aspects related to milk origin, manufacturing steps, ripening processes and microbial communities are expected to determine the type and concentration of such biopeptides. Valine–proline–proline tripeptide (VPP) and isoleucine–proline–proline tri-peptide (IPP) carry clinically documented benefits in the reduction of mild hypertension. They are encrypted in the casein primary structure and released during ripening (Hayes et al. 2007).

Bioactive peptides may be released from bovine, goat or sheep milk proteins (Raynal-Ljutovac et al. 2008).

In general, the microbial ecology of cheeses is poorly understood. Anti-microbial substances are believed to be produced during cheese ripening, helping regulate the microbial population dynamics. Anti-listerial bacteriocins are commonly produced by LAB and include nisin—an authorised preservative coded E234 in the European Union), which may contribute to the safety and general quality of cheese. Moreover, bacteriocins and other antimicrobial compounds have been frequently observed in probiotic strains, and their anti-microbial properties are thought to be present in vivo. Cottage or fresh cheese can also be produced from different types of milk (small ruminants and buffala are the most valued). Pasteurised milk is used, and the cheese manufacturing only involves protein coagulation and removal of the water phase. Whey proteins are sometimes used in the manufacture of cottage cheese. Fresh cheese has a higher water content and a shorter shelf-life than ripened cheese.

Although no clear definition of functional food exists, the concept refers to a food that beneficially affects one or more target functions in the body, beyond adequate nutritional effects, in a way that is relevant to either an improved state of health and well-being and/or reduction of risk of disease (EUFIC 2015). Some foods considered to be functional are actually natural whole foods for which new scientific information about their health qualities can be used to proclaim benefits (Phelan et al. 2009). Traditional cheeses are included in this category and have much to exploit, as many bioactive peptides are encrypted within the primary casein structure (Choi et al. 2012; Korhonen and Pihlanto 2006).

In short, cheese is manufactured from the milk of several ruminants. The Mediterranean basin is characterized by a wide variety of landscapes with distinct topographies, influenced by the Mediterranean Sea and the Atlantic Ocean (Fig. 1.1), with plains, deserts and many mountains, particularly in southern Europe. These mountain areas have been exploited since ancient times for the grazing of sheep and goats. Most artisanal cheeses are obtained from seasonal ovine and caprine milks, with peak production in spring and the lowest production in fall/autumn.

6.5.1 Gorgonzola (Italy)

The first evidence of Gorgonzola[4] production dates back to 879 AC in Gorgonzola, near Milan. It is made of pasteurised cow milk inoculated with *Lactobacillus* sp. and *Streptococcus* sp. starter cultures, together with *Penicillium roqueforti*. It was awarded with the PDO EC No: IT/PDO/117/0010/12.4.2002 and later amended (EC 2008).

According to those regulations, Gorgonzola is a large cylindrical cheese produced in three sizes (6–13 kg) with flavour variations from a slight to a pronounced tangy taste, called 'dolce' (sweet) and 'piccante' (hot), respectively. It is produced in several provinces of Italy from pasteurised whole cow milk. The rind is grey and/or pink and inedible. The paste is soft, white to pale yellow with blue-green mould (marbling) veins (EC 2008).

***Gorgonzola**. Gorgonzola is originally a large cylindrical cheese, although many sizes and shapes are available in modern markets, with flavour variations from a slight to a pronounced tangy taste, called 'dolce' (sweet) and 'piccante' (hot), respectively. It is a typical Italian product. Visibly, the most distinctive feature is the presence of a grey and/or pink inedible rind, while the paste is soft, white to pale yellow with blue-green mould (marbling) veins. Its properties depend mostly on the production process and fermentative features of starter cultures, including Penicillium roqueforti. Photo reprinted with kind permission from M. Barone*

According to the above-mentioned European Council (EC) regulation, during the curdling phase, milk enzymes are added and the milk is inoculated with a starter of *Penicillium* sp. and yeasts; calf rennet is used as a coagulant at 28–36 °C. The

[4] Further information can be found at the following web address: www.gorgonzola.com/gorgonzola.asp?id=16.

curd is pressed into a special apparatus, and the resulting wheel is dry salted for several days at 18–24 °C. Maturation lasts from 50 to 80 days, at a temperature between 2 and 7 °C and moisture of 85–99 %. Holes are made in the paste on several occasions to provide the necessary oxygen for the development of the desirable *Penicillium* strains of *P. roqueforti*, *P. glaucum* and *P. weidemannii*, which produce the 'marbling' effect.

Microbial communities that cover the cheese surface are believed to be cellar-dependent and relevant to, not only flavour and texture, but also to food safety. The dominance of Gram-positive bacteria (non-pathogenic, coagulase negative), namely from genus *Staphylococcus*, *Enterococcus*, *Lactobacillus*, *Leuconostoc* and *Carnobacterium*, has been registered (Fontana et al. 2010). Cocolin and colleagues observed that LAB, namely lactococci and lactobacilli, dominated the microbial ecology of the rinds, with enterococci present within the rinds at different frequencies among samples. These authors also reported the presence of *Carnobacterium* sp. and *Staphylococcus* sp., while *Debaryomyces hansenii* seems to be the predominant yeast in Gorgonzola rinds (Cocolin et al. 2009). According to these authors, marked differences in microbial characterisation can result from the analytical approach (culture-based methods vs. molecular biology methods).

In terms of flavour, the most important odorants in Gorgonzola are reported to be 2-heptanone, 2-nonanone (monoketones), 1-octen-3-ol (also known as amylvinyl-carbinol), 2-heptanol, ethyl hexanoate (ethyl caproate) and 2-methylanisole (Moio et al. 2000).

6.5.2 Queso de Murcia (Spain)

Murcia cheese[5] is made from the milk of the Murciana breed of goat, in that region of southern Spain. It has been produced and consumed for centuries, fresh or ripe, either with ham and wine or cooked in tomato sauce. PDO status was applied to three distinct variants—'Queso de Murcia fresco' (fresh Murcia cheese), 'Queso de Murcia Curado' (cured Murcia cheese) and 'Queso de Murcia al Viño' (Murcia cheese with wine) by EC No: ES-PDO-0105-01062-05.11.2012 and later amended (EC 2013).

According to the above regulation and amendments, Murcia cheese is made of the whole raw milk of Murcia goats with 4.7–5 % fat, although pasteurised milk may be used.

The manufacture of fresh Murcia cheese starts with filtration through a cloth, followed by the addition of animal rennet and heating to 32–35 °C for coagulation for 30–45 min. The curd is then coarsely sliced in grains with diameters of about 10 mm, and heated to about 40 °C before the grains are pressed and moulded.

[5] Further information can be found at the following web address: http://www.quesosdemurcia.com/.

Cheeses weighing 1–2 kg, 5–8 cm in height, and 7–9 cm in diameter are obtained and salted for 10 h (by immersion in cold brine) and kept at a temperature below 4 °C to drain excess whey.

The final product is white, of light 'chévre' flavour, with a few small holes or no holes, exhibiting flower-type marks resulting from curd compression in moulds, and no rind. Fresh Murcia cheese has about 45 % total fat, 32 % proteins, pH $\geq$6.5 and a minimum dry extract of 40 %. It is intended to be consumed fresh and kept under refrigeration.

Cured Murcia cheese undergoes the same first two steps as fresh cheese, but coagulation occurs at 30–34 °C for 40–60 min and curd is sliced in smaller pieces of 5 mm diameter. The curd temperature is raised about 5 °C to mix and press it more firmly, before introducing the paste into the molds, where curd is continuously pressed for 2–4 h until the desirable pH is reached. Cheeses are salted by immersion in cold brine for about 20 h. The ripening phase takes about 60 days at 9–13 °C and at least 80 % relative moisture. Cheeses are turned regularly, and their rinds are sanitised whenever necessary.

The final product has a height of 7–9 cm and a diameter of 12–18 cm, with a light beige to ochre rind, and a white or yellowish compact paste with very few or no holes. 'Queso de Murcia curado' has about 45 % total fat, 32 % proteins, pH $\geq$5 and a minimum dry extract of 55 %.

In the case of Murcia cheese with wine, the same steps as for cured Murcia cheese are followed to obtain the curd, with the difference that the curd is washed in water to remove about 15 % of whey before being heated to about 40 °C, pressed in molds and salted in brine. Before and during maturation, cheeses are immersed in the red wine of the region for a variable time span according to the degree of ripening, giving the rind a reddish-wine colour. Following the conditions described above, maturation lasts for 30–45 days, depending on the size of the cheese. The paste is white, compact but creamy and elastic, with low salt and a slightly acidic flavour when compared with 'Queso de Murcia curado'. Size, shape and physico-chemical characteristics are the same as above.

Proteolysis is one of the main occurrences during cheese ripening and a measure of the state of maturation. Abellan and co-workers observed that the total free amino acid concentration increased during ripening, as expected. Ile, Val, Ala, Phe, Gaba, Arg and Lys represent more than 50 % of the total free amino acid content at 60 days. Consequently, it is plausible that VPP and IPP are present (Abellan et al. 2012). As mentioned above, these bioactive peptides carry clinically documented benefits in the reduction of mild hypertension.

Sánchez-Macías and colleagues studied the ripening process in goat cheeses, particularly the lipolytic and proteolytic profiles and their interdependence. These authors observed that β-casein was the most abundant protein in all cheeses and at all ripening times, but its degradation seemed to be dependent on fat content. Degradation rates were higher for α(S1)-casein and intact α(S2)-casein, and α(S1)-casein levels were higher for low-fat milk at the end of the maturation period (Sánchez-Macías et al. 2011). The free fatty acid concentration per 100 g of cheese was higher in full-fat cheese than in reduced- and low-fat cheese.

Proteolysis and lipolysis are thus linked phenomena that contribute not only to the final texture but also to improve the nutraceutical properties of cheese.

6.5.3 Queijo Serra da Estrela (Portugal)

'Queijo Serra da Estrela'[6] is a cured cheese that exists in two varieties: soft paste and hard paste. The latter is designated 'Queijo Serra da Estrela velho' (meaning 'old mountain cheese') because of its longer ripening process. Both variants were awarded the PDO status by reg. EC No EC: PT/PDO/117/0213/16.05.2002 and later amended (EC 2007a).

The origin of this cheese is lost in time. It is referred to in Roman texts and is also noted as being one of the food items taken by Portuguese sailors on sea expeditions during the fifteenth and sixteenth centuries. Currently, it is mainly consumed as a table cheese.

'Queijo Serra da Estrela' (in both varieties of soft or hard paste) is made with the whole raw milk of 'Bordaleira Serra da Estrela' or 'Churra Mondegueira' sheep breeds; coagulation is achieved by means of the flower of *Cynara cardunculus*, a common plant in that mountain region. The high levels of proteolytic enzymes in thistle, or the cardoon's flowers are responsible for the effective clotting of milk.

The cheese is obtained by adding salt and thistle flower to the filtered raw milk followed by slow agitation to promote coagulation. Traditionally, the quality of the cheese depended on the low ambient temperatures of mountain winters, reflected in the cold hands of women manually working the curd.

The curd is cut into coarse pieces, filtered through cotton cloths and gently pressed to slowly drain the whey. The curd is then poured into molds, where it continues to be pressed to release more whey. Straps are then opened and the still soft cheeses are wrapped with a white cloth and shaped with an adjustable stainless steel ring and gently pressed to release more whey.

Maturation involves two distinct phases. The first one occurs at low temperatures (ideally about 6 °C and no higher than 14 °C) and high moisture levels (80–95 %), in natural or industrial chambers, where a natural lactic acid fermentation takes place. After about 30 days, the surface of the cheese is washed and it then undergoes the second ripening stage, a drying phase lasting for a few days in a cold dry chamber (about 15 °C).

The final product should look like a low cylinder ranging in size from 0.7 to 1.7 kg, 11–20 cm in diameter and 3–6 cm in height. It has a soft yellowish rind and a white to pale yellow very soft, butter-like paste with no holes. The flavour is slightly acidic but not intense. On a dry weight basis, 'Queijo Serra da Estrela' has 26–33 % proteins, 45–60 % fat, 61–69 % water and 5–6.5 % ash. On a fresh weight basis, 'Queijo Serra da Estrela' provides 167 kcal/100 g from 21 % proteins, 0.2 % mono- and

[6] Further information can be found at the following web address: http://confrariadoqueijoserradaestrela.com/index.php.

disaccharides (mostly lactose), 1.6 % organic acids (including lactic and isobutyric acids) and 27 % total fat from SFA (14.2 %), MUFA (6.5 %), PUFA (0.9 %), trans fats (1.1 %), linoleic acid (0.8 %) and cholesterol (88 mg/100 g) (INSA 2015). A large portion of the TFA are probably CLA, which are believed to have a beneficial role in human health. The differences between TFA of dairy and industrial origin have been acknowledged and are a subject of scientific debate (Sect. 6.4).

'Queijo Serra da Estrela' is a relevant source of carotenes (80 mg/100 g), vitamin A (240 μg of retinol eq.), vitamin D (0.21 μg, mostly cholecalciferol) and B-complex vitamins. It also provides similarly high amounts of sodium (Na: 706 mg/100 g) and Ca (700 mg/100 g). As with other cheeses, salt is added during manufacturing.

During the manufacture of 'Queijo Serra da Estrela Velho', after the lactic acid fermentation step and rind wash, a paste of paprika in olive oil is spread over the cheese, and the pieces are kept in the maturation chamber for at least an extra 90 days.

In the final product, the rind is brownish or reddish and rough; the paste is white and hard and appears dry and brittle. The flavour is more intense ('picante'), with a salty taste, than the soft paste variety. According to its PDO requirements and due to its lower moisture content, 'Queijo Serra da Estrela Velho' contains 36–43 % proteins, >60 % fat and 7–8 % ash on a dry weight basis.

On a fresh weight basis, 'Queijo Serra da Estrela velho' provides 389 kcal/100 g: proteins (25.5 %), mono- and disaccharides (mostly lactose, 0.2 %), organic acids (1.0 %) and total fat (31.5 %: SFA 16.6 %, MUFA 7.6 %, PUFA 1.1 %, TFA 1.3 %, which may include CLA, as the total determined linoleic acid was 1.0 %) (INSA 2015). Moreover, the Portuguese National Institute of Health (Instituto Nacional de Saúde Doutor Ricardo Jorge [INSA]) database reported 81 mg/100 g of cholesterol similarly to the less ripened variety. 'Queijo Serra da Estrela velho' contains fewer carotenes (10 mg/100 g) and is richer in vitamin A (285 μg/100 g of retinol eq.); the levels of other vitamins are similar to those in the less ripened variety. The balance between Ca and Na differs, with 792 mg/100 g Na and 815 mg/100 g of Ca. The moisture differences do not totally explain these figures. This nutritional profile is somewhat similar to that of other cheeses of the same type, mainly found in other regions of Portugal and Spain. The soft paste 'Queijo Serra da Estrela' is the most well-known variety and preferred by a growing number of consumers, due to its exquisite flavour and texture.

Different casein profiles are obtained with different coagulants used for cheese making, the differences being more evident throughout maturation. In the case of cardoon's flower, the enzymatic extract is a mix of cinarases or cyprosins (which are glycoprotein proteases of the aspartic acid type), cardosins (similar to chymosin), pepsin-like proteases and other unidentified proteases (Roa et al. 1999). Higher proteolytic rates were observed in cheeses coagulated with *Cynara* sp. when compared with other rennets of animal origin (Roa et al. 1999; Roseiro et al. 2003).

The initial hydrolysis of caseins is caused chiefly by residual coagulant, resulting in the formation of large- and intermediate-sized peptides that are subsequently degraded by the rennet itself and by enzymes contributed by microorganisms.

Cynara cardunculus enzymes have been noted as having a stronger hydrolytic activity towards β-casein, as well as a higher residual activity and stability during maturation than other coagulants, even those of vegetable origin (Roa et al. 1999; Roseiro et al. 2003). Roseiro and co-workers (2003) registered the presence of specific peptides from the action of this particular vegetable coagulant on ovine caseins, as well as the release of short peptide sequences from casein (Hayes et al. 2007). As mentioned in Sect. 6.5, some of these short peptide sequences may display relevant bioactivities.

The pH is crucial in cheese ripening because of its influence on proteolytic activity. Initial pH values are contributed via the lactic acid fermentation by the native microflora of 'Serra da Estrela' cheese, mainly of *Lactococcus* and *Lactobacillus* genera. Tavaria and colleagues (2002) attributed the free amino acid degradation and consequent production of volatiles to the activity of *Lactobacillus rhamnosus*, *L. plantarum*, *Leuconostoc lactis*, *Lactococcus lactis* ssp. *lactis*, and *Enterococcus faecium*, which they isolated during cheese ripening.

Extensive lipolysis also takes place during ripening, as free fatty acids are the dominant type of compounds in the mature cheese. In addition, the concentrations of acetic, isobutyric and isovaleric acids were found to steadily increase with maturation, contributing to flavour (Tavaria et al. 2004).

6.5.4 Feta (Greece)

The PDO's Feta[7] is the most well-known Greek cheese and has been produced since ancient times. Homer's *The Odyssey* describes how the Cyclops prepares a cheese from sheep milk, which is, believed to be Feta. According to the documents provided to the EC by Greece,[8] Feta has been produced in much the same way for thousands of years. The unique climate of the designated areas of the Greek mountains is believed to give feta cheese its distinct taste, because of the strict use of sheep and goat milk that graze in the area. Feta cheese is still a key component of Greek's cuisine. It has a place in almost every meal, at any time of the day. In the classic MD, cheese often took the place of meat, being cheaper and more easily available.

Thus, according to Reg. EC No CE: EL/PDO/0017/0427 15.10.2002 (EC 2002), feta cheese is traditionally prepared from pure sheep milk or from a mixture of not less than 70 % sheep milk and not more than 30 % goat milk. Goats and sheep have been bred with a strong constitution that can adapt to the difficult mountainous

[7] Further information can be found at the following web address: http://www.cheesenet.gr/english-html/cheeses/feta.htm.

[8] Reference: document EL/PDO/0017/0427 at in http://ec.europa.eu/agriculture/quality/door/registeredName.html?denominationId=876.

conditions, with most sheep belonging to the Zackel breed. The cheese is made of raw or pasteurised milk, with a minimum of 6 % fat and a pH value of at least 6.5.

LAB and 10–20 % sodium chloride (NaCl) are added to the milk and allowed to ferment at 34–36 °C for 20 min to 1 h. When it reaches the necessary acidity, rennet is added to coagulate the caseins. In about 1 h, a compact curd is obtained, which is then cut into 1- to 2-cm cubes, gently stirred and left to rest for about 10 min.

Feta*. Feta is a white, rindless cheese ripened and preserved in brine. It is a typical and traditional Greek cheese made exclusively from pure sheep milk or from a mixture of sheep and goat's milk (local production), it is distinguished by its soft to semi-hard structure, with scratches and holes of a mechanical nature. Interestingly, Feta is a long-life product that, similar to wines, continues to develop as it ages. Photo reprinted with kind permission from M. Barone*

The curd is then transferred to cylindrical molds with many small holes, if the cheese is to be ripened and preserved in barrels; and rectangular molds if tins are to be used. This stage, during which the biological acidification of the curd takes place, is particularly important for feta's quality. Whey is allowed to drain for about 24 h at 16–18 °C. During this time, molds are upturned 2–3 times and the curd surface is salted to help remove excess whey. The next day, the curd is removed from molds and cut in rectangular shapes. It is placed temporarily in open barrels or containers in layers, and salted at a proportion of 3 % weight/weight. After 2–3 days, the pieces of cheese are finally placed in wooden barrels or metal containers, and brine of approximately 7 % NaCl is added. Containers are placed in ripening chambers at 16–18 °C and at least 85 % relative humidity, where they remain for about 15 days. The second ripening stage takes place in refrigerated rooms at 2–4 °C and a relative humidity of at least 85 % if wooden containers are used;

this stage lasts for about 2 months, at which point the cheese should have acquired its rich smell and pleasant taste. Some producers claim that wooden barrels are more traditional and transmit a unique flavour. Containers are then shipped to market, where the cheese is cut and sold directly from the container; alternatively, blocks of standardised weight are packaged in sealed plastic cups with some brine.

Feta is a white, rindless cheese ripened and preserved in brine. It is distinguished by its soft to semi-hard structure, with scratches and holes of a mechanical nature. It has a slightly acidic and salty taste and the resulting properties of mild lipolysis. Feta has a pH around 4, a moisture content of about 50 %, 25–26 % fat, 16–17 % total proteins, about 3 % NaCl and traces of lactose. The interesting thing about feta cheese is that it continues to develop as it ages, much like fine wine. It will taste different depending on the number of months it has been allowed to age.

The main phenomena in cheese making impacting its quality are the physical responses of the protein matrix to salting and cooling, namely in water retention. Although working with a distinct milk mixture, brine temperature was found to be the factor that most affected cheese weight and volume, due to the expansion of the protein matrix in the cold (at 3 °C, curd expands 20–30 %) (McMahon et al. 2009).

Given the impact of LAB on quality, a wide variety of strains have been isolated from feta and evaluated for their biochemical properties. Xanthopoulos and colleagues (2000) isolated and characterised 32 *L. plantarum* strains with distinct acidifying and proteolytic activities, suggesting the possibility of selecting strains with specific biotechnologically interesting properties as starters.

Ripening has been attributed to proteolysis, lipolysis and to the production of aroma by viable LAB and yeast consortia. The role of *Lactobacillus paracasei* subsp. *paracasei*, *Debaryomyces hansenii* and *Saccharomyces cerevisiae* in the feta-ripening process has been discussed (Bintsis et al. 2003). Esterase activity was less frequent in lactobacilli than in yeast, which also showed higher caseinolytic activity than lactobacilli, which, in turn, were found to preferentially degrade β-caseins. Esterases from yeasts preferentially degraded short-chain fatty acids. Thus, lactobacilli and yeasts contribute differently to the ripening and flavour development of feta. Volatile compounds that were detected and believed to be of microbial origin were ethanol, acetate, acetone, acetaldehyde, acetoin and diacetyl (Sarantinopoulos et al. 2002).

As with other traditional cheeses (namely from Portugal and Spain), enterococci were found not only to be part of the normal mixed population but also to play critical roles in ripening, positively affecting taste, aroma, colour and structure of the cheese. These populations may include *Enterococcus faecium* strains. Selected *E. faecium* strains showed a pronounced proteolytic activity, enhancing the degradation of α- and β-caseins with the release of peptides and a corresponding increase in water-soluble nitrogen fraction (Sarantinopoulos et al. 2002).

6.5.5 Mozzarella di Bufala Campana (Italy)

The buffalo was probably introduced to Italy in the seventh century, and was a familiar sight in the countryside, as it was widely used as a draught animal in ploughing both compact and watery terrains. Mozzarella from buffala's milk[9] became widespread throughout the south of Italy from the second half of the eighteenth century. It is a fresh, stringy textured cheese with a porcelain-white colour, an extremely thin rind and delicate taste. When cut, it produces a white watery fluid with a milky aroma. It is molded in a typical round shape. This creamy cheese from Southern Italy has won fans all over the world.

Mozzarella was awarded its PDO status by Reg. N. CE: IT/PDO/117/0014/20.09.2002, later amended (EC 2007b). It is a fresh variety of cheese made from whole buffala's milk (raw or pasteurised), from seven provinces in central-south Italy: Caserta and Salerno provinces, and part of Benevento, Naples, Frosinone, Latina and Rome. The milk is required to have a minimum content of 7.2 % fat and 4.2 % proteins, which makes it unsuitable for human consumption as is.

Mozzarella. *Mozzarella is a fresh cheese produced from buffala's milk, and traditionally used on pizzas. Mozzarella may contain beneficial lactic acid bacteria, as well as some bioactive peptides, released from the breakdown of casein, with protective effects against coronary heart disease. Photo reprinted with kind permission from T. N. Wassermann*

The first step in cheese manufacturing is the acidification of milk through the addition of whey from previous batches from the same facility. LAB are most probably responsible for the pH drop that facilitates coagulation. Calf rennet is then

[9] Further information can be found at the following web address: http://www.mozzarelladibuffala.org/.

added, and milk is kept at 33–39 °C for a maturation period of about 5 h. After the milk curdles, most of the whey is drained. The curd is then cut into small pieces, ground up in a sort of primitive mill and reduced to a crumble; the curd is put into a mould and immersed in boiling water, where it is stirred until it takes on a rubbery texture. The cheesemaker kneads pieces of curd, like a baker kneading bread, until a smooth, shiny paste is obtained, a strand of which is pulled out and lopped off, forming the individual mozzarella (the word 'mozzare' in Italian means to lop off), in pieces of about 10–800g. These pieces are in turn put into cold water and are then allowed to soak in brine. The cheese absorbs as much salt as is necessary to take on the correct consistency and can be consumed the next day. It should show an onion-like leafy but elastic structure and, according to the corresponding PDO regulation, Mozzarella should have a minimum of 52 % fat (on a dry weight basis) and a maximum of 65 % water.

Commercial fresh Mozzarella cheese is stored dry or in water without additional salt. The cheese has a shelf life of 4–6 weeks, as it may develop an off-flavour, and it loses textural integrity by 4 weeks, potentially due to bacterial growth. Microbial growth was found to be inevitable, even in the presence of 2 % NaCl and adequate refrigeration conditions (Baruzzi et al. 2012).

Devirgiliis and co-workers (2008) suggest that the traditional procedures necessary for the manufacture of this typical cheese, such as high-temperature treatments, led to a final product with low bacterial counts and lower biodiversity. Nevertheless, although its complexity has been recognized, the microbiota involved in mozzarella production has never been thoroughly assessed. Ercolini and colleagues (2001) described the microbiome of whey extracted from Mozzarella cheese, by using culture-dependent and pure molecular biology methods (all bacteria were identified via the sequencing of 16S rDNA). According to these authors, the dominant species in the samples directly analysed by molecular biology methods were *Streptococcus thermophilus*, *Lactococcus lactis*, *Lactobacillus delbrueckii* and *L. crispatus*, whereas *Lactobacillus fermentum* and *Enterococcus faecalis* were found instead in the cultivable community. Later, the most abundant species in the natural starter whey were found to be *Streptococcus thermophilus*, *L. delbrueckii* and *Lactobacillus helveticus*, suggesting thermophilic LAB may be primarily responsible for the fermentation (Ercolini et al. 2012). The presence of the psychrotrophic spoilage bacteria *Acinetobacter* sp. and *Pseudomonas* sp. was also reported. Thus, bacterial growth will eventually led to cheese spoilage; however, before that, bacteria may have beneficial effects, such as contributing to aroma and to casein hydrolysis. The presence of bioactive peptides in 'Mozzarella di bufala campana' has been registered, with some of the health-beneficial properties referred to in Sect. 6.5 (De Simone et al. 2009).

6.6 Yoghurt

The word yoghurt derives from Turkish; yoghurt itself is thought to originate in central Asia as an accidental phenomenon of milk coagulation by adventitious bacteria to a product of pleasant acidic taste. The milk of different mammals was sometimes used to obtain this dairy product, which is generally safer than milk.

Yoghurt. *Yoghurt is a fermented food, a class that has recently gained notice for its health-promoting properties. Yoghurt is also the first identified source of probiotics, with proven positive impact in gut microbiota, thus reinforcing immunity and the perception of wellness. Photo reprinted with kind permission from T. N. Wassermann*

Nowadays, yoghurt is created via an industrially controlled food process and made of pasteurised cow milk to which selected strains of *Lactobacillus delbrueckii* subsp. *bulgaricus* and *Streptococcus salivarius* var. *thermophilus* are added. Further lactobacilli and bifidobacteria are sometimes added during or after culturing, mainly for their probiotic properties. The fermentation takes place at about 42 °C for 4–8 h. The addition of extra milk solids regulates the texture, from firm to creamy or liquid. The accumulation of lactic acid causes the coagulation of milk caseins and lowers the pH to about 4.5, improving the preservation and safety of yoghurt, which generally is not suitable for the growth of pathogens.

Greek-style yoghurt is strained extensively to remove much of the liquid whey, leaving the remaining lactose and other sugars, thus giving a thicker consistency to the product.

Industries adapt their products according to market demand; yoghurt is most commonly flavoured or has fruit or jam added, and is packed in individual portions. Large amounts of sugar—or other sweeteners for low-calorie yoghurts—are often used in commercial yoghurt. Some yoghurts contain added starch, pectin (found naturally in fruit), and/or gelatin to artificially create thickness and creaminess at a lower cost.

Until 1900, yoghurt was a staple food in the diets of western Asia, Caucasus, India and border regions, as in Turkey and Greece. By this time, Elie Metchnikoff (a Nobel laureate) was the director of 'Institut Pasteur'. He promoted important studies on the identification and characterisation of yoghurt microorganisms and introduced the term 'probiotic' (Anukam and Reid 2007).

As a consequence, dairy industries have been developing and promoting a wide range of healthy foods, from probiotic acidified milk to Greek-style yoghurt, which, with roughly the same amount of energy, can deliver up to double the protein while cutting sugar content by half, and it is more satiating than regular yoghurt for the same portion.

6.6.1 Regular Plain Yoghurt

Regular plain yoghurt provides 54–63 kcal from: proteins (4.2–5.2 %), sugars (5.0–7.0 %, some lactose plus added sucrose) and total fat (1.6–1.8 %). Lipid fraction primarily comprises SFA (1.0 %) from C4 to C18, with a predominance of C14:0 to C18:0; MUFA (0.4 %), mostly C18:1 (oleic acid); and PUFA (44 mg to 0.1 g/100 g), C18:2 and C18:3, which are believed to be conjugated isomers of rumenic, linolenic (0.1 %) and rumelenic acids. This type of yoghurt also contains 6.0 mg/100 g of cholesterol. Yoghurt is a good source of retinol (14–30 μg/100 g), riboflavin (0.21–34 μg/100 g), vitamin B_{12} (0–0.56 μg/100 g), and Ca (118–183 mg/100 g). It also supplies some trace minerals, such as Se (3.3 μg/100 g) and F (12 μg/100 g) (INSA 2015; USDA 2015).

As seen before, when comparing compositional data obtained from the American and the Portuguese databases, for the same type of food product, in some cases no variation in the concentration of individual components is found, as is the case of cholesterol in regular plain yoghurt. However, for the same type of product, significant differences in the average concentration values for other components can be observed, as happens with the vitamins retinol (14 and 30 μg/100 g), riboflavin (0.21 and 34 μg/100 g) and B_{12} (0 and 0.56 μg/100 g) (INSA 2015; USDA 2015). These differences can be a result of many variations such as cattle feeding, handling practices and climate.

At the date of the current publication, the composition of Greek-type yoghurt is only available from the USDA (2015) database. The most traded flavour of a major global dairy company was chosen and subsequently described here.

6.6.2 Strawberry Greek-Type Yoghurt (Oikos)

Strawberry Greek-type yoghurt provides 106 kcal from: proteins (8.25 %), sugars (about 10 %: sucrose 3.7 %, fructose 4.1 %, lactose 2.3 %, galactose 0.5 % and dextrose 0.3 %), dietary fibre (1 %) and total fat (2.92 %). As before, lipid fraction is mainly composed of SFA (1.83 %) from C4 to C20, with a predominance of C14:0 and C16:0; MUFA (0.75 %), mostly C18:1 (OA); and PUFA (0.13 g/100 g), including 18:2 CLA, C18:3 ALA and 22:5 n-3 (docosapentaenoic acid [DPA]).

This type of yoghurt contains more cholesterol (13.0 mg/100 g) and almost no vitamin A, riboflavin (0.23 μg/100 g) or vitamin B_{12} (0.56 μg/100 g); it also provides less Ca (86 mg/100 g) but more Se (9.5 μg/100 g) (USDA 2015).

Yoghurt is a classic example of a probiotic food. Even if no other cultures are added, starter cultures *S. salivarius* var. *thermophilus* and *L. delbrueckii* ssp. *bulgaricus* (independent of the strain) are recognised as enhancing lactose digestion in lactose-intolerant individuals (FAO/WHO 2002).

Elie Metchnikoff is regarded as the grandfather of modern probiotics, in hypothesizing that "the dependence of the intestinal microbes on the food makes it possible to adopt measures to modify the flora in our bodies and to replace the harmful microbes by useful microbes" (Anukam and Reid 2007). Molecular and genetic studies, with strong contributions by Fuller, Klaenhammer, Reids and Salminem, have helped to uncover the mechanistic basis for the beneficial activities of probiotics; however, recent findings turned the Metchnikoff hypothesis into a topic of debate, by showing that permanent changes in gut microbiota are difficult or even unlikely to occur in adults (Sect. 6.2).

Yet, probiotic dairy products are targeted at healthy individuals who care about their long-term well-being. These products include fermented milk and yoghurt, with one or more strains for which a health benefit is normally marketed. It is a growing business segment, based on the concept that introducing beneficial bacteria into the human intestines will improve human health.

At the date of the present publication, neither the US Food and Drug Administration nor the EFSA have approved any health claim for food probiotics. Nevertheless, probiotic microorganisms (e.g. specific strains of *Saccharomyces boulardii* and *Lactobacillus plantarum*) are used in the medical treatment of a variety of GI conditions, such as infectious diarrhoea and diarrhoea associated with antibiotic use and irritable bowel syndrome. Probiotic formulations are also being proposed to prevent tooth decay and to prevent or treat other oral health problems such as gingivitis and periodontitis.

Probiotic strains with scientific evidence for health benefits are reported to be mainly added to dairy products and are as follows (Anukam and Reid 2007):

(a) Genus *Lactobacillus* (*L. rhamnosus GG*, *L. rhamnosus* GR-1, *L. reuteri* RC-14, *L. casei* DN114001, *L. acidophilus* LA-1, *L. reuteri* SD2112, *L. plantarum* 299v, *L. casei* Shirota, *L. acidophilus* LB, *L. rhamnosus* HN001, *L. salivarius* UCC118, *L. acidophilus* NCFM, *L. fermentum* VRI003, *L. johnsonii Lj*-1, *L. paracasei* F19)
(b) Genus *Bifidobacterium* (*B. lactis* Bb 12, *B. infantis* 35624, *B. breve* strain Yakult, *B. animalis* DN 117-001, *B. lactis* HN019, *B. longum* BB536)
(c) *Lactococcus lactis* L1A

The above list does not include enterococci, although evidence is building in this direction. *Enterococcus* sp. is a complex genus encompassing opportunistic pathogens as well as commensal bacteria, including potential probiotics (Cuív et al. 2013). They are members of the order Lactobacillales, which also includes

Lactobacillus, and are present in numerous cheeses, where they are believed to play important roles, as mentioned in Sect. 6.2.

In a recent study, piglets were fed with a supplement of the probiotic *E. faecium* NCIMB 10415 and exhibited a reduction in diarrhoea caused by *E. coli* and a greater increase in body mass. These findings were attributed to sterical interference between the *E. faecium* probiotic and the pathogen to binding sites (although other mechanisms may also be involved), causing a reduction in the adhesive properties of the pathogen and thus to its decreased virulence (Bednorz et al. 2013).

It is necessary to note that the approval process through the relevant authorities is complex and involves several types of studies, including risk assessment and epidemiological studies.

Regardless, it is well established that probiotic usage should be complemented by the ingestion of prebiotics, which are mainly oligosaccharides (as oligofructose) that improve probiotic viability, both in the dairy product (to which these compounds are most often added) and in the gut. Inulin is a term applied to a heterogeneous blend of fructose-based polymers widely distributed in nature as plant storage carbohydrates. Oligofructose is an anologue of inulin, with a polymerisation degree >10. The criteria used for classifying a food component as a prebiotic are its resistance to digestion, hydrolysis, the ability to be fermented by colonic microbiota and, most importantly, selective growth stimulation of one or a limited number of bacteria in the human colon. This last criterion makes the distinction between a prebiotic and a normal dietary fibre (Ziesenitz et al. 2012).

In a Mediterranean cohort study, Martinez-Gonzalez and colleagues (2014) found that yoghurt consumption was inversely associated with the incidence of overweight/obesity, especially among participants who consumed more fruit. The effect was more pronounced in individuals who consumed more than seven servings/week of total and whole-fat yoghurt, which may be partially explained by the maintenance of the short-term alterations provoked in gut microbiota.

In short, dairy products play an important role in diet as relevant sources of protein, calcium, vitamins and bioactive compounds.

Milk is regarded as nearly or almost a complete food and has been consumed in large quantities by children and adults, even in South Europe and North Africa, where the prevalence of hypolactasia is high and worsens with ageing. Cheese, butter and yoghurt do not pose such a problem, since they have a naturally reduced concentration of lactose.

Some milk proteins can be allergens or cause intolerance. Once again, individuals with these intolerances may find plausible alternatives in dairy products such as ripened cheese (in which proteolysis not only produces aroma, but also alters the protein profile).

The composition of dairy products reflects their ruminant origin. Major bioactive compounds are CLA isomers. Cheese is the main contributor to total CLA's daily intake among the Portuguese population (Martins et al. 2007). Again, and according to Martins and co-workers (2007), cheese is the primary contributor of the most important CLA isomers (c9,t11 and t7,c9) to which beneficial health effects have been attributed (besides being precursors of eicosanoids). On the other hand, intake of the t10,c12 isomer of dairy origin by the Portuguese population is negligible. As noted above, this particular isomer may have some adverse effects, particularly in obese individuals.

Moreover, cheese frequently carries bioactive peptides (e.g. VPP, IPP), with well-known health-beneficial properties that may contribute to creating more added value, particularly among small producers.

Thus, the transmission of more information to consumers about the alternatives to milk could decrease milk consumption, seeing it replaced with cheese and yoghurt. Yoghurt is demanded by health-conscious consumers, since all yoghurt may provide probiotic bacteria. The rate of approval of health claims for specific strains is increasing. Nevertheless, the efficacy of these probiotics depends on the host's enterotype and individual specificities, on the viability of yoghurt's bacteria and on regular intake, as the effects are most probably not permanent. Further insights into diet/microbiome relationships could eventually have an impact on nutritional guidelines for both healthy individuals and patients with chronic intestinal diseases and metabolic diseases such as obesity and diabetes.

References

Abellan A, Cayuela JM, Pino A, Martinez-Cacha A, Salazar E, Tejada L (2012) Free amino acid content of goat's milk cheese made with animal rennet and plant coagulant. J Sci Food Agric 92(8):1657–1664. doi:10.1002/jsfa.5528

Alkanani AK, Hara N, Lien E, Ir D, Kotter CV, Robertson CE, Wagner BD, Frank DN, Zipris D (2014) Induction of diabetes in the RIP-B7.1 mouse model is critically dependent on TLR3 and MyD88 pathways and is associated with alterations in the intestinal microbiome. Diabetes 63(2):619–631. doi:10.2337/db13-1007

Anukam KC, Reid G (2007) Probiotics: 100 years (1907-2007) after Elie Metchnikoff's Observation. In: Méndez-Vilas A (ed) Communicating Current Research and Educational Topics and Trends in Applied Microbiology, vol 1, Microbiology series. Formatex, Badajoz

Arumugam M, Raes J, Pelletier E, Le Paslier D, Yamada T, Mende DR, Fernandes GR, Tap J, Bruls T, Batto JM, Bertalan M, Borruel N, Casellas F, Fernandez L, Gautier L, Hansen T, Hattori M, Hayashi T, Kleerebezem M, Kurokawa K, Leclerc M, Levenez F, Manichanh C, Nielsen HB, Nielsen T, Pons N, Poulain J, Qin J, Sicheritz-Ponten T, Tims S, Torrents D, Ugarte E, Zoetendal EG, Wang J, Guarner F, Pedersen O, de Vos WM, Brunak S, Doré J, MetaHIT Consortium (additional members), Weissenbach J, Ehrlich SD, Bork P (2011) Enterotypes of the human gut microbiome. Nature 473(7346):174–180. doi:10.1038/nature09944

Banwo K, Sanni A, Tan H (2012) Technological properties and probiotic potential of Enterococcus faecium strains isolated from cow milk. J Appl Microbiol 114(1):229–241. doi:10.1111/jam.12031

Baruzzi F, Lagonigro R, Quintieri L, Morea M, Caputo L (2012) Occurrence of non-lactic acid bacteria populations involved in protein hydrolysis of cold-stored high moisture Mozzarella cheese. Food Microbiol 30(1):37–44. doi:10.1016/j.fm.2011.10.009

Bednorz C, Guenther S, Oelgeschläger K, Kinnemann B, Pieper R, Hartmann S, Tedin K, Semmler T, Neumann K, Schierack P, Bethe A, Wieler LH (2013) Feeding the probiotic Enterococcus faecium strain NCIMB 10415 to piglets specifically reduces the number of Escherichia coli pathotypes that adhere to the gut mucosa. App Environ Microbiol 79(24): 7896–7904. doi:10.1128/AEM.03138-13

Berg JM, Tymoczko JL, Stryer L (2002) Section 26.4: Important derivatives of cholesterol include bile salts and steroid hormones. In: Berg JM, Tymoczko JL, Stryer L (eds) Biochemistry, 5th edn. W. H. Freeman, New York

Bintsis T, Vafopoulou-Mastrojiannaki A, Litopoulou-Tzanetaki E, Robinson RK (2003) Protease, peptidase and esterase activities by lactobacilli and yeast isolates from Feta cheese brine. J Appl Microbiol 95(1):68–77. doi:10.1046/j.1365-2672.2003.01980.x

Brinkmann CR, Heegaard CW, Petersen TE, Jensenius JC, Thiel S (2011) The toxicity of bovine α-lactalbumin made lethal to tumor cells is highly dependent on oleic acid and induces killing in cancer cell lines and noncancer-derived primary cells. FEBS J 278(11):1955–1967. doi:10.1111/j.1742-4658.2011.08112.x

Brouwer IA, Wanders AJ, Katan MB (2010) Effect of animal and industrial trans fatty acids on HDL and LDL cholesterol levels in humans—a quantitative review. PLoS One 5(3):e9434. doi:10.1371/journal.pone.0009434

Burcelin R (2012) Regulation of metabolism: a cross talk between gut microbiota and its human host. Physiology 27(5):300–307. doi:10.1152/physiol.00023.2012

Chassaing B, Koren O, Goodrich JK, Poole AC, Shanthi Srinivasan S, Ley RE, Gewirtz AT (2015) Dietary emulsifiers impact the mouse gut microbiota promoting colitis and metabolic syndrome. Nature 519(7541):92–96. doi:10.1038/nature14232

Chen J, He X, Huang J (2014) Diet effects in gut microbiome and obesity. J Food Sci 79: R442–R451

Choi J, Sabikhi L, Hassan A, Anand S (2012) Bioactive peptides in dairy products. Int J Dairy Technol 65(1):1–12. doi:10.1111/j.1471-0307.2011.00725.x

Christoffersen TE, Jensen H, Kleiveland CR, Dorum G, Jacobsen M, Lea T (2012) In vitro comparison of commensal, probiotic and pathogenic strains of Enterococcus faecalis. Br J Nutr 108(11):2043–2053. doi:10.1017/S0007114512000222

Cocolin L, Nucera D, Alessandria V, Rantsiou K, Dolci P, Grassi MA, Lomonaco S, Civera T (2009) Microbial ecology of Gorgonzola rinds and occurrence of different biotypes of Listeria monocytogenes. Int J Food Microbiol 133(1–2):200–205. doi:10.1016/j.ijfoodmicro.2009.05.003

Cuív PO, Klaassens ES, Smith WJ, Mondot S, Durkin AS, Harkins DM, Foster L, McCorrison J, Torralba M, Nelson KE, Morrison M (2013) Draft genome sequence of Enterococcus faecalis PC1.1, a candidate probiotic strain isolated from human feces. Genome Announc 1(1):e00160-12. doi:10.1128/genomeA.00160-12

De Simone C, Picariello G, Mamone G, Stiuso P, Dicitore A, Vanacore D, Chianese L, Addeo F, Ferranti P (2009) Characterisation and cytomodulatory properties of peptides from Mozzarella di Bufala Campana cheese whey. J Pept Sci 15(3):251–258. doi:10.1002/psc.1093

De Vos WM, De Vos EA (2012) Role of the intestinal microbiome in health and disease: from correlation to causation. Nutr Rev 70(Suppl 1):S45–S56. doi:10.1111/j.1753-4887.2012.00505.x

Delgado A, Brito D, Peres C, Noe-Arroyo F, Garrido-Fernández A (2005) Bacteriocin production by Lactobacillus pentosus B96 can be expressed as a function of temperature and NaCl concentration. Food Microbiol 22(6):521–528. doi:10.1016/j.fm.2004.11.015

Delgado A, Arroyo López FN, Brito D, Peres C, Fevereiro P, Garrido-Fernández A (2007) Optimum bacteriocin production by Lactobacillus plantarum 17.2b requires absence of NaCl and apparently follows a mixed metabolite kinetics. J Biotechnol 130(2):193–201. doi:10.1016/j.jbiotec.2007.01.041

Devirgiliis C, Caravelli A, Coppola D, Barile S, Perozzi G (2008) Antibiotic resistance and microbial composition along the manufacturing process of Mozzarella di Bufala Campana. Int J Food Microbiol 128(2):378–384. doi:10.1016/j.ijfoodmicro.2008.09.021

Doell D, Folmer D, Lee H, Honigfort M, Carberry S (2012) Updated estimate of trans fat intake by the US population. Food Addit Contam Part A Chem Anal Control Expo Risk Assess 29(6):861–874. doi:10.1080/19440049.2012.664570

Duca FA, Sakar Y, Lepage P, Devime F, Langelier B, Doré J, Covasa M (2014) Replication of obesity and associated signaling pathways through transfer of microbiota from obese-prone rats. Diabetes 63(5):1624–1636. doi:10.2337/db13-1526

EC (2000) Directive 2000/13/EC of the European parliament and of the council of 20 March 2000 on the approximation of the laws of the Member States relating to the labelling, presentation and advertising of foodstuffs, Ingredients referred to in Article 6(3a), (10) and (11), Annex IIIa. Off J Eur Comm L 109:29–42

EC (2002) Commission Regulation (EC) No 1829/2002 of 14 October 2002 amending the Annex to Regulation (EC) No 1107/96 with regard to the name 'Feta'. Off J Eur Comm L 277:10–14

EC (2007a) Publication of an amendment application pursuant to Article 6 (2) of Council Regulation (EC) No 510/2006 on the protection of geographical indications and designations of origin for agricultural products and foodstuffs. Off J Eur Union C127:10–14

EC (2007b) Publication of an amendment application pursuant to Article 6(2) of Council Regulation (EC) No 510/2006 on the protection of geographical indications and designations of origin for agricultural products and foodstuffs. Off J Eur Union C90:5–9

EC (2008) Publication of an amendment application pursuant to Article 6(2) of Council Regulation (EC) No 510/2006 on the protection of geographical indications and designations of origin for agricultural products and foodstuffs. Off J Eur Union C111:51–55

EC (2013) Publication of an amendment application pursuant to Article 50(2)(a) of Regulation (EU) No 1151/2012 of the European Parliament and of the Council on quality schemes for agricultural products and foodstuffs. Off J Eur Union C329:4–15

Economist (2014) Farming in the Netherlands: Polder and wiser, 23 Aug 2014. The Economist Newspaper Limited, London

EFSA (2004) Opinion of the Scientific Panel on Dietetic Products, Nutrition and Allergies on a request from the Commission related to the presence of trans fatty acids in foods and the effect on human health of the consumption of trans fatty acids (Request N° EFSA-Q-2003-022). EFSA J 81:1–49. doi:10.2903/j.efsa.2004.81

Enattah NS, Sahi T, Savilahti E, Terwillerger JD, Peltonen L, Jarvela I (2002) Identification of a variant associated with adult-type hypolactasia. Nat Genet 30(2):233–237. doi:10.1038/ng826

Ercolini D, Moschetti G, Blaiotta G, Coppola S (2001) The potential of a polyphasic PCR-dGGE approach in evaluating microbial diversity of natural whey cultures for water-buffalo Mozzarella cheese production: bias of culture-dependent and culture-independent analyses. Syst Appl Microbiol 24(4):610–617. doi:10.1078/0723-2020-00076

Ercolini D, De Filippis F, La Storia A, Iacono M (2012) 'Remake' by high-throughput sequencing of the microbiota involved in the production of water buffalo mozzarella cheese. Appl Environ Microbiol 78(22):8142–8145. doi:10.1128/AEM.02218-12

EUFIC (2015) The basics 06/2006: functional foods. European Food Information Council, Brussels, available at http://www.eufic.org/article/en/expid/basics-functional-foods/. Accessed 4 Nov 2015

FAO (2015a) Statistics Division. Trade. http://faostat3.fao.org/browse/T/. Accessed 16 Mar 2015

FAO (2015b) Dairy production and products. Milk composition. Food and Agriculture Organization, Rome, available at http://www.fao.org/agriculture/dairy-gateway/milk-and-milk-products/milk-composition/en/#.Vjo9HPl_Oko. Accessed 4 Nov 2015

FAO/WHO (2002) Guidelines for the evaluation of probiotics in food. Report of a Joint FAO/WHO Working Group on Drafting Guidelines for the Evaluation of Probiotics in Food. London, Ontario, Canada, 30 April and 1 May 2002. Food and Agriculture Organization of the United Nations, Rome, and World Health Organization, Geneva. Available at http://www.who.int/foodsafety/fs_management/en/probiotic_guidelines.pdf. Accessed 26 Nov 2015

FIL-IDF, International Dairy Federation (2005) Trans fatty acids: scientific progress and labelling In Bulletin of the International Dairy Federation, 393/2005. IDF, Brussels

Fontana C, Cappa F, Rebecchi A, Cocconcelli PS (2010) Surface microbiota analysis of Taleggio, Gorgonzola, Casera, Scimudin and Formaggio di Fossa Italian cheeses. Int J Food Microbiol 138(3):205–211. doi:10.1016/j.ijfoodmicro.2010.01.017

Gayet-Boyer C, Tenenhaus-Aziza F, Prunet C, Marmonier C, Malpuech-Brugère C, Lamarche B, Chardigny JM (2014) Is there a linear relationship between the dose of ruminant trans-fatty acids and cardiovascular risk markers in healthy subjects: results from a systematic review and meta-regression of randomised clinical trials. Br J Nutr 112(12):1914–1922. doi:10.1017/S0007114514002578

Gerber PJ, Steinfeld H, Henderson B, Mottet A, Opio C, Dijkman J, Falcucci A, Tempio G (2013) Tackling climate change through livestock—a global assessment of emissions and mitigation opportunities. Food and Agriculture Organization of the United Nations, Rome, available at http://www.fao.org/3/i3437e.pdf. Accessed 4 Nov 2015

Giorgetti G, Brandimarte G, Fabiocchi F, Ricci S, Flamini P, Sandri G, Trotta MC, Elisei W, Penna A, Lecca PG, Picchio M, Tursi A (2015) Interactions between innate immunity, microbiota, and probiotics. J Immunol Res 2015:501361. doi:10.1155/2015/501361

Gophna U (2011) The guts of dietary habits. Science 334(6052):45–46. doi:10.1126/science.1213799

Hayes M, Ross RP, Fitzgerald GF, Stanton C (2007) Putting microbes to work: dairy fermentation, cell factories and bioactive peptides. Part I: overview. Biotechnol J 2(4):426–434. doi:10.1002/biot.200600246

INSA (2015) Tabela da Composição de Alimentos (TCA). http://www.insa.pt/sites/INSA/Portugues/AreasCientificas/AlimentNutricao/AplicacoesOnline/TabelaAlimentos/PesquisaOnline/Paginas/PorPalavraChave.aspx. Accessed 5 Nov 2015

Kelly D, Mulder IE (2012) Microbiome and immunological interactions. Nutr Rev 70(Suppl 1):S18–S30. doi:10.1111/j.1753-4887.2012.00498.x

Kieffer D, Martin R, Marco M, Kim E, Keenan M, Dunn T, Adams S (2014) Resistant starch significantly alters gut microbiota and liver metabolome in mice fed a high fat diet. FASEB J 28(1 suppl):822.13

Kim HJ, Park KS, Lee SK, Min KW, Han KA, Kim YK, Ku BJ (2012) Effects of pinitol on glycemic control, insulin resistance and adipocytokine levels in patients with type 2 Diabetes mellitus. Ann Nutr Metab 60(1):1–5. doi:10.1159/000334834

Koba K, Yanagita T (2014) Health benefits of conjugated linoleic acid (CLA). Obes Res Clin Pract 8(6):e525–e532

Korhonen H, Pihlanto A (2006) Bioactive peptides: production and functionality. Int Dairy J 16(9):945–960. doi:10.1016/j.idairyj.2005.10.012

Logan AC (2015) Dysbiotic drift: mental health, environmental grey space, and microbiota. J Physiol Anthropol 34(1):23–40. doi:10.1186/s40101-015-0061-7

Mao XY, Cheng X, Wang X, Wu SJ (2011) Free-radical-scavenging and anti-inflammatory effect of yak milk casein before and after enzymatic hydrolysis. Food Chem 126(2):484–490. doi:10.1016/j.foodchem.2010.11.025

Martinez-Gonzalez MA, Sayon-Orea C, Ruiz-Canela M, de la Fuente C, Gea A, Bes-Rastrollo M (2014) Yoghurt consumption, weight change and risk of overweight/obesity: the SUN cohort study. Nutr Metab Cardiovasc Dis 24(11):1189–1196. doi:10.1016/j.numecd.2014.05.015

Martins SV, Lopes PA, Alfaia CM, Ribeiro VS, Guerreiro TV, Fontes CMGA, Castro MF, Soveral G, Prates JAM (2007) Contents of conjugated linoleic acid isomers in ruminant-derived foods and estimation of their contribution to daily intake in Portugal. Br J Nutr 98(6):1206–1213. doi:10.1017/S0007114507781448

McMahon DJ, Motawee MM, McManus WR (2009) Influence of brine concentration and temperature on composition, microstructure, and yield of feta cheese. J Dairy Sci 92(9):4169–4179. doi:10.3168/jds.2009-2183

Michaëlsson K, Wolk A, Langenskiöld S, Basu S, Lemming EW, Melhus H, Byberg L (2014) Milk intake and risk of mortality and fractures in women and men: cohort studies. BMJ 349:g6015. doi:10.1136/bmj.g6015

Mills S, Ross RP, Hill C, Fitzgerald GF, Stanton C (2011) Milk intelligence: mining milk for bioactive substances associated with human health. Int Dairy J 21(6):377–401. doi:10.1016/j.idairyj.2010.12.011

Moio L, Piombino P, Addeo F (2000) Odour-impact compounds of Gorgonzola cheese. J Dairy Res 67(2):273–285

Nakamura T, Aizawa T, Kariya R, Okada S, Demura M, Kawano K, Makabe K, Kuwajima K (2013) Molecular mechanisms of the cytotoxicity of human α-lactalbumin made lethal to tumor cells (HAMLET) and other protein-oleic acid complexes. J Biol Chem 288(20): 14408–14416. doi:10.1074/jbc.M112.437889

Ostaff MJ, Stange EF, Wehkamp J (2013) Antimicrobial peptides and gut microbiota in homeostasis and pathology. EMBO Mol Med 5(10):1465–1483. doi:10.1002/emmm.201201773

Palmnäs MS, Cowan TE, Bomhof MR, Su J et al (2014) Low-dose aspartame consumption differentially affects gut microbiota-host metabolic interactions in the diet-induced obese rat. PLoS One 9(10), e109841

Phelan M, Aherne A, FitzGerald RJ, O'Brien NM (2009) Casein-derived bioactive peptides: biological effects, industrial uses, safety aspects and regulatory status. Int Dairy J 19(11): 643–654. doi:10.1016/j.idairyj.2009.06.001

Pihlanto A (2006) Antioxidative peptides derived from milk proteins. Int Dairy J 16(11): 1306–1314. doi:10.1016/j.idairyj.2006.06.005

Rajilic-Stojanovic M, de Vos WM (2014) The first 1000 cultured species of the human gastrointestinal microbiota. FEMS Microbiol Rev 38(5):996–1047. doi:10.1111/1574-6976.12075

Rammer P, Groth-Pedersen L, Kirkegaard T, Daugaard M, Rytter A, Szyniarowski P, Høyer-Hansen M, Povlsen LK, Nylandsted J, Larsen JE, Jäättelä M (2010) BAMLET activates a lysosomal cell death program in cancer cells. Mol Cancer Ther 9(1):24–32. doi:10.1158/1535-7163.MCT-09-0559

Raynal-Ljutovaca K, Lagriffoulb G, Paccardb P, Guillet I, Chilliard Y (2008) Composition of goat and sheep milk products: An update. Small Ruminant Res 79:57–72

Roa I, Lopez MB, Mendiola FJ (1999) Residual clotting activity and ripening properties of vegetable rennet from Cynara cardunculus in La Serena cheese. Food Res Int 32(6): 413–419. doi:10.1016/S0963-9969(99)00098-8

Roseiro LB, Gómez-Ruiz JA, García-Risco M, Molina E (2003) Vegetable coagulant (Cynara cardunculus) use evidenced by capillary electrophoresis permits PDO Serpa cheese authentication. Lait 83:343–350. doi:10.1051/lait:2003017

Rubio RC, Collado MC, Laitinen K, Salminen S et al (2012) The human milk microbiome changes over lactation and is shaped by maternal weight and mode of delivery. Am J Clin Nutr 96: 544–551

Sánchez-Macías D, Morales-Delanuez A, Moreno-Indias I, Hernández-Castellano LE, Mendoza-Grimón V, Castro N, Argüello A (2011) Lipolysis and proteolysis profiles of fresh artisanal goat cheese made with raw milk with 3 different fat contents. J Dairy Sci 94(12):5786–5793. doi:10.3168/jds.2011-4423

Sandrini S, Aldriwesh M, Alruways M, Freestone P (2015) Microbial endocrinology: host-bacteria communication within the gut microbiome. J Endocrinol 225(2):R21–R34. doi:10.1530/JOE-14-0615

Sarantinopoulos P, Kalantzopoulos G, Tsakalidou E (2002) Effect of Enterococcus faecium on microbiological, physicochemical and sensory characteristics of Greek Feta cheese. Int J Food Microbiol 76(1–2):93–105. doi:10.1016/S0168-1605(02)00021-1

Scherr C, Ribeiro JP (2010) Gorduras em laticínios, ovos, margarinas e óleos: implicações para a aterosclerose. Arq Bras Cardiol 95(1):55–60

Schnorr SL, Candela M, Rampelli S, Centanni M, Consolandi C, Basaglia G, Turroni S, Biagi E, Peano C, Severgnini M, Fiori J, Gotti R, De Bellis G, Luiselli D, Brigidi P, Mabulla A, Marlowe F, Henry AG, Crittenden AN (2014) Gut microbiome of the Hadza hunter-gatherers. Nat Commun 5:3654. doi:10.1038/ncomms4654

Selhub EM, Logan AC, Bested AC (2014) Fermented foods, microbiota, and mental health: ancient practice meets nutritional psychiatry. J Physiol Anthropol 33:2. doi:10.1186/1880-6805-33-2

Silva SV, Pihlanto A, Malcata FX (2006) Bioactive peptides in ovine and caprine cheeselike systems prepared with proteases from Cynara cardunculus. J Dairy Sci 89(9):3336–3344. doi:10.3168/jds.S0022-0302(06)72370-0

Sokol H, Pigneur B, Watterlot L, Lakhdari O, Bermúdez-Humarán LG, Gratadoux JJ, Sébastien Blugeon S, Chantal Bridonneau C, Jean-Pierre Furet JP, Corthier G, Grangette C, Vasquez N, Pochart P, Trugnan G, Thomas G, Blottière HM, Doré J, Marteau P, Seksik P, Langella P (2008) Faecalibacterium prausnitzii is an anti-inflammatory commensal bacterium identified by gut microbiota analysis of Crohn disease patients. Proc Natl Acad Sci U S A 105: 16731–16736. doi:10.1073/pnas.0804812105

Suez J, Korem T, Zeevi D, Zilberman-Schapira G, Thaiss CA, Maza O, Israeli D, Zmora N, Gilad S, Weinberger A, Kuperman Y, Harmelin A, Kolodkin-Gal I, Shapiro H, Halpern Z, Segal E, Elinav E (2014) Artificial sweeteners induce glucose intolerance by altering the gut microbiota. Nature 514(7521):181–186. doi:10.1038/nature13793

Suzuki TA, Worobey M (2014) Geographical variation of human gut microbial composition. Biol Lett 10:20131037. doi:10.1098/rsbl.2013.1037

Szabó I, Wieler LH, Tedin K, Scharek-Tedin L, Taras D, Hensel A, Appel B, Nöckler K (2009) Influence of a probiotic strain of Enterococcus faecium on Salmonella enterica serovar Typhimurium DT104 infection in a porcine animal infection model. Appl Environ Microbiol 75(9):2621–2628. doi:10.1128/AEM.01515-08

Tavaria FK, Dahl S, Carballo FJ, Malcata FX (2002) Amino acid catabolism and generation of volatiles by lactic acid bacteria. J Dairy Sci 85(10):2462–2470. doi:10.3168/jds.S0022-0302(02)74328-2

Tavaria FK, Ferreira ACS, Malcata FX (2004) Volatile free fatty acids as ripening indicators for Serra da Estrela cheese. J Dairy Sci 87(12):4064–4072. doi:10.3168/jds.S0022-0302(04)73548-1

Turpin W, Humblot C, Thomas M, Guyot JP (2010) Lactobacilli as multifaceted probiotics with poorly disclosed molecular mechanisms. Int J Food Microbiol 143(3):87–102. doi:10.1016/j.ijfoodmicro.2010.07.032

Tyburczy C, Major C, Lock AL, Destaillats F, Lawrence P, Brenna JT, Salter AM, Bauman DE (2009) Individual trans octadecenoic acids and partially hydrogenated vegetable oil differentially affect hepatic lipid and lipoprotein metabolism in golden Syrian hamsters. J Nutr 139(2): 257–263. doi:10.3945/jn.108.098004

USDA (2015) Agricultural Research Service National Nutrient Database for Standard Reference Release 27, Software v.2.2.4, The National Agricultural Library. http://ndb.nal.usda.gov/ndb/foods. Accessed 5 May 2015

USDA and USDHHS (2010) Dietary guidelines for Americans, 7th edn. U.S. Government Printing Office, Washington, DC, available at www.dietaryguidelines.gov. Accessed 30 Oct 2015

Velasquez-Manoff M (2015) Gut microbiome: the peacekeepers. Nature 518:S3–S11. doi:10.1038/518S3a

Voreades N, Kozil A, Weir TL (2014) Diet and the development of the human intestinal microbiome. Front Microbiol 5:494. doi:10.3389/fmicb.2014.00494

Waworuntu R, Hain H, Chang Q, Thiede L, Hanania T, Berg B (2014) Dietary prebiotics improve memory and social interactions while reducing anxiety when provided early in life to normally developing rodents. FASEB J 28(1 Suppl):637.5

WHO-ROEM (2012) Promoting a healthy diet for the WHO Eastern Mediterranean Region: user-friendly guide. WHO Regional Office for the Eastern Mediterranean, Cairo, available at http://applications.emro.who.int/dsaf/emropub_2011_1274.pdf?ua=1. Accessed 30 Oct 2015

Xanthopoulos V, Hatzikamari M, Adamidis T, Tsakalidou E, Tzanetakis N, Litopoulou-Tzanetaki E (2000) Heterogeneity of Lactobacillus plantarum isolates from feta cheese throughout ripening. J Appl Microbiol 88(6):1056–1064. doi:10.1046/j.1365-2672.2000.01056.x

Yamaguchi M, Takai S, Hosono A, Seki T (2014) Bovine milk-derived α-lactalbumin inhibits colon inflammation and carcinogenesis in azoxymethane and dextran sodium sulfate-treated mice. Biosci Biotechnol Biochem 78(4):672–679. doi:10.1080/09168451.2014.890034

Ziesenitz S, Eldridge A, Antoine J-L, Coxam V, Flynn A, Fox K, Gray J, Macdonald I, Maughan R, Samuels F, Sanders T, Tomé D, van Loveren C, Williamson G (2012) Healthy lifestyles diet, physical activity and health. ILSI Europe, International Life Sciences Institute, Belgium, available at http://www.ilsi.org/Europe/Publications/ILSIcm11-004_Diet08.pdf. Accessed 4 Nov 2015

Fish, Meat and Other Animal Protein Sources

7

Abstract

The Mediterranean dietary pattern is based on a reduced intake of foods of animal origin, which provide essential amino acids (those the human body is unable to synthesize), as well as certain minerals and vitamins that are difficult to find or have a reduced bioavailability in vegetables. In this regard, fish is preferable to meat as a protein source, since the associated fats are mainly long-chain polyunsaturated fatty acids, with a predominance of n-3, as in sardines, anchovies and codfish. The Mediterranean Sea is less rich in fish than the Atlantic Ocean. The Atlantic coast is important to Portugal, Spain and Morocco, and Portugal and Spain are among the largest consumers of fish and seafood in the world. It is noteworthy that the composition of meat is highly dependent on how the animal was fed and grown and on the butcher's cut. In this regard, a lean steak from cows raised in the open air via extensive systems can be less deleterious to health than a piece of broiler chicken raised under an intensive system. In certain cases, total fat can be lower in beef (e.g. lean steak) than in a piece of chicken. Ancient meat preservation methods have been developed mainly for pork and rely on seasoning, smoking, fermentation and drying steps, aiming at reducing pH and water activity, which, in combination with spices, restricts microbial growth. Nowadays, some preservatives are allowed, even in 'Protected Designation of Origin' products as *chorizo, salami* and *jamon iberico*. The intake of large portions of meat is viewed by many as related to a high socioeconomic status, but excess meat is deleterious to health because of the accumulation of acidic deamination products such as uric acid, which becomes more difficult to eliminate with age, leading to gout, urate kidney stones and other ailments. Moderation of animal protein consumption and an increase in the ratio of vegetables to animal proteins may confer health-protective effects and result in a balanced diet.

A.M. Delgado et al., *Chemistry of the Mediterranean Diet*,
DOI 10.1007/978-3-319-29370-7_7

7.1 Fish, Meat and Other Animal Protein Sources: An Introduction

Proteins are the primary structural components of cells, especially muscle. Along with peptides, they play a multitude of crucial functions in the human body, from enzymes to signalling molecules (many important neurotransmitters are primary or secondary amines, derived from amino acids). Moreover, the immune system mainly relies on protein interactions (e.g. immunoglobulins [Igs], surface cell proteins).

Body proteins are synthesised as needed, after the polymerisation of 20 standard amino acids, which in turn are obtained from food or synthesized *de novo*. Some non-standard amino acids (obtained by post-translational modifications) also have important functions and are incorporated into some proteins, such as collagen or myosin.

Humans do not have the complete pathways to synthesise nine amino acids (phenylalanine, valine, threonine, tryptophan, methionine, leucine, isoleucine, lysine and histidine). They are therefore essential or indispensable and must be supplied by the diet. Under some circumstances, the synthesis of some other amino acids can be limited.[1]

The inclusion of animal protein sources (fish, shellfish, seafood, meat, dairy and eggs) in the diet ensures that all the necessary amino acids are provided in balanced proportions; for this reason, proteins of animal origin are called 'complete proteins'. On the other hand, most vegetable protein sources (e.g. cereals, pulses and nuts) generally lack one or more of the indispensable amino acids (limiting amino acids); for this reason, they are called 'incomplete proteins'. In most food cultures, several traditional dishes combine, for instance, cereals and pulses, which result in a higher-quality protein intake as the limiting amino acids of these foods are different and therefore complement each other. Examples of such dishes combine maize and beans, wheat (such as pasta) and chickpeas, black beans and rice. Another way to overcome this issue is the introduction, in the diet of a large variety of vegetable foods supplemented with some animal protein sources.

In a balanced diet, it is then necessary to observe the requirements and bioavailability of indispensable amino acids (NAP 2015), as shown in Table 7.1 (percentage values). Animal proteins are a rich source of sulphur-containing amino acids, such as cysteine and methionine, the metabolism of which generates sulphuric acid, as well as lysine, arginine and histidine, the metabolism of which yields hydrochloric acid, increasing the acid load of the organism. On the other hand, the catabolism of vegetable foods generally originates bicarbonate and inorganic anions. The bicarbonate ion, a base with buffering capacity, can reduce the net rate of endogenous acid accumulation (Frassetto et al. 2000; Sellmeyer et al. 2001). In other words, vegetables contain significant amounts of base

[1] Further information on proteins and their functions in the human body can be found in chapters 5, 6, 7, 8 and 22 of Nelson DL, Cox MM. (2012) Lenhinger—Principles of Biochemistry. W. H. Freeman and Company, New York and Basingstoke.

Table 7.1 Estimated average requirements of indispensable amino acids, expressed as percentage values of the total recommended intake of dietary protein (NAP 2015)

Amino acid	Minimum % proportion
Histidine	6.27
Isoleucine	8.71
Leucine	19.16
Lysine	17.77
Methionine & cysteine	8.71
Phenylalanine & tyrosine	16.38
Threonine	9.41
Tryptophan	2.44
Valine	11.15

precursors that may help neutralise acidic metabolites in the body (e.g. from protein digestion).

With aging, the glomerular filtration rate is reduced and the kidney's ability to excrete this dietary acid load decreases. Thus, otherwise healthy individuals develop progressively increasing blood acidity and decreasing plasma bicarbonate as they age. Because urinary excretion of acid is insufficient, other homeostatic systems, such as bone, buffer the excess dietary acid load (Sellmeyer et al. 2001). Over decades, the magnitude of daily positive acid balance may be sufficient to induce osteoporosis, as bone is the only internal reservoir capable of supplying a base to the systemic circulation over the lifetime (Frassetto et al. 2000).

The accumulation of acidic metabolites has been correlated with health issues, namely the incidence of hip fracture in the elderly. Renal net acid excretion correlates positively with animal protein intake and negatively with vegetable protein intake (Frassetto et al. 2000; Sellmeyer et al. 2001).

Not all proteins are easily digested. In the human gut, most globular proteins from animal sources are almost completely hydrolysed to amino acids, but some fibrous proteins, such as keratin, are only partially digested. On the other hand, some proteins from plant foods are enclosed in cellulose husks, thus impairing their enzymatic breakdown.

In developed countries, protein availability exceeds average population requirements, and measured food intake also reveals that consumption, mainly from animal sources, is often much higher than internationally recommended intake levels. Excessive proteins are digested and converted into their amino acid residues, the carbon backbones of which are used to produce energy via the Krebs cycle. Some amino acids are converted into ketone bodies and others to glucose or both. The use of proteins to produce energy, instead of fats and carbohydrates, occurs in starvation (when body proteins are consumed), in untreated diabetes and in cases of excess protein in the diet.

The link between the Krebs and urea cycles ensures that the N-moieties (obtained by deamination) are converted to urea and uric acid and eventually eliminated. High blood concentrations of uric acid can lead to gout. Excess of urea and uric acid are also commonly associated with medical conditions such as diabetes and the formation of ammonium acid urate kidney stones. On the other hand, the accumulation of ketone bodies in the blood can alter blood pH (which can be life threatening) and may be related to an imbalanced lipid metabolism and/or diabetes.

In the case of the Mediterranean diet (MD), the daily pattern of animal protein consumption is generally low. Traditional Mediterranean dishes usually do not include animal foods as the main ingredient but instead as a source of taste.

The composition, health and nutritional aspects of the main sources of animal protein are reviewed. Animal sources of protein also contain fats. The drawbacks as well as the relevance of certain fats and vitamins of animal origin to a balanced diet are also discussed in this chapter.

7.2 Seafood

In the Mediterranean basin, particularly in the countries that this work refer to (Fig. 1.1), a large variety of fish is consumed, particularly pelagics but also shellfish, cephalopods and crustaceans. Since freshwater fish is not so commonly consumed in these countries, only marine fish and seafood are the objects of this review.

The large Atlantic coasts of Portugal, Spain and Morocco, and the richer variety of fishes found in the Atlantic Ocean, influence local patterns of fish consumption, although with common features. The most prominent species consumed in the above-mentioned countries (Fig. 1.1) are tuna (*Thunnus alalunga*), mackerel (*Scomber scombrus*), aquaculture produce such as gilt-head bream (*Sparus aurata*) and Mediterranean seabass (*Dicentrarchus labrax*), as well as smaller species such as sardine (*Sardina pilchardus*) and anchovy (*Engraulis encrasicolus*). Cod (*Gadus morhua*), a demersal fish, is very popular in all these areas, although it is captured in the North Sea. The intake of species such as eel (*Anguilla anguilla*), a migrating snake-like fish that is captured in fresh water, sole (and related flat fish species such as Dover sole), or cephalopods such as squid (*Loligo forbesii*, *Loligo vulgaris* and related species), cuttlefish (*Sepia officinalis*) and octopus (*Octopus vulgaris*) are noteworthy.

The most widely consumed species of shellfish, or bivalves, in the Mediterranean region are clams, mussels (*Mytilus edulis*, *M. galloprovincialis* and related species) and cockle (*Cerastoderma edule*), among many other species, such as razor clams and others for which the English name is hard to find, as in the case of 'tellina' or 'conquilha'.

Crustaceans, such as different types of crabs (*Cancer pagurus*) and shrimps (*Crangon crangon*, as well as *Penaeus* sp. and *Palaemon* sp.) have been also consumed in the area. Lobster is not so widely available in the area.

A larger variety of seafood is available in Portugal and Spain because of their Atlantic coastline. Both countries are at the top of the world ranking of countries of per capita seafood consumption (Banks 2012).

In general, fish is known to have a balanced essential amino acid composition, and it is recommended for a healthy diet. However, the free amino acid composition changes considerably during storage and subsequent processing (if any). A possible issue, involving fish proteins (but easily avoided with adequate preservation methods) is the accumulation of biogenic amines (organic bases derived from the degradation of free amino acids), including histamine. Histamine is produced from the decarboxylation of histidine (a common amino acid in fish) by the enzyme histidine decarboxylase. The reaction is favoured by temperature and by the release of cellular contents (which ensures the presence of the enzyme in sufficient amounts). Thus, correct processing and refrigeration prevents the formation of histamine, a potent allergen that can cause food poisoning.

Besides protein, fish and seafood also provide relevant amounts of long chain (n3) polyunsaturated fatty acids (PUFA) playing several important physiological roles in the body (DiLeone 2011; Galli and Calder 2009; Jump 2002; Mozaffarian and Wu 2012; Sanders 2009; Stark et al. 2008). These essential fatty acids include (Fig. 5.5) eicosapentaenoic acid (EPA: 20:5, n-3), docosapentaenoic acid (DPA: 22:5, n-3) and docosahexaenoic acid (DHA: 22:6, n-3).

Dietary patterns change over time, and a westernisation of the MD has been observed in the area. One concern is the increase of the ratio of n-6 to n-3 fatty acids in the diet. According to Stark and colleagues (2008), n-6 and n-3 fatty acids are metabolised by the same set of enzymes thus, competing for the same active sites. According to the same authors, metabolites of n-3 fatty acids are anti-inflammatory and anti-arrhythmic, while metabolites of n-6 fatty acids may result in the accumulation of pro-thrombotic and pro-inflammatory compounds (Galli and Calder 2009). The preference for fish may mitigate this issue, given its high n-3 PUFA content.

EPA and DHA are reported to modulate a variety of relevant biological pathways (Mozaffarian and Wu 2012). In human studies, both fatty acids lower triglyceride levels and collagen-stimulated platelet aggregation; they seem to favourably affect cardiac diastolic filling, arterial compliance and to have some effect against inflammation and oxidative stress (Mozaffarian and Wu 2012). Less is known about the effects of DPA on biological pathways or physiological risk factors; a few observational studies suggest potential benefits for inflammation.

Moreover, eicosanoids are chemical messengers derived from 20-carbon PUFA that play critical roles in immune and inflammatory responses. Both 20-carbon omega-6 fatty acids (arachidonic acid) and 20-carbon omega-3 fatty acids (EPA) can be found in cell membranes. DPA can be synthesised endogenously from EPA (Mozaffarian and Wu 2012).

EPA and DHA are metabolised by cytochrome P450 enzymes to generate mono-epoxides derived from (n-3) fatty acids (MEFA). An EPA-derived MEFA was referred to display potent anti-inflammatory activity, in human bronchi and MEFA derived from either EPA or DHA were similarly potent at achieving mesenteric microvessel dilation (Mozaffarian and Wu 2012).

Generally, dietary guidelines converge on a recommended minimum consumption of 250–500 mg/day of combined EPA+DHA for an average adult (Smit et al. 2009). Despite the alternative sources of essential fatty acids and protein, in the context of the MD, fish should be preferred to meat, with two or more servings per week (Bach-Faig et al. 2011).

7.2.1 Tuna (*Thunnus alalunga*)

Tuna or albacore is a pelagic fish that can be found in the open waters of all tropical and temperate oceans, including the Mediterranean Sea. It has migration habits and a lifespan of 11–12 years, reaching reproductive maturity at around 5–6 years. The increasing world demand for tuna, preserved and traded by the canning industry, are rapidly decreasing the population numbers of albacore tuna.

Like other fish captured in the wild, tuna accumulates methylmercury, a very toxic compound that undergoes biomagnification along the food chain. For safety reasons, a moderate consumption of tuna and similar predator fishes should be observed. Tuna is consumed primarily as canned food. Canned tuna is presented as cooked fish meat in vegetable oil, olive oil or in water. The proportion of salt may vary. Tuna is typically gutted by hand, and then pre-cooked for prescribed times. The fish is processed again to separate lateral dark meat, filleted and canned, usually in oil. Cans are sealed and heated for 2–4 h to ensure sterility.

The data presented below refer to the most commonly found commodity: canned tuna in vegetable oil with added salt.

Canned tuna supplies 186–214 kcal/100 g from: proteins (24.3–26.5 %) and fat (about 8–13 %). No carbohydrates or organic acids are found. Fat is mainly PUFA (7.1 %, mostly conjugated linoleic acid [CLA]), and monounsaturated fatty acids [MUFA: 3.8 %). Saturated fatty acids (SFA) account for 0.9–1.3 % of the global composition, trans-fatty acids account for 0.2 % and cholesterol for 31–41 mg/100 g (INSA 2015; USDA 2015).

Tuna is a source of the omega-3 fatty acids: EPA (0.07 g/100 g), DPA (0.005 %) and DHA (0.18 %), as well as the fat-soluble vitamins retinol (5–23 μg/100 g), α-tocopherol (1.9–2.3 mg/100 g) and vitamin D (0.4 μg/100 g). Most water-soluble vitamins are absent, except for some of B-complex vitamins such as niacin (9.8–11.7 mg/100 g), B_6 (0.23–0.43 mg/100 g) and B_{12} (2.2–2.4 μg/100 g). Tuna also contains relevant micronutrients such as phosphorus (P: 202 mg/100 g), manganese (Mn: 0.016 mg/100 g), selenium (Se: 60.1 μg/100 g) and fluorine (F: 31 μg/100 g) (INSA 2015; USDA 2015).

7.2.2 Sardine (*Sardina pilchardus*)

Sardines are small epipelagic fish that have a cylindrical shape, are about 10–20 cm long and sometimes migrate along the coast in shoals. Sardines are forage for large pelagic fish and present a lower risk of accumulating persistent pollutants such as mercury compounds because they are low in the food chain; thus, they are safer for human consumption. Sardine is an important fishery species, and the countries with the largest catches in 1999 were Morocco and Spain (FAO 2015b). Sardine is consumed fresh, frozen and canned.

7.2.2.1 Fresh Sardines

Fresh sardines are consumed all over the Mediterranean area and are most popular in Portugal, Spain, Italy and Greece. Sardines have been captured in coastal areas since ancient times and consumed during summer, mainly grilled or fried. Sardines that are not consumed fresh are canned, generally in a seasoned sauce containing olive oil and tomato paste. Canned sardines are exported and/or consumed during winter.

The fishing season for sardines usually starts in June, which in Portugal overlaps 'Santos Populares', a series of popular festivals that take place outdoors. Sardines are the symbol of Lisbon festivities, 'Santo António' (the city's patron saint). Sardines are grilled on charcoal in downtown streets, where eating and dancing takes place, attracting many tourists and artists that dare to face the strong smell and flavour. In early summer, sardines are at their best because the higher fat content contributes to their taste and flavour. The fresh sardine season ends with summer.

Consumption of sardines grilled on charcoal, served with boiled potatoes and salad (green pepper, onion, tomato and lettuce), generously seasoned with olive oil, in the evening at a terrace near the sea, is a pleasure shared mainly by Portuguese and Spanish populations, and is a good portrait of the MD and way of life.

Since fresh sardines are not commonly eaten worldwide, only the Portuguese Food Information Resource (PortFir) database (INSA 2015) could be used in this review. Data correspond to consumers' preferences for raw fresh high-fat sardines. Comparisons are made to the same category of grilled and fried fish.

Raw sardine supplies 221 kcal/100 g, whereas grilled sardines contain 211 kcal/100 g and fried sardines contain 247 kcal/100 g. Raw sardines have a total fat content of 16.4 %, slightly below the protein content of 18.4 %; no carbohydrates are referred. Total fats are 4.7 % SFA, 4.0 % MUFA, 5.6 % PUFA, 0.5 % trans-fatty acids (TFA), also including 0.5 % linoleic acid and 20 mg/100 g cholesterol. Sardines are a good source of fat-soluble vitamins, particularly vitamin A (47 μg/100 g of retinol eq.), vitamin D (21 μg/100 g), α-tocopherol (0.66 mg/100 g) and the water-soluble B vitamins niacin (4.4 mg/100 g), B_6 (0.57 mg/100 g) and B_{12} (10 μg/100 g). Sardines have a high P content (314 mg/100 g). When grilled, sardines lose water and fat. This cooking method results in a decrease of SFA, MUFA (the larger loss) and PUFA but slightly concentrates cholesterol, the level of which increases from 20 to 43 mg/100 g; vitamin levels may result a little higher too. In the case of fried sardines it should be noted that the raw fish is coated with batter (a mix of wheat flour with milk/water and eggs) before frying. Batter residues contribute to the frying oil degradation. Nowadays, olive oil is rarely used for frying, given its current high price; soy, sunflower or corn oils are often used instead, and thus, the accumulation of degradation products is more likely, and this cooking method should not be preferred.

In short, fresh sardines are a good source of protein, supplying a wide variety of amino acids and are a well-known fatty fish, containing essential fatty acids but low levels of cholesterol.

7.2.2.2 Canned Sardines

Canned sardines (in olive oil, drained) supply 186 kcal/100 g and a total fat content of 9 %, which is significantly less energy and fat than their fresh cooked counterparts. The quantity of protein is higher (26.3 %) and the SFA levels are much lower (1.5 %). The predominance of MUFA should be noted, as well as cholesterol levels that are higher (60 mg/100 g) than in (cooked) fresh sardines. Moreover, the industrial processing of sardines significantly reduces its vitamin content, with the exception of vitamin B_{12} (13 μg/100 g) (INSA 2015).

7.2.3 Anchovy (*Engraulis encrasicolus*)

The European anchovy is a fish with a slender elongated body, oval in cross-section and common in the coastal areas of Europe and the Mediterranean basin. It is a pelagic coastland fish forming large schools, similar to sardines. It has a tendency to extend into more northern waters in the summer, generally moving into the surface layers, retreating and descending in winter. Anchovies feed on planktonic organisms and reach about 12–15 cm (FAO 2015c). They are consumed as fresh or canned foods or after traditional preservation involving brining and maturation. Anchovies have been brined since the times of the Roman Empire, when they formed the base of a very popular fermented fish sauce called 'garum'. The traditional method of preserving anchovies involves evisceration, followed by brining and a maturation stage until fish flesh turns deep grey, after which anchovies are industrially canned or preserved in olive oil and capers (*Capparis spinosa*) in small jars. This results in a characteristically strong flavour for many dishes, although anchovies became world famous because of their use in pizza. No data on composition for this type of preserve were available in the food databases used in the current publication.

7.2.3.1 Fresh Anchovies

Fresh anchovies supply 131 kcal/100 g from: proteins (20.35 %) and total fat (4.84 %). SFA, MUFA and PUFA proportions are quite close, with a slight predominance of PUFA (USDA 2015).[2] Anchovies contain 1.3 % SFA (mainly 16:0, 14:0 and 18:0), 1.2 % MUFA (mainly 18:1 and 16:1) and 1.6 % PUFA, mainly the essential fatty acids EPA (0.54 %), DHA (0.91 %) and DPA (0.029 g/100 g), and also include cholesterol (60 mg/100 g). Fresh anchovies are also a good source of fat-soluble

[2] Report 15001: fish, anchovy, European, raw (*Engraulis encrasicholus*).

vitamins, particularly vitamin A (15 μg/100 g of retinol) and α-tocopherol (0.57 mg/100 g), as well as water-soluble B vitamins: niacin (14.0 mg/100 g), B_6 (0.14 mg/100 g) and B_{12} (0.62 μg/100 g). Anchovies have high levels of Ca (147 mg/100 g) and P (174 mg/100 g), as well as the micronutrients Mn and Se.

7.2.3.2 Canned Anchovies

Compared with fresh anchovies, processed anchovies supply more energy (210 kcal/100 g) from a slightly higher concentration of proteins (28.9 %) and total lipids (9.71 %). The relative proportions of SFA, MUFA and PUFA are similar, resulting in an increased concentration of every component, including EPA (0.76 %), DPA (0.04 %), DHA (1.29 %) and cholesterol (85 mg/100 g). Processing affects vitamin content by decreasing vitamin A and E, while concentrating niacin (19.9 mg/100 g), vitamin B_6 (0.2 mg/100 g) and vitamins B_{12} and D_3 (cholecalciferol). In particular, vitamin D_3 can reach 0.62 μg/100 g, whereas it was most likely below the detection level in fresh anchovies. Concentration of Ca (232 mg/100 g) and P (252 mg/100 g) seems to increase with processing but not in the same proportions (USDA 2015).[3]

7.2.4 Codfish (*Gadus morhua*)

Cod is a demersal, voracious fish, feeding largely on other fish and various invertebrates. The species *Gadus morhua* is found on both sides of the North Atlantic. Although some groups of small cod are relatively stationary, individuals or groups may undertake astonishingly long migrations. For some reason, larger fish are found in colder waters, at 0–5 °C (FAO 2015d). A North Pacific species of cod, *G. macrocephalus*, is very similar in appearance to the Atlantic form but does not have the same economic value. *Gadus morhua* is captured in Norway, dried, salted and, ideally, exposed to the sun.

Cod livers are processed to make cod liver oil, used as a food supplement as a source of fat-soluble vitamins (A, D, E) and essential n-3 fatty acids (EPA and DHA).

Cod has been an important economic commodity in international markets since the Viking period, as Vikings travelled with dried cod and soon developed a dried cod market in southern Europe, enduring the black plague of the fourteenth century, wars and crisis. By the fifteenth century, Portuguese and Spanish peoples started sailing to Norway to fish for cod. Dry salted cod is still an important Norwegian trade commodity, although ecological organizations (World Wildlife Fund and Greenpeace) have placed *Gadus morhua* on the list of endangered species. Efforts have been directed towards the development of effective and sustainable cod farming systems (FAO 2015d).

[3] Report 15002: fish, anchovy, European, canned in oil, drained solids (*Engraulis encrasicholus*).

Dry salted codfish is consumed in all countries referred in the present review, from Portugal to Croatia. It is locally known as 'bacalhau' (Portugal), 'bacalao' (Spain), 'baccalá' (Italy), 'bakaliáros' (Greece) or 'bakalar' (Croatia). It is cooked in many different ways after being desalted in water, and is appreciated for its succulent white and flaky meat. In Portugal, its presence is mandatory in both Christmas Eve and Christmas lunch meals. Unlike sardines, which are known as a fat fish, codfish has a low fat content.

Merged data from the Portuguese National Institute of Health (Instituto Nacional de Saúde Doutor Ricardo Jorge [INSA]) and the US Department of Agriculture (USDA) food composition databases are discussed below, although records diverge at the desalting step. The INSA database refers to dry salted Atlantic cod (*Gadus morhua*) at the stage it is normally cooked in South Europe (after being desalted in water and with 76.2 % water content), while the USDA database (2015) refers to the same fish but not ready for consumption (16.1 % water, and with the salt added for preservation). Thus, the primary composition data were retrieved from INSA and complemented with relevant information from the USDA database (mainly in a qualitative way).

Thus, uncooked codfish supplies 80 kcal/100 g from: proteins (19 %) and total lipids (0.4 %; equal fractions of SFA [16:0, 18:0, 14:0], MUFA [mainly 18:1] and PUFA [18:2, 20:4, traces of 18:3 and 18:4, and relevant quantities of EPA, DPA and DHA]), as well as cholesterol (52 mg/100 g). Fat-soluble vitamins are present in fair amounts (retinol [4 μg/100 g], vitamins $D_2 + D_3$ [4.5 μg] and α-tocopherol [0.28 mg/100 g]), as are the water-soluble B vitamins thiamine (0.047 mg/100 g), riboflavin (0.068 mg/100 g), niacin (0.76 mg/100 g), B_6 (0.072 mg/100 g) and B_{12} (0.95 μg/100 g). Relevant minerals are Ca (33 mg/100 g), P (116 mg/100 g), Mg (23 mg/100 g) and Mn and Se (INSA 2015; USDA 2015).

7.2.5 Cephalopods

Cephalopods (squid, cuttlefish and octopus) are members of the molluscan class *Cephalopoda*. These exclusively marine animals are characterised by bilateral body symmetry, a prominent head and a set of arms or tentacles (muscular hydrostats) modified from the primitive molluscan foot. They are called inkfish because of their ability to squirt ink (with the exception of a few species). Cephalopods play a major role in the trophic web of all marine ecosystems, both as predators and as prey. Cephalopods are found in oceans worldwide (except the polar seas), and they cannot tolerate fresh water (only one species was detected in brackish waters). Like most molluscs, cephalopods use haemocyanin—a copper-containing protein—rather than haemoglobin to transport oxygen. As a result, their blood is colourless when deoxygenated and turns blue when exposed to air (Jereb and Roper 2010). Cephalopods are balanced food protein sources (containing all the essential amino acids). Its flesh contains lower amounts of essential fatty acids than most fishes. It is also lower in some vitamins than the so-called fat fishes, and cholesterol and sodium levels are generally higher.

Cephalopods are mainly fished by Southeast Asian and Mediterranean countries, which are also traditionally the major consumers (Jereb and Roper 2010), with Portugal, Spain and Italy registering the highest per capita intake values in 2011 (FAO 2015e).[4]

7.2.5.1 Squid (*Loligo forbesii, L. vulgaris,* and Related Species)

Loligo spp. includes a large variety of edible species, and this neritic cephalopod genus is distributed worldwide, particularly in western Atlantic and Mediterranean areas (Jereb and Roper 2010). According to the same authors, adults undergo vertical migrations, living close to the bottom during the day, and ascending towards the surface at night. Squids grow rapidly and may significantly vary in size. *Loligo* spp. is a voracious and opportunistic predator, feeding on a wide spectrum of prey only limited by size. Mature squids are encountered almost throughout the year, but two peak spawning periods are generally observed: the first and more important in spring, and the second, less intense, in late summer and fall (Jereb and Roper 2010). Squid is commonly known in Portugal, Spain and Italy as 'lula' or 'calamar/calamari' and marketed either fresh, frozen, or processed into canned or pre-cooked dishes. Squids are cooked in a variety of ways (stuffed with tentacles and 'chouriço/chorizo'[5] and cooked in tomato sauce, grilled, cooked with vegetables, etc.), although the most popular form is in rings previously seasoned, battered, and fried.

According to the INSA and USDA databases (accessed in 2015), raw squid, mixed species,[6] supplies 92 kcal/100 g, mainly from proteins (15.6 %) and total lipids (1.38 %; the sum of SFA [0.2–0.4 %; mostly 16:0 and traces of 14:0 and 18:0], MUFA [0.1 %; mainly 18:1 and 20:1] and PUFA [0.4–0.52 %; mainly EPA and DHA] as well as cholesterol [140–233 mg/100 g]). According to the INSA database (2015) no trans-fatty acids or linoleic acid were found. When merging data from the INSA and USDA databases, it can be noted that the fat-soluble vitamins retinol (10 μg/100 g) and α-tocopherol (1.2 mg/100 g) are present in fair and constant amounts, whereas the levels of vitamins $D_2 + D_3$ (0–3.5 μg/100 g) vary widely. Water-soluble B vitamins are thiamine (0.047–0.071 mg/100 g), riboflavin (0.016–0.068 mg/100 g), niacin (0.76–1.0 mg/100 g), B_6 (0.053–0.072 mg/100 g) and B_{12} (0.95–1.1 μg/100 g). Ascorbic acid can be detected in highly variable amounts (0–4.7 mg/100 g). Relevant minerals are Ca (18–32 mg/100 g), P (221–261 mg/100 g) and Mg (33–49 mg/100 g) (INSA 2015; USDA 2015). The following minerals are also worth mentioning: copper (Cu: 1.9 mg/100 g), Mn (0.035 mg/100 g) and Se (44.8 μg/100 g) (USDA 2015).

[4] Food supply quantity (from food balance sheets, 2011 statistical data), referring to the total amount of the commodity available for human consumption during 1 calendar year (FAO 2015d, 2015e).

[5] Dry fermented sausage: a traditionally processed pork meat that is seasoned and fermented prior to a smoking 'cure'.

[6] Analysed species may differ between databases. It is probable that PortFir (INSA 2015) includes just *Loligo* sp. while the USDA database specifically notes that *Ommastrephidae* sp. is also included.

7.2.5.2 Cuttlefish (*Sepia officinalis*)

Cuttlefish are found predominantly on sandy to muddy bottoms of the coastline to about 200 m in depth; larger individuals are encountered in the deeper part of the range. In the western Mediterranean, in early spring, large individuals leave the deeper water, where they have spent the winter, to migrate into shallower water (males precede females by about a week). The maximum mantle length is 45 cm, weighing up to 4 kg in temperate waters. This group is followed by a succession of smaller cuttlefish, arriving in shallow waters throughout the summer. In fall/autumn, the gradual descent begins. Common cuttlefish is usually marketed fresh and frozen, and is a highly valued food item, particularly in Italy and Spain (FAO 2015f). Cuttlefish is similarly cooked in Portugal and North Spain, e.g. in its own ink, with beans and 'chouriço' ('feijoada de choco') or seasoned and fried, resembling fish fingers.

According to the INSA and USDA databases, raw cuttlefish, mixed species (*Sepiidae* sp.) supplies 79 kcal/100 g from: proteins (16.2 %) and total lipids (0.7 %; being the sum of SFA 0.12–0.4 %, mostly16:0 and traces of 18:0; MUFA 0.08 %, mainly 18:1 and 20:1; and PUFA 0.13–0.52 %, mainly 20:4, EPA and DHA), as well as cholesterol (112–233 mg/100 g). No trans-fatty acids or linoleic acid are found (INSA 2015). Fat-soluble vitamins are present in fair and constant amounts: retinol (10 μg/100 g) and α-tocopherol (1.2 mg/100 g). Vitamin $D_2 + D_3$ content varies widely (0–3.5 μg/100 g). Water-soluble B vitamins are also found: thiamine (0.047–0.071 mg/100 g), riboflavin (0.016–0.068 mg/100 g), niacin (0.76–1.0 mg/100 g), B_6 (0.053–0.072 mg/100 g) and B_{12} (0.95–1.1 μg/100 g). Ascorbic acid can be found at highly variable levels (0–4.7 mg/100 g). Relevant minerals are Ca (18–32 mg/100 g), P (221–261 mg/100 g) and Mg (33–49 mg/100 g) (INSA 2015; USDA 2015). In addition, the following minerals are also detected: Cu (1.9 mg/100 g), Mn (0.035 mg/100 g) and Se (44.8 μg/100 g) (USDA 2015).

7.2.5.3 Octopus (*Octopus vulgaris*)

Octopus is found worldwide, mostly in temperate and tropical waters as it becomes inactive below 7 °C. It is a medium- to large-sized animal, chunky in appearance and commonly found in shallow waters in a variety of habitats (e.g. rocks, coral reefs and grass beds) (FAO 2015g). The pattern of octopus consumption is similar to that of other cephalopods: it is consumed in all countries of the present study, with the most probable exception of Morocco. Small- and medium-size animals are preferred for the softness of the meat. Octopus can be cooked in a variety of ways: as a stew, baked, grilled, fried, incorporated in rice dishes (e.g. the portuguese 'arroz de polvo') or as an appetiser salad, consisting of cooked and sliced pieces of octopus immersed in a mixture of olive oil, vinegar, onion and aromatic herbs. Octopus is most popular in Portugal and in Spain.

According to the INSA and USDA databases (accessed in 2015), raw octopus, *Octopus vulgaris*, supplies 73–82 kcal/100 g from: proteins (14.9–15.6 %) and total lipids (1.04–1.20 %, being the sum of SFA 0.23–0.3 %, 14:0, 16:0 and 18:0; MUFA 0.10–0.16 %, mainly 16:1 and 18:1; and PUFA 0.24–0.60 %, mainly 18:4, 20:4, EPA and DHA), as well as cholesterol (48–64 mg/100 g). No trans-fatty acids or linoleic acid are registered in the PortFir database (INSA 2015). α-tocopherol (0.73–1.2 mg/100 g), and retinol (3.0–4.5 μg/100 g), which is present in variable amounts. Vitamin D is absent, and vitamin K is present only in very small amounts (0.1 μg/100 g). Water-soluble B vitamins are thiamine (0.02–0.03 mg/100 g), riboflavin (0.04 mg/100 g) and niacin (1.3–2.1 mg/100 g). Again, divergent values for some vitamins and minerals are found when merging data from the INSA and USDA databases, as is the case of vitamins B_6 (0.07–0.36 mg/100 g), B_{12} (1.3–2.0 μg/100 g), ascorbic acid (0–5.0 mg/100 g), Ca (32–53 mg/100 g), and P (186–261 mg/100 g); also relevant are Mg (30–49 mg/100 g), Fe (5.3 mg/100 g), Cu (0.43 mg/100 g), Mn (0.025 mg/100 g) and Se (44.8 μg/100 g) (USDA 2015).

When compared with other cephalopods, octopus provides a higher vitamin content and less cholesterol for the same edible portion.

7.2.6 Bivalves

Bivalves have been consumed by humans at least since the time of the Roman Empire judging by the empty shells found in archaeological sites.

The circulatory system relies on haemoglobin as the oxygen-carrying protein, with iron (Fe) in its active centre instead of Cu, as in the case of cephalopods and crustaceans. Bivalves are filter feeders, passing large quantities of water through their gills and filtering out organic particles, including chemical pollutants and microbial pathogens. These are retained in tissues and become concentrated in their liver-like digestive glands. Another issue, closely monitored by health authorities, is the presence of biotoxins produced by microalgae ingested by bivalves. To address the growing trend of consumption and to reduce food safety issues, aquaculture systems have been improving cultivation techniques that take advantage of bivalves' sedentary habits, and contributing to the lowering of market prices. Generally, juveniles are transferred from hatcheries (in controlled tanks) and reared off the seabed in suspended rafts, on floating trays or cemented to ropes. Here, they are largely free from bottom-dwelling predators (such as starfish and crabs), are less exposed to contaminated debris, and can be harvested by hand when they reach a suitable size.

Oysters, scallops, clams, ark clams, mussels and cockles are the most commonly consumed bivalves, and are included as ingredients in main courses (such as 'carne de porco à alentejana', with clams and pork) or appetisers (generally cooked with wine and herbs) accompanied with bread that is dipped in sauce.

None of the food databases accessed in the current work provided adequate information about the species of clams consumed in the area (namely blue mussel, Mediterranean mussel, Atlantic surf clam and common cockle). For this reason, no detailed food composition is provided, but no major differences from other bivalves are expected.

7.2.6.1 Mussels (*Mytilus edulis, M. galloprovincialis* and Related Species)

Mytilus edulis and *M. galloprovincialis* are two closely related species, commonly known simply as mussels. Mussels have a solid shell and are approximately triangular in outline. The shell exhibits fine concentric lines, of different tones of purple, blue and brown on an almost black periostracum. The interior is pearl-white with a wide border of purple or dark blue. Mussels, in general, are common in Europe and in the Mediterranean basin, living on all coasts attached to rocks and piers from intertidal areas to 40 m deep, sometimes building large groups of individuals. The diet of mussels consists of phytoplankton and detritus filtered from the surrounding water. The species dimensions are greatly influenced by its biotope: intertidal shells often remain smaller (5–6 cm long) than deep-water shells, which easily measure 9 cm (FAO 2015h, 2015i).

Mussels (raw, *Mytilus* sp.) supply 69–86 kcal/100 g: protein (11.9–12.1 %), starch (2 %) and total lipids (1.50–2.24 %, being the sum of SFA 0.3–0.42 %, 14:0, 16:0 and 18:0; MUFA 0.3–0.51 %, mainly 16:1, 18:1 and 20:1; and PUFA 0.50–0.61 %, mainly 20:4, EPA and DHA) as well as cholesterol (28–40 mg/100 g) (INSA 2015; USDA 2015). No trans-fatty acids or linoleic acid have been registered (INSA 2015).

α-Tocopherol (0.55–0.74 mg/100 g) and γ-tocopherol (0.02 mg/100 g) are present in fairly stable amounts, when merging data from INSA and USDA databases, whereas the concentration of retinol (48–360 μg/100 g) and ascorbic acid (0–8.0 mg/100 g) vary widely. Vitamin D is absent, and vitamin K is present only in very small amounts (0.1 μg/100 g). Mussels contain the B vitamins thiamine (0.10–0.16 mg/100 g), riboflavin (0.14–0.21 mg/100 g), niacin (1.2–1.6 mg/100 g), B_6 (0.05–0.08 mg/100 g) and B_{12} (12.0–19.0 μg/100 g). Relevant minerals are Ca (26–56 mg/100 g), P (197–240 mg/100 g), Mg (34–36 mg/100 g), Mn (3.4 mg/100 g) and Se (44.8 μg/100 g). Like other molluscs, mussels provide quite a high level of sodium (290 mg/100 g) (INSA 2015; USDA 2015).

7.2.6.2 Cockle (*Cerastoderma edule*)

Cockle is a common designation for bivalves that have an almost globular very thick shell with radiating ribs and very fine irregular concentric lines. Growth stages are prominent, and may include other species related to *Cerastoderma edule* (the most common in the area). The shell is dirt white or pale yellow and white inside. Cockles can jump by bending and straightening their foot, although (like other molluscs) they are quite sedentary. Preferred habitats are sandy bays that are intertidal to only a few meters deep, with some arrival of fresh water, where they live just under the bottom surface on sand, mud or gravel. They are relatively common along the Atlantic coast but quite rare in the Mediterranean. Consequently, they are more commonly consumed in Portugal and Spain, particularly the small subspecies (3–4 cm) (FAO 2015j).

Cockle shell *(on scallop shells). Shellfish or bivalves, like other fish, provide a protein with the correct balance in amino acids and are simultaneously an important source of unsaturated long-chain fatty acids n-3, which play important physiological roles in the organism. Photo reprinted with kind permission from T. N. Wassermann*

Cockles (raw, mixed species) supply 60–79 kcal/100 g from: proteins (10.5–13.5 %), some available carbohydrates (2.7 % of starch) and total lipids (0.7 %, being the sum of SFA 0.1–0.3 %, MUFA 0.1–0.3 % and PUFA 0.3–0.50 %) (INSA 2015; USDA 2015). No trans-fatty acids or linoleic acid are registered (INSA 2015).

Cockles have an average content of 30 mg/100 g cholesterol. Fat-soluble vitamins are absent but water-soluble B vitamins are found in variable amounts: thiamine (0.005–0.16 mg/100 g), riboflavin (0.11–0.20 mg/100 g), niacin (1.7–3.2 mg/100 g), B_6 (0.04 mg/100 g) and B_{12} (41.0 µg/100 g). Relevant minerals are Ca (30–56 mg/100 g), P (159 mg/100 g) and Mg (58 mg/100 g). Like other molluscs, cockles have a relatively high sodium content (376 mg/100 g) (INSA 2015; USDA 2015).

7.2.7 Crustaceans

Crustaceans are highly valued worldwide and moderately consumed in the countries that have been mentioned throughout this text. Crustaceans include species of crab, shrimp and lobster belonging to the class *Malacostraca*. Besides an exoskeleton, crustaceans have an open circulatory system, with haemocyanin as the oxygen-carrying metal-protein.

The most popular crustaceans are crabs and shrimps, consumed as appetisers or as main course ingredients (as in the well-known 'paella'). As opposed to lobsters, crabs and shrimps are farmed on a large scale, which makes them more easily available and cheaper. Lobster is still consumed, in the area, as a delicacy.

Crabs are gathered in the wild but are mostly cultured. They are benthic organisms, living on a wide range of sand, gravel and rocky bottoms, usually at a depth of 6–40 m. The species *Cancer pagurus* is found from the Eastern Atlantic (northern Morocco) extending along to the north coast of the Mediterranean (Marseille, Napoli). It is the most common edible species in the area. Its body is broadly oval, with large and smooth pincers, without spinules and four pairs of similar legs. The carapace exhibits a homogenous reddish brown pale colour and is about 20 cm long.

No detailed information is available from databases on the crab species normally found and consumed in the area. Therefore, the discussion on crustaceans is only based on data for shrimp/prawn, raw, mixed species, mostly from the family *Penaeidae*. The most common shrimp in the area is *Crangon crangon*, which is distributed along the Atlantic coast of Europe, including Portugal and Morocco, as well as in the Mediterranean. Like other shrimps, *C. crangon* prefer shallow coastal, marine or slightly brackish waters (0–20 m), where it lives near the bottom. A larger quantity of edible species is available in the Atlantic coast than in the Mediterranean Sea. Nowadays, farmed prawns (mainly of genus *Penaeus*) are usually preferred for their low prices (FAO 2015k).

Thus, shrimp, raw, mixed species (including those from aquaculture) supply 77–83 kcal/100 g from: proteins (13.6–17.6 %), starch (0.3–0.9 %) and total fat (0.6–1.0 %, being the sum of SFA 0.1 %, mostly 16:0), MUFA 0.1 %, conjugated *cis–trans* isomers of 16:1 and 18:1; and PUFA 0.3 %, mainly 18:2, n-6; 20:4; EPA, DHA and some DPA). Shrimps also contain cholesterol (126–154 mg/100 g) (INSA 2015; USDA 2015). According to INSA (2015), *Penaeus* sp. contain linoleic acid (0.1 %) and no trans-fatty acids, while the USDA database (2015) registers trans isomers (0.018 g/100 g, of which 0.011 g corresponds to *trans*-monoenoic acids). By overlapping data from both databases, it can be deduced that *trans*-monoenoic acids most probably account for CLA.

Fat-soluble vitamins found in shrimp are retinol (0–54 μg/100 g), α-tocopherol (0.70–1.32 mg/100 g), γ-tocopherol (0.12 mg/100 g) and δ-tocopherol (0.03 mg/100 g). Shrimps also contain water-soluble B vitamins: thiamine (0.02–0.03 mg/100 g), riboflavin (0.010–0.015 mg/100 g), niacin (1.78–2.0 mg/100 g), pantothenic acid (0.31 mg/100 g), B_6 (0.05–0.16 mg/100 g) and B_{12} (1.1–2.1 μg/100 g). Relevant minerals are Ca (54–87 mg/100 g), P (150–244 mg/100 g) and Mg (22–30 mg/100 g).

In short, fish and seafood provide a balanced source of animal protein (containing all essential amino acids) with low energy. Fat content is variable but generally rich in essential fatty acids, namely omega-3, and a wide range of vitamins. With the exception of some cephalopods (squid and cuttlefish) and crustaceans, fish and seafood is low in cholesterol and contains no TFA.

7.3 Meat

Meat has been part of the food habits of humankind since the origin of the species, as a protein and energy source, and providing the essential amino acids (Table 7.1), vitamins and minerals. Meat is generally a good source of iron present in a haemoglobin centre and a major source of vitamin B_{12}, essential fatty acids (EPA, DHA and DPA) and other bioactive fats such as CLA. Meat is a very satiating food, which may compel an individual to eat larger portions than necessary. Excess animal protein consumption is deleterious to health, as mentioned above and further explained below.

Roughly, 'red meat' includes all meat from mammals (beef, lamb, pork, veal and goat), whereas 'white meat' refers mainly to poultry (turkey, duck, goose, chicken and rabbit). Animals with 'white' meat have muscles with fast-twitch fibres that enable quick bursts of activity (such as fleeing from predators) at the expense of locally stored glycogen. On the other hand, 'red meat animals' have muscles with slow-twitch fibres. These muscles are used for extended periods of activity such as standing or walking, and need a consistent energy source such as fat reserves. As the muscles of red meat animals perform aerobic respiration, the concentration of myoglobin is higher than in the muscles of white meat animals, which mainly work under anaerobiosis (breaking glycogen down to lactic acid). International organisations have been advising consumers to reduce the intake of red meat, particularly processed meats, due to the risk of coronary disease and bowel cancer (American Heart Association 2015; NHS 2015). Red meat can be part of a healthy diet, if lean parts and small portions are preferred and consumed with vegetables. In the context of the MD, Bach-Faig and colleagues (2011) recommend, in increasing order of preference, lower intakes of processed red meat, such as 'jamon' (<1 serving/week), red meat (<2 servings/week), white meat (about 2 servings/week) and 2–4 eggs/week.

On average, white meat contains less saturated fat and less cholesterol than red meat, but this only applies to the average of whole carcasses, as in fact such content depends on the type of each piece of meat. Both total fat and the composition of the lipid fraction depend on many factors, from the feed and breed of the animals to the butcher's cut, even the cooking method.

Table 7.2 shows the lipid profile of different types of raw meat, referring to the portion ready to be cooked (the figures should be regarded only as a guidance; data were obtained from different sources and correspond to commonly consumed butcher cuts). Roughly, total fat and saturated fat are commonly within the same range of values for white and red meats. However, in commonly consumed butcher cuts, total fat tends to be lower in the case of beef and higher in chicken and in pork. Cholesterol levels are in the same range (INSA 2015; USDA 2015) but slightly higher in chicken (the skin is probably the main contributor). According to Table 7.2, chicken and pork generally contain a lower concentration of TFA than beef (even the lean cuts). Nevertheless, TFA of ruminant origin may include *trans-*

Table 7.2 Lipid profile for the main types of consumed meat (chicken [poultry], beef [ruminants], pork). Mean values from distinct breeds and geographical locations (EU and USA) expressed as percentage of edible portion

Type of fat	Chicken (a)	Beef (fat trimmed) (b)	Pork (fat trimmed) (c)
Total fat, %	14.4	4.0	10.2
SFA	3.6	1.8	3.51
MUFA	5.9	2.2	3.9
PUFA	2.7	0.33	1.65
Cholesterol	0.009	0.006	0.007
TFA	0.1	0.2	0.1
Linoleic acid/ALA	2.5	0.1	0.36
EPA	0.01	0.002	$<10^{-4}$
DPA	0.01	0.012	$<10^{-4}$
DHA	0.03	0.001	$<10^{-4}$

Displayed results have been calculated by the elaboration of single data points from different sources (Alfaia et al. 2007; INSA 2015; Martins et al. 2007; Marušić et al. 2013; USDA 2015). Total fat includes SFA, MUFA, PUFA, cholesterol, and TFA; PUFA includes ALA, EPA, DPA and DHA; (a) whole carcass, meat and skin, without gizzards; (b) Beef, grass-fed, lean only, raw; (c) pork ribs with bone, raw

isomers of CLA and other compounds potentially beneficial to human health (Alfaia et al. 2007).

Chicken contains about ten times more essential fatty acids (ALA, EPA, DPA and DHA) than beef and far more than pork's meat. On the other hand, pork does not contain EPA or DHA but is richer than beef, in linoleic acid.

Although marginal, some uncommon types of meat are consumed in the area, such as rabbit, frog, snail and small birds (quail, pheasant). Horse meat seems to be poorly accepted by many consumers in the Mediterranean area and for that reason it is not discussed in the current work. Farmed animals intended for meat consumption are regularly inspected by a veterinarian to ensure not only their health but also hygiene and comfort. When slaughtered, every carcass is again inspected for compliance with requirements for human consumption.

7.3.1 Poultry (Chicken, Turkey, Duck)

Although chicken and turkey are categorised as white meat animals, their legs, which support their body weight, rely on muscles that provide endurance work. Consequently, they contain larger amounts of myoglobin than breast meat; this enables leg muscles to use oxygen more efficiently for respiration. Ducks, which use chest muscles for long flights, have dark meat all over the body. Chicken and turkeys are categorised as white meat because most muscles perform anaerobically and hence contain low concentrations of myoglobin, which is responsible for the meat pigmentation. Thus, both types of meat—white and dark (or red)—can be found within poultry.

The chicken (*Gallus gallus domesticus*) is one of the most common and widespread domestic animals. Since ancient times, humans have kept chickens close to their homes as a source of food, consuming both their meat and their eggs. Chickens are omnivores, eating seeds, insects, small lizards and even small snakes or young mice. Chickens are gregarious birds with a communal approach to the incubation of eggs and raising of young; they may live for 5–10 years, depending on the breed.

However, the majority of poultry for human consumption is raised using intensive-farming techniques; a smaller proportion is raised with an organic approach, relying on different feed and keeping the animals in the open air for some periods of time. Chickens intensively farmed for meat take less than 6 weeks to reach slaughter size, while organic meat chicken will usually be slaughtered at about 14 weeks of age. Chicken farmed for their eggs can produce more than 300 eggs/year. After 12 months of laying, egg-laying ability usually starts to decline, and the animals are then slaughtered and used in processed foods or as pet food ingredients.

Thus, broiler chicken, meat and skin, raw, supplies 201–215 kcal/100 g from: proteins (18.6–19.6 %) and total fat (13.6–15.1 %) (INSA 2015; USDA 2015).

Chicken contains all fat-soluble vitamins in fair amounts: retinol (20–41 μg/100 g), α-tocopherol (0.20–0.30 mg/100 g), vitamins $D_2 + D_3$ (0.20–0.62 μg/100 g) and K (1.5 μg/100 g). The water-soluble vitamins are ascorbic acid (0–1.6 mg/100 g), folates (6–9.4 mg/100 g), thiamine (0.06–0.12 mg/100 g), riboflavin (0.120–0.25 mg/100 g), niacin (6.6–6.8 mg/100 g), pantothenic acid (0.91 mg/100 g), B_6 (0.30–0.35 mg/100 g) and B_{12} (0.31–0.80 μg). Relevant minerals are P (147–176 mg/100 g) and Mg (20–23 mg/100 g) (INSA 2015; USDA 2015), Mn (0.019 mg/100 g) and Se (14.4 μg/100 g) (USDA 2015).

Another detected substance in chicken meat is betaine[7] (7.8 mg/100 g), a non-protein amino acid and a compatible solute that protects cells against osmotic stress and may promote water retention. Betaine is ubiquitous in nature, occurring in plant foods as well as in animal foods and there are no reported toxic effects associated to it (NCBI 2015). Duck and turkey meat share most of the characteristics of chicken meat.

7.3.2 Ruminants (Bovine, Lamb and Goat)

Cattle (*Bos taurus*) is raised for milk or meat production. Domestic cows naturally feed on grasses, stems and other herbaceous plant material. An average cow can

[7] Also named 2-(trimethylazaniumyl)acetate.

consume about 70 kg of grass in an 8-h day and may reach a body mass of more than 600 kg. The lifespan of domestic cows may exceed 20 years (EOL 2015).

Cows, like lambs (*Ovis aries*) and goats (*Capra hircus*) are herbivorous ruminants, meaning that their digestive system breaks down cellulose and other relatively indigestible plant material. Ruminants have a four-chambered stomach, including a rumen where specialised bacteria take part in digestion. The process of digestion is very slow (taking 70–100 h in cows), and therefore extracts the most nutrients from plant materials (EOL 2015).

The meat of young cows is known as veal, and that of adults is called beef. Cattle may be raised by intensive or extensive systems, which have an impact on the lifespan, feeding, handling and the final quality of the meat. In extensive systems, cows are kept outdoors most of the time, feeding on pasture but with access to shelter. Alternatively cattle can be confined in intensive systems, although in the observance of international guidelines for minimising stress and improving the wellbeing of animals. In intensive farming systems, cattle for meat are raised to slaughter in about 3–16 months, thus significantly reducing their lifespan. Extensive systems are less productive in the short-term but produce the best-quality meat while providing an increased lifespan and animal welfare. These systems are more adequate in mild climates, as in the Mediterranean basin, where they are common for small ruminants.

Domestic sheep (*Ovis aries*) are extremely versatile and exist in a wide variety of habitats worldwide, ranging from temperate mountain forests to desert conditions. Their natural life expectancy is about 22–23 years, and their average weight is about 35–40 kg (Reavill 2000).

Domestic goats (*Capra hircus*) have similar habits to sheep but a slight higher body mass and shorter life expectancy of 15 years (Mileski 2004). Roughly, the meat composition of large and small ruminants has an identical balance of compounds, but contents may significantly differ according to breed, climate, breeding systems, etc. Data on the composition of beef from two distinct databases are presented below as an example.

Thus, regarding meat composition (INSA 2015; USDA 2015), Beef, grass-fed, strip steaks, lean only, raw, supply 117–122 kcal/100 g from: proteins (20.9–23.1 %) and total fat (2.7–4.3 %).

Beef does not contain retinol; other fat-soluble vitamins are α-tocopherol (0.04–0.22 mg/100 g), vitamins $D_2 + D_3$ (0.10–0.40 μg/100 g) and vitamin K (0.9 μg/100 g). Regarding water-soluble vitamins, ascorbic acid is absent, but beef contains folates (13–16 μg/100 g), thiamine (0.05–0.10 mg/100 g), riboflavin (0.12–0.16 mg/100 g), niacin (4.6–6.7 mg/100 g), pantothenic acid (0.68 mg/100 g) and vitamins B_6 (0.51–0.65 mg/100 g) and B_{12} (1.18–2.0 μg/100 g). Relevant minerals are P (169–212 mg/100 g) as well as the microelements Mn (0.009 mg/100 g), Se (21.1 μg/100 g) (INSA 2015; USDA 2015). Betaine is quantified at a concentration of 7.6 mg/100 g (USDA 2015).

Martins and co-workers (2007) determined the average contents of CLA isomers in the most consumed ruminant meats in Portugal. Regarding the most relevant isomers, the concentration of the well-recognized beneficial isomer c9,t11 CLA was found to be significantly higher in extensively produced beef (78.4 ± 6.3 mg/g) and lamb meat (77.3 ± 6.3 mg/g) than in intensively produced beef (59.9 ± 13.7 mg/g). On the other hand, the meat concentration of t10,c12 CLA (which is not so desirable, as explained in Sect. 6.2) is generally higher in intensively produced beef than in beef from extensive systems (Martins et al. 2007), and it is also significantly lower in lamb meat (extensive system).

7.3.3 Pork

Pork is the name of the meat from the domestic pig (*Sus domesticus*), a voracious omnivore animal of which many subspecies exist, ranging from 1 to 2 m in length and 50 to 350 kg in weight. The colour of its skin may vary from pale pink to black, depending on the breed. Domestic pigs are farmed primarily for their meat, which is consumed fresh or processed (e.g. bacon, ham, etc.). Pork has been a very popular meat in the western world, including southern Europe, particularly due to the diversity of end products that can be obtained: smoked ham, chorizo, salami, mortadella, etc. Consumption of pork is forbidden for both Muslim and Jewish people for religious reasons.

Breeds from northern European countries, of large size and of white pinky skin, are generally exploited in intensive husbandries, whereas others, such as 'cerdo iberico' (of dark skin), are preferably raised by extensive systems. This is the way to obtain meat of the proper quality to use in the production of 'chorizo', smoked ham ('jamon') and other valuable 'Protected Designation of Origin' (PDO) products. The natural lifespan of pigs (in the wild) is about 15 years, but an adult may reach the necessary size and weight to slaughter in less than a year in intensive farming systems. Juveniles (suckling pigs) are slaughtered a few weeks after birth, when the tender meat (called 'leitão') is most appreciated.

Pork, if fat trimmed, can be leaner than chicken although higher in cholesterol and SFA than other meats, whereas it is higher in thiamine (vitamin B_1).

Thus, for example, pork ribs with bone, fat trimmed, raw (about 9 % separable fat), supply 172–221 kcal/100 g from: protein (19.8–20.9 %) and total fat (9.8 %) (INSA 2015; USDA 2015).

Pork meat contains negligible amounts of retinol (0–3 μg/100 g), and vitamin K is absent, but other fat-soluble vitamins are provided in fair amounts: α-tocopherol (0.11–0.2 mg/100 g) and cholecalciferol (0.7–0.9 μg/100 g). Regarding water-soluble vitamins, ascorbic acid is absent as are folates, but pork meat is generally richer in thiamine (0.50–0.74 mg/100 g) and riboflavin (0.20–0.33 mg/100 g) than meat from other animals; the concentration of the other B vitamins is as follows: niacin (3.5–6.5 mg/100 g), pantothenic acid (0.86 mg/100 g), B_6 (0.40–0.46 mg/100 g) and B_{12} (0.54–1.0 μg/100 g). Relevant minerals are P (165–193 mg/100 g) and Se (33.3 μg/100 g) (INSA 2015; USDA 2015). Betaine, which may be correlated with water retention in meat, is quantified at a concentration of 2.8 mg/100 g (USDA 2015), lower than typical values from broiler chicken and beef.

7.3.4 Traditionally Processed Meat

Meat-preserving methods aim to decrease water activity and pH and may additionally avoid microbial growth by adding spices and compounds from wood fumes (in large-scale production, synthetic food preservatives may be added, within the allowed limits). In the Mediterranean basin, techniques for preparing cured sausages (chorizo) and dry smoked meat (jamon) have been developed locally, resulting in distinctive regional products. The technique for producing cured sausages evolved from ancient recipes involving filling pig intestines with meat by-products and drying with smoke. The manufacture and preservation methods were most probably improved by Roman legions to obtain an easily transported food resource with extended shelf-life at room temperature. The Romans also made other types of sausage products, 'circelli', 'tomacinae', 'butuli', which were eaten during annual orgiastic festivals and sacrifices.

Preservation methods include curing (by adding salt and spices), drying and (in many cases) smoking; these systems give the best results in mild climates. Meat is a complex raw material in which an alteration in a single factor results in a series of interrelated changes and processes. Traditional know-how and scientific knowledge together are determinant to quality and safety, and hence the creation of successful products. Nowadays, industries also include preservatives such nitrites and benzoates, as happens with cooked ham and bacon.

Pork is more often used in meat preserves, although beef sausages have been produced in Muslim countries such as Egypt (Rabie et al. 2010; Rabie and Toliba 2013).

Maturation of pork meat (e.g. ham and 'chorizo') involves lipolysis and proteolysis processes, conducted by endogenous enzymes or enzymes of microbial origin. The biochemical reactions that take place during the maturation of many of these traditional products is far from being understood from a scientific viewpoint, with quality and safety mostly relying on local knowledge. The extent of proteolytic and lipolytic reactions depends on the breed and feeding system of pigs (impacting the patterns of muscle proteolytic and lipolytic enzymes), processing techniques and, ultimately, the temperature of chambers (Alfaia et al. 2004; Cava et al. 1999).

However, many aspects of modern meat technology, combined with traditional processing practices, have been implemented, particularly in the smoked ham and sausage-producing segment of the meat industry. Current dry sausage and smoked ham production methods are continuously evolving, and changes based on technological advances are in sight.

7.3.4.1 'Chouriço/Chorizo' (Dry Fermented Sausages) and Similar Products

'Chouriço/chorizo' (dry fermented sausages) and similar products are fermented and cured spicy and smoked pork sausages, stuffed into dried and pre-treated intestines or bladder of the pork. Many different types of this kind of 'sausage' exist, depending on the ingredients and spices, which affect the cure and smoking periods. Fermented pork sausages of different sizes and flavours are most popular in Portugal, Spain and Italy. The spices included in the sausage are a particular feature that distinguishes them from other meat derivatives. They not only contribute to the aroma but also play a key role in the selection of adequate microbial communities responsible for the fermentation step. Salt, garlic, paprika and 'pimento' are spices commonly used in these products.

These sausages have traditionally been produced in the autumn and matured (by exposure to smoke) over the winter. The right firmness and flavour were normally attained after 2–3 months. The flavour of dry fermented sausages results from chemical, biochemical and microbiological reactions that take place alongside ripening, as well as compounds contributed by the smoke in products that undergo this step.

With reference to 'chorizo', a microbial fermentation step is relevant to quality, as opposed to the desirable low microbial activity in cured ham.

Lactic acid bacteria (LAB) are well known for their fastidious nutritional requirements but wide tolerance to environmental factors, contributing to the aroma and safety of many fermented foods. They play a key role in the first maturation step of 'chorizo' by producing lactic and acetic acids and later contributing to proteolysis and lipolysis. Species of genera *Lactobacillus* and *Pediococcus* are predominant (Santos et al. 1997).

Ansorena and colleagues (2000) registered a stepwise increase in acids as a consequence of LAB activity, followed by the accumulation of long-chain fatty acids (C16 and C18), resulting from the hydrolysis of triglycerides by both microbial and endogenous lipases. The next phase (after 21 days of maturation) is the accumulation of short-chain fatty acids ($C < 6$), which contribute to the typical organoleptic characteristics (Ansorena et al. 2000) and were noted as playing a relevant role in maintaining a healthy gut microbiota (Sect. 6.1). Esters and aldehydes were also reported by the same authors, as maturation products, some of which have been detected in red pepper (e.g. 2,4-decadienal and pentadecanal).

Bacterial proteolysis causes the accumulation of benzeneacetaldehyde (2-phenyl-acetaldehyde) from the oxidation of phenylalanine, as well as 2-methylbutanoic acid and 3-methylbutanoic acid, which result from microbial degradation of the amino acids, isoleucine and leucine, respectively. Thus, 2-phenyl-acetaldehyde communicates a green odour and sweet, floral taste with a spicy nuance. On the other hand, 2-methyl- and 3-methylbutanoic acid communicate a sweaty odour to the 'chorizo de Pamplona' (Ansorena et al. 2000).

In Portuguese 'chouriço' from Alentejo, the primary group of volatiles by the end of maturation are hydrocarbons, alcohols and sulphur compounds (Partidário et al. 2011). At least some of these sulphur compounds may be organosulphur compounds from garlic (most of them are under investigation for their potential health-promoting properties).

Many of these traditional products have been granted PDO or similar status by the EC. This status implies the mandatory use of specific techniques, with meat from local pork breeds raised in extensive farming systems and fed on pasture and acorn. Procedures, bags (intestine and bladder) and their treatments prior to use, temperatures of the maturation chambers and the type of wood used to produce fumes, are all regulated and stated in each PDO or similar EU-issued protocol. Industrial processes have been developed to shorten the maturation period, aiming to retain its uniqueness while allowing small sausages to be ready to market in about 4–6 weeks.

Among many examples of such products are 'chouriço de carne de Estremoz e Borba' (PT/PGI/0005/0159), 'chosco de Tineo' (ES-PGI-0005-0696) and 'salama da sugo' with IT-PGI-0005-01114 (EC 2002, 2010, 2014).

Fuet jamon*. Fuet jamon is a thin dry-cured sausage from Catalonia, Spain, obtained from minced pork meat and usually seasoned with garlic and black pepper. Unlike other Spanish sausage varieties such as chorizo, paprika is not used for fuet production. According to experts, this product may be defined as similar to Italian salami. Photo reprinted with kind permission from M. Barone*

7.3.4.2 'Jamón/Presunto' (Smoked Ham)

High-quality smoked hams are manufactured from Iberian pigs that are raised in an extensive system in the regions near the south border between Portugal and Spain (Alentejo, Andalucia and Granada) and mainly fed on pasture and acorn. A second (mountainous) region near the north border between Tràs-os-Montes and Castilla-Leon is also relevant (jamón serrano).

Relevant products within this category, holding PDO or similar status, are as follows (EC 1996, 2000, 2005, 2007, 2010, 2014):

(a) 'Jamón de Huelva' (ES/PDO/0005/0009)
(b) 'Jamón de Trevélez' (ES/PGI/0005/0309)
(c) 'Jamón de Serón' (ES/PGI/0005/01052), from the southern Iberian Peninsula
(d) 'Presunto de Vinhais/Presunto bísaro de Vinhais' (PT/PGI/0005/0456) and 'Lacón Gallego' (ES/PGI/0005/0104), both from the northern region of the Iberian Peninsula

The smoked ham preparation method follows some common steps, with specificities according to traditional recipes and the microclimate of the region. According to the methodology described in PDO documentation (ES/PDO/0005/0009) for 'Jamón de Huelva' (commonly known as 'jamón ibérico'), only the rear extremities larger than 7 kg are used. The first step is the salting with sodium chloride (NaCl) and nitrates, followed by a curing step at 0–5 °C and 70–90 % relative humidity for a variable time depending on the size of the piece (about 1 day/kg weight). The excess salt on the surface is washed out with water, and the hams are kept in chambers at 3–7 °C and 70–90 % relative humidity for 30–60 days to equilibrate salt content inside the piece to about 1 % and are then hung to dry for about 6 months followed by maturation in 'bodegas' until obtaining the correct sensory characteristics. The external aspect is determined by typical moulds, which give ham a white to blue-greyish colour. The meat in the interior is purple rose, non-fibrous, elastic, and of greasy consistency, with thin adipose veins distributed in the muscle.[8]

Iberian ham ripening involves little microbial activity, with the predominance of *Micrococcaceae*, which is well adapted to the salty environment (Alfaia et al. 2004). Similar to what was described for cheese, these proteolytic reactions yield peptides and free amino acids, which are correlated with flavour development in aged dry-cured ham, e.g. by contributing to the formation of aromatic compounds through pathways such as Strecker degradation or Maillard reaction (Alfaia et al. 2004). Free amino acids can also be decarboxylated into biogenic amines; although this should be avoided as they spoil the aroma and are dangerous to public health if maximum legal levels are approached.

Proteolysis also releases short peptides and non-protein amino acids, such as creatine, creatinine, glutathione, carnosine, carnitine and taurine (Marušić et al. 2013). The antioxidant and anti-hypertensive functions of short bioactive peptides have been described in Sect. 6.5. Moreover, Marušić et al. (2013) observed that carnosine, creatinine and serine were associated with antioxidant activity, while cysteine, glutathione and carnosine were associated with anti-hypertensive activity; all of them are present in dry cured ham.

[8] Adapted from ES/PDO/0005/0009-27.01.1998, available at: http://ec.europa.eu/agriculture/quality/door.

Lipolytic reactions are important phenomena that occur during ripening. Triglycerides are broken down, and some fatty acids may undergo several types of reactions. Thus, long-chain saturated aldehydes (C4–C10) can be formed, negatively affecting aroma and hence ham quality. These compounds were found to decrease with ripening in the case of Iberian pigs fed on acorn and pasture (Cava et al. 1999), but that is not the case in pigs confined and fed on industrial food. On the other hand, and according to the same authors, volatile aldehydes, such as pentanal and acetaldehyde are only predominant in 'jamón ibérico de bolota'.

Prolonged ageing does not always result in an improved taste with dry cured hams (Cava et al. 1999; Alfaia et al. 2004).

7.4 Eggs

The term 'eggs' generally means chicken eggs; however, quail eggs are also consumed, mostly as a delicacy. Eggs comprise an egg white (mostly albumin) and an egg yellow (yolk), of more complex composition.

Fresh raw eggs supply 143–149 kcal/100 g from: proteins (12.6–13.0 %), glucose (0.37 %) and total fat (9.51–10.0 %, being the sum of SFA 2.7–3.1 %, predominantly 16:0 and 18:0; MUFA 3.7–3.9 %, predominantly 16:1c and 18:1c; PUFA 1.9–2.1 %; TFA 0.04 %) and cholesterol (372 mg/100 g) (INSA 2015; USDA 2015).

Eggs are important sources of Ca and P and are rich in B vitamins, including B_{12} (0.89–1.0 μg/100 g), also containing retinol (160 μg/100 g), carotenoids (503 μg/100 g of lutein + zeaxanthin) and tocopherols (α-, β-, δ- and γ- in the concentrations of 1.05, 0.01, 0.06 and 0.5 mg/100 g, respectively). Eggs also supply cholecalciferol (1.7–2.0 μg/100 g) and vitamin K (0.3 μg/100 g) (INSA 2015; USDA 2015).

Nowadays, eggs are used in the dry or pasteurised liquid form for most industrial applications. These heating and drying processes somehow affect the flavour and texture of the final product also altering its nutritional value. Thus, the bioavailability of some vitamins seems to be reduced, as is the case of retinol and tocopherols (only α-tocopherol remains after the drying process). The lipid fraction also suffers some alterations (mainly in the concentrations of essential fatty acids) but the proportions of SFA/MUFA/PUFA do not significantly change.

We have drawn statistical data on total protein availability for human consumption from the United Nations (UN) Food and Agriculture Organization 'Food Balance Sheet' (FBS[9]) (FAO 2015e) for the countries on the UN Educational, Scientific and Cultural Organization (UNESCO) MD representative list (Portugal, Spain, Morocco, Italy, Greece, Cyprus and Croatia), and carried out some simple calculations. Results are presented in Table 7.3; the explanation about the FBS (Sect. 1.2) should also be noted. In the year 2011, the lowest value for total protein availability was registered in Cyprus (78.9 g/capita/day) while the highest value

[9] See Sect. 1.2 for information on FBS methodologies and meanings.

Table 7.3 Most relevant protein sources (g protein/capita/day) for the countries that integrate the representative list of UNESCO for the Mediterranean diet in the year 2011

Country	Fish and seafood	Meat	Dairy (except butter)	Eggs	Pulses	Total[a]
Portugal	15.2	31.2	16.8	2.8	2.0	111.2
Spain	12.9	30.8	15.4	4.3	5.2	103.3
Morocco	4.1	12.3	5.0	1.9	4.6	95.6
Italy	7.0	29.7	17.9	3.6	2.9	109.9
Greece	5.3	26.1	26.2	2.8	2.8	111.4
Cyprus	6.2	25.0	12.5	2.2	1.9	78.9
Croatia	6.0	17.5	18.5	2.8	0.6	82.4

Values were collected and/or calculated from the United Nations Food and Agriculture Organization Food Balance Sheets (accessed 2015) and refer to the year 2011. Food consumption per person is the amount of food, in terms of quantity, of each commodity and its derived products for each individual in the total population. Figures are shown for food items

[a]The difference in the values between total available protein (last column) and the sum of protein from discriminated items (previous columns) accounts for miscellaneous sources (not shown)

(111.4 g/capita/day) was registered in Greece. The main source of protein (in this representative list of countries) is meat, despite the almost threefold variation in its availability (from 12.3 g/capita/day in Morocco to 31.2 g/capita/day in Portugal). Dairy products are the second source of proteins (Table 7.3). The availability of fish varies as much as that of meat, and it is not necessarily associated with total protein intake. It is noteworthy that Portugal and Spain were amongst the world's largest consumers of fish and seafood in 2011 (FAO 2015a), showing total protein intakes of 111.2 g/capita/day (Portugal) and 103.3 g/capita/day (Spain), for which fish and seafood account for 14 and 12 % of total protein intake, respectively. In the remaining countries in our work, the contribution of fish to total protein intakes ranges from 4 to 7 % (Table 7.3). Nowadays, pulses seem to play a much less important role as a protein source than in the early 1960s, particularly in Croatia, where the availability of pulses is marginal (0.6 g/capita/day).

As can be deduced from that set of statistical data, there has been a remarkable increase in per capita meat consumption in all countries that make up the UNESCO representative list for 'Mediterranean Diet'.

By closely comparing data from the FBS over time (FAO 2015e), it can be observed that pork was the first choice in Spain and Portugal in 1961, closely followed by bovine meat. It should be recalled that pork meat can be more easily preserved via traditional methods such as smoking and curing than other types of meat. This pattern was not reflected in the other countries. In 1961, bovine was the most available type of meat in Italy. It can be deduced from market availability figures that Greeks consumed mostly the meat of small ruminants (lamb and goat), followed by bovine. In the same decade, bovine was preferred in Cyprus and Morocco, followed by the meat of small ruminants; no data for the 1960s are available from Croatia, thus impairing comparisons.

The equivalent dataset from 2011 (FAO 2015e) indicates that pork now seems to be the preferred meat in all countries, (highlighted in Fig. 1.1) with the exception of Morocco, the reasons for which are probably religious. Italy, Greece and Croatia show similar patterns of availability for consumption (data from 2011), with pork as the preferred meat, followed by bovine and poultry at quite similar proportions (FAO 2015e). Moreover, and still according to the same source, the availability of poultry for consumption has significantly increased. In 2011, poultry was the first choice in Morocco, followed by ruminant meat, in similar proportions. In Portugal and Spain, poultry displaced bovine meat as the second choice. Poultry is also the second choice in Cyprus.

These trends may indicate alterations in the pattern of meat consumption in the region.

With the development of agriculture and the increasing industrialisation of their products, intensive poultry breeding has turned out to be a good investment, mainly due to the short life cycle of the intensively farmed animals (about 1 month from egg to slaughter). The 'bovine spongiform encephalopathy' crisis may have discouraged the consumption of ruminant meat, while the economic crisis (quite severe in south Europe since 2009) may have triggered a demand for cheaper meats.

On the other hand, as can be observed in Table 7.3, pulses are currently far from counterbalancing animal protein intake, but the increase of public awareness about their health benefits may alter this deleterious increasing trend in meat consumption. One of the objectives of the International Year of Pulses (FAO 2015l).

In short, whether 'excess' dietary protein intake adversely affects bone in humans is currently a controversial subject in nutrition. Although not entirely conclusive, studies by both Frassetto and colleagues (2000) and Sellmeyer and co-workers (2001) support the hypothesis that "excessive dietary protein from foods with high potential renal acid load (e.g. animal foods) adversely affects bone, unless buffered by the consumption of alkali-rich foods (e.g. vegetable foods)." In other words, moderating animal protein intake and increasing the vegetable-to-animal protein ratio may confer a protective effect against osteoporosis, among other ailments (Frassetto et al. 2000; Sellmeyer et al. 2001).

Within animal protein sources, fish is known to have a balanced essential amino acid composition, and a desirable lipid profile, which is why it is recommended within a balanced healthy diet. Nevertheless, it should be noted that the characteristics of each type of meat, and consequently their potential benefits and deleterious effects will always strongly depend on the frequency and quantity of consumed meat, and is not dissociable from the other components of the MD.

Pulses may partially or totally substitute meat when combined with other protein sources, providing satiety and improving gut's health, along with other benefits.

References

Alfaia CP, Castro ML, Martins SI, Portugal AP, Alves SP, Fontes CM, Bessa RJ, Prates JA (2007) Influence of slaughter season and muscle type on fatty acid composition, conjugated linoleic acid isomeric distribution and nutritional quality of intramuscular fat in Arouquesa-PDO veal. Meat Science 76:787–95. doi:10.1016/j.meatsci.2007.02.023

Alfaia CM, Castro MF, Reis VA, Prates JM, Almeida IT, Correia AD, Dias MA (2004) Changes in the profile of free amino acids and biogenic amines during the extended short ripening of Portuguese dry-cured ham. Food Sci Technol Int 10(5):297–304. doi:10.1177/1082013204047597

American Heart Association (2015) Eat more chicken, fish and beans. American Heart Association, Dallas. Available at http://www.heart.org/HEARTORG/GettingHealthy/NutritionCenter/HealthyEating/Eat-More-Chicken-Fish-and-Beans-than-Red-Meat_UCM_320278_Article.jsp. Accessed 4 Nov 2015

Ansorena A, Astiasarán I, Bello J (2000) Changes in volatile compounds during ripening of chorizo de Pamplona elaborated with Lactobacillus plantarum and Staphylococcus carnosus. Food Sci Technol Int 6(6):439–447. doi:10.1177/108201320000600602

Bach-Faig A, Berry EM, Lairon D, Reguant J, Trichopoulou A, Dernini S, Medina FX, Battino M, Belahsen R, Miranda G, Serra-Majem L (2011) Mediterranean diet pyramid today. Science and cultural updates. Public Health Nutr 14(12A):2274–2284. doi:10.1017/S1368980011002515

Banks J (2012) Grocery consumers in the recession. In: Ryder J, Ababouch L, Balaban M (eds) Proceedings of the second international congress on seafood technology on sustainable, innovative and healthy seafood, 10–13 May 2010, Anchorage. Food and Agriculture Organization of the United Nations, Rome. Available at http://www.fao.org/docrep/015/i2534e/i2534e.pdf. Accessed 4 Nov 2015

Cava R, Ruiz J, Ventanas J, Antequera T (1999) Oxidative and lipolytic changes during ripening of Iberian hams as affected by feeding regime: extensive feeding and alpha-tocopheryl acetate supplementation. Meat Sci 52(2):165–172. doi:10.1016/S0309-1740(98)00164-8

DiLeone RJ (2011) Neuroscience gets nutrition. Nat Neurosci 14:271–272. doi:10.1038/nn0311-271

EC (1996) Publication of an application for registration pursuant to Article 6 (2) of Regulation (EEC) No 2081/92 on the protection of geographical indications and designations of origin. Off J Eur Comm C246:9–11

EC (2000) Publication of an application for registration pursuant to Article 6(2) of Regulation (EEC) No 2081/92 on the protection of geographical indications and designations of origin. Off J Eur Comm C264:2–4

EC (2002) Publication of an application for registration pursuant to Article 6(2) of Council Regulation (EEC) No 2081/92 on the protection of geographical indications and designations of origin. Off J Eur Comm C102:2–3

EC (2005) Publication of an application for registration pursuant to Article 6(2) of Regulation (EEC) No 2081/92 on the protection of geographical indications and designations of origin. Off J Eur Union C51:2–4

EC (2007) Publication of an application pursuant to Article 6(2) of Council Regulation (EC) No 510/2006 on the protection of geographical indications and designations of origin for agricultural products and foodstuffs. Off J Eur Union C236:10–12

EC (2010) Publication of an application pursuant to Article 6(2) of Council Regulation (EC) No 510/2006 on the protection of geographical indications and designations of origin for agricultural products and foodstuffs. Off J Eur Union C166:8–12

EC (2014) Publication of an application pursuant to Article 50(2)(a) of Regulation (EU) No 1151/2012 of the European Parliament and of the Council on quality schemes for agricultural products and foodstuffs. Off J Eur Union C 101:10–14

EOL, Encyclopedia of Life (2015). Bos Taurus, cattle. http://eol.org/pages/328699. Accessed 15 May 2015

FAO (2015a) Statistics Division. Trade. http://faostat3.fao.org/browse/T/. Accessed 16 Mar 2015

FAO (2015b) Fisheries and Aquaculture Department. Species Fact Sheets: Sardina pilchardus (Walbaum, 1792). FAO, Rome. http://www.fao.org/fishery/species/2910/en. Accessed 26 Feb 2015

FAO (2015c) Fisheries and Aquaculture Department. Species Fact Sheets: Engraulis encrasicolus (Linnaeus, 1758). FAO, Rome. http://www.fao.org/fishery/species/2106/en. Accessed 26 Feb 2015

FAO (2015d) Fisheries and Aquaculture Department. Species Fact Sheets: Gadus morhua (Linnaeus, 1758). FAO, Rome. http://www.fao.org/fishery/species/2218/en. Accessed 26 Feb 2015

FAO (2015e) Statistics Division. Food Balance. Food supply—livestock and fish primary equivalent. FAO, Rome. http://faostat3.fao.org/download/FB/. Accessed 14 May 2015

FAO (2015f) Fisheries and Aquaculture Department. Species Fact Sheets: Sepia officinalis (Linnaeus, 1758). FAO, Rome. http://www.fao.org/fishery/species/2711/en. Accessed 26 Feb 2015

FAO (2015g) Fisheries and Aquaculture Department. Species Fact Sheets: Octopus vulgaris (Lamarck, 1798). FAO, Rome. http://www.fao.org/fishery/species/3571/en. Accessed 26 Feb 2015

FAO (2015h) Fisheries and Aquaculture Department. Species Fact Sheets: Mytilus edulis (Linnaeus, 1758). FAO, Rome. http://www.fao.org/fishery/species/2688/en. Accessed 26 Feb 2015

FAO (2015i) Fisheries and Aquaculture Department. Species Fact Sheets: Mytilus galloprovincialis (Lamarck, 1819). FAO, Rome. http://www.fao.org/fishery/species/3529/en. Accessed 26 Feb 2015

FAO (2015j) Fisheries and Aquaculture Department. Species Fact Sheets: Cerastoderma edule (Linnaeus, 1758). FAO, Rome. http://www.fao.org/fishery/species/3535/en. Accessed 26 Feb 2015

FAO (2015k) Fisheries and Aquaculture Department. Species Fact Sheets: Crangon crangon (Linnaeus, 1758). FAO, Rome. http://www.fao.org/fishery/species/3435/en. Accessed 26 Feb 2015

FAO (2015l) Observances. International years. 2016 International Year of Pulses (A/RES/68/231). FAO, Rome. Available at http://www.fao.org/pulses-2016/en/. Accessed 2 Mar 2015

Frassetto LA, Todd KM, Morris RC Jr, Sebastian A (2000) Worldwide incidence of hip fracture in elderly women: relation to consumption of animal and vegetable foods. J Gerontol Med Sci A 55(10):M585–M592. doi:10.1093/gerona/55.10.M585

Galli C, Calder PC (2009) Effects of fat and fatty acid intake on inflammatory and immune responses: a critical review. Ann Nutr Metab 55(1–3):123–139. doi:10.1159/000228999

INSA (2015) Tabela da Composição de Alimentos (TCA). http://www.insa.pt/sites/INSA/Portugues/AreasCientificas/AlimentNutricao/AplicacoesOnline/TabelaAlimentos/PesquisaOnline/Paginas/PorPalavraChave.aspx. Accessed 5 Nov 2015

Jereb P, Roper CFE (eds) (2010) Cephalopods of the world. An annotated and illustrated catalogue of cephalopod species known to date, Volume 2. Myopsid and oegopsid squids, vol 2, FAO species catalogue for fishery purposes. FAO, Rome

Jump DB (2002) Dietary polyunsaturated fatty acids and regulation of gene transcription. Curr Opin Lipidol 13(2):155–164. doi:10.1097/00041433-200204000-00007

Martins SV, Lopes PA, Alfaia CM, Ribeiro VS, Guerreiro TV, Fontes CMGA, Castro MF, Soveral G, Prates JAM (2007) Contents of conjugated linoleic acid isomers in ruminant-derived foods and estimation of their contribution to daily intake in Portugal. Br J Nutr 98 (6):1206–1213. doi:10.1017/S0007114507781448

Marušić N, Aristoy M-C, Toldrá F (2013) Nutritional pork meat compounds as affected by ham dry-curing. Meat Sci 93(1):53–60. doi:10.1016/j.meatsci.2012.07.014

Mileski A (2004) 'Capra hircus' (On-line), Animal Diversity Web. Regents of the University of Michigan, Ann Arbor. Available at http://animaldiversity.org/accounts/Capra_hircus/. Accessed 4 Nov 2015

Mozaffarian D, Wu JHY (2012) (n-3) Fatty acids and cardiovascular health: are effects of EPA and DHA shared or complementary? J Nutr 142(3):614S–625S. doi:10.3945/jn.111.149633

NAP (2015) Dietary reference intakes for energy, carbohydrate. Fiber, fat, fatty acids, cholesterol, protein, and amino acids (2002/2005). www.nap.edu. Accessed 12 May 2015

NCBI (2015) PubChem compound Database. CID 247. National Center for Biotechnology Information (NCBI), Bethesda. Available at http://pubchem.ncbi.nlm.nih.gov/compound/betaine. Accessed 4 Nov 2015

NHS (2015) Your health, your choices. Live Well. Red meat and the risk of bowel cancer. National Health Service (NHS), London. Available at http://www.nhs.uk/livewell/goodfood/pages/red-meat.aspx. Accessed 4 Nov 2015

Partidário AM, Roseiro C, Santos C (2011) Effect of processing in volatiles from a Portuguese traditional dry-fermented ripened sausage 'chouriço grosso Borba-Estremoz PGI'. Food Sci Technol Int 17(1):15–22. doi:10.1177/1082013210367512

Rabie MA, Toliba AO (2013) Effect of irradiation and storage on biogenic amine contents in ripened Egyptian smoked cooked sausage. J Food Sci Technol 50(6):1165–1171. doi:10.1007/s13197-011-0444-7

Rabie MA, Siliha H, el-Saidy S, el-Badawy A-A, Malcata FX (2010) Effects of γ-irradiation upon biogenic amine formation in Egyptian ripened sausages during storage. Innovat Food Sci Emerg 11(4):661–665. doi:10.1016/j.ifset.2010.08.007

Reavill C (2000) 'Ovis aries' (on-line), Animal Diversity Web. Regents of the University of Michigan, Ann Arbor. Available at http://animaldiversity.org/accounts/Ovis_aries/. Accessed 4 Nov 2015

Sanders TAB (2009) Fat and fatty acid intake and metabolic effects in the human body. Ann Nutr Metab 55(1–3):162–172. doi:10.1159/000229001

Santos EM, González-Fernández C, Jaime I, Rovira J (1997) Identification and characterization of lactic acid bacteria isolated from traditional chorizo made in Castilla-León. Food Sci Technol Int 3(1):21–29. doi:10.1177/108201329700300103

Sellmeyer DE, Stone KL, Sebastian A, Cummings SR (2001) A high ratio of dietary animal to vegetable protein increases the rate of bone loss and the risk of fracture in postmenopausal women. Am J Clin Nutr 73(1):118–122

Smit LA, Mozaffarian D, Willett W (2009) Review of fat and fatty acid requirements and criteria for developing dietary guidelines. Ann Nutr Metab 55(1–3):44–55. doi:10.1159/000228995

Stark AH, Crawford MA, Reifen R (2008) Update on alpha-linolenic acid. Nutr Rev 66 (6):326–332. doi:10.1111/j.1753-4887.2008.00040.x

USDA (2015) Agricultural Research Service National Nutrient Database for Standard Reference Release 27, Software v.2.2.4, The National Agricultural Library. http://ndb.nal.usda.gov/ndb/foods. Accessed 5 May 2015

Infusions and Wines

8

Abstract

Typical drinks of the Mediterranean diet, associated with convivial meals, are wine (in non-Muslim countries)—consumed during main courses—and infusions and coffee, generally consumed after the meal. Many herbal infusions of wild weeds are often related to folk medicine. None of these drinks provide major nutrients in significant amounts, and their effects on health result from the presence of minor components. Coffee is a very popular drink, consumed all over the world in many different ways: brewed, filtered or prepared using particular techniques. Italians popularised the 'espresso' coffee, produced from a blend of selected proportions of 'Robusta' and 'Arabica' grains. Coffee became popular for its stimulant properties attributed to caffeine, which is thought to improve mental acuity, particularly in the elderly, and recent studies suggest that moderate coffee consumption may reduce the risk of stroke. On the other hand, tea (the infusion of the leaves of *Camellia sinensis*) provides a wide range of bioactive compounds. Tea is rich in catechins and derivatives as well as in other flavonoids; it also provides two stimulant compounds: caffeine and theobromine (in lower amounts than coffee). While black tea is predominant in Europe, green tea is preferred in the south of the Mediterranean Sea; the difference between both types of tea relies on the existence of an oxidation step during manufacturing of black tea. Regular tea consumption has been associated with a decreased rate of some non-communicable diseases and may also be helpful in weight control. Herbal infusions, sometimes called medicinal teas, have been used since immemorial times, mostly for curative purposes. The most popular herbal infusions, in the countries referred in the current work, are chamomile (*Chamaemelum nobile*), lemon balm and balm mint (variant of *Melissa officinalis*), lemon verbena (*Aloysia triphylla*) and lemon peel (*Citrus limon*). Their antioxidant and anti-inflammatory properties have been attributed to different flavonoids and essential oils. In its turn, wine is one of the oldest fermented drinks, admired for its flavour and health-promoting effects, particularly in the improvement of cardiovascular health. Wine making and

A.M. Delgado et al., *Chemistry of the Mediterranean Diet*,
DOI 10.1007/978-3-319-29370-7_8

consumption is a mixture of science, culture and art. A wine is chosen to complement a meal and should be consumed in moderate amounts.

8.1 Infusions and Wines: An Introduction

Since water is an essential molecule for life, human beings need to keep themselves hydrated. Thus, an instinctive need to drink is regulated by the hypothalamus in response to subtle changes in the body's electrolyte levels, and also as a result of changes in the volume of circulating blood. Although drinking is a primary need, over time different cultures have developed distinct drinks associated with ceremonies, rituals and social events. Wines, infusions, teas and coffee are among the most ancient examples.

Spirits (distilled alcoholic beverages), beer, pasteurised fruit juices and derivatives, and more recently carbonated and soft drinks have been developed by food industries. As with food habits, drink options are affected by cultural aspects, including religious beliefs, social habits, rituals and health concerns.

Typical drinks of the Mediterranean diet (MD) associated with convivial meals are wine—consumed with main courses—and infusions and coffee, which are generally consumed after the meal.

8.2 Coffee

Coffee is a drink obtained from the beans of plants from the family *Rubiaceae*, of the genus *Coffea* and the species *C. arabica* (cv *Typica*) and *C. canephora* (cv *Robusta*). Although other species can be consumed, variants of *Robusta* and *Arabica* cultivars account for the majority of international coffee trades (ICO 2015). Current *Arabica* cultivars have been developed in Brazil, Colombia, Central America, Jamaica and India, whereas *Robusta* is grown in West and Central Africa, South-East Asia and to some extent in Brazil (ICO 2015). Coffee has been selectively bred to improve yield, cup quality, disease resistance etc. In this sense, the Portuguese developed *Híbrido de Timor*, a natural hybrid of *Arabica* and *Robusta* with improved characteristics, such as cup quality and resistance to plagues. Icatu is a similar hybrid obtained in Brazil. In 1985, Kenya also launched a hybrid, Ruiru 11 (ICO 2015).

Coffee (espresso). *Coffee is prepared from ground roasted beans of the species Coffea arabica and Coffea canephora and is the main source of caffeine, a purine that stimulates central nervous system activity and is thought to contribute to improved mental acuity in the elderly. Coffee contains many volatile compounds that are important for its aroma, although the physiological function of most is unknown. Photo reprinted with kind permission from T. N. Wassermann*

Coffee cultivation first took place in Southern Arabia, and evidence of coffee drinking appears in the middle of the fifteenth century in the Horn of Africa and Yemen. The stimulant properties of coffee were discovered in the middle ages and are attributed to caffeine (ICO 2015).

According to the International Coffee Organization (ICO 2015), the first coffee houses opened in Mecca and quickly spread throughout the Arab world, thriving as places where cards and other games were played, and gossip was exchanged.

Despite some Arab resistance, by the late 1600s the Dutch were growing coffee at Malabar, in India, and later in 1699 also in Indonesia. Within a few years, the Dutch colonies had become the main suppliers of coffee to Europe, where it was sold by Venetian traders.

Although coffee houses have been repeatedly forbidden in ancient times on both sides of the Mediterranean Sea, nowadays they exist worldwide in different versions according to the local culture but retaining its essence of a convivial place. *Caffé Florian*, dated from 1720 and located at Piazza San Marco, Venice, is the oldest coffee house in Europe. In Morocco and other Muslim countries, coffee houses (kahwa) keep their original essence as places where (mostly) men meet to gossip and to play games while drinking coffee or tea and smoking 'shisha' (water pipe).

The flavour and aroma of coffee is created during the roasting process. Green coffee beans are heated at high temperatures for several minutes. This pyrolysis process removes the moisture, causing chemical changes and altering the cellular structure of the bean. Starches and proteins are broken down to some extent into their reactive monomers, and aromatic compounds are precipitated mainly in the organic phase, which is known as 'caffeol'. The precursors of aroma molecules are sucrose, chlorogenic acid, proteins and carbohydrates, which are present in the aqueous and lipid phases of beans and undergo chemical reactions during the roasting process (mainly Maillard reactions). Aroma volatile compounds are (by definition) small molecules, mostly polar and often reactive. Over 850 of these compounds have been identified in coffee, a number that increases each year and includes ketones, acids, phenols, furans, pyrans, thiophenes, pyrroles, oxazoles, thiazoles and pyridines (Fisk et al. 2012). Volatile and water-soluble compounds are all important to the aroma of coffee, which can be damaged by moisture, light and oxygen.

In addition to the blend and the roasting processes, coffee's aroma varies according to many other factors: geographical origin of the beans, blend of cultivars, differences in ageing before roasting, packaging and storage and variables introduced by the consumer. Nowadays, blending, roasting and packaging are carefully controlled industrial processes.

Coffee is a very popular drink and it is consumed all over the world in many different ways: brewed, filtered or prepared using particular techniques. In Morocco, Croatia, Greece and Cyprus, although with different local names, coffee is prepared with some slight variation of the Turkish recipe, inscribed in 2013 in the United Nations' Representative List of the Intangible Cultural Heritage of Humanity. Turkish coffee is prepared from roasted beans that are ground to a fine powder, to which cold water and sugar are added in a special coffee pot, and brewed slowly on a stove to produce the desired foam (UNESCO 2015). This infusion is served very hot, in small cups, accompanied by a glass of cold water. Turkish coffee plays an important role in social occasions such as engagement ceremonies and holidays, and the grounds left in the empty cup are often used to tell a person's fortune.

On the other hand, Italians popularised the 'espresso' coffee, produced from a blend of selected proportions of *Robusta* and *Arabica* grains, coarsely ground at the time and brewed in special machines that combine steam and hot water to produce a creamy foam. Different dilution versions of the original espresso coffee have been developed worldwide. As the preparation of coffee involves an extraction of compounds by boiling water, the extent of the extraction ('large' vs. 'short' coffee) influences the composition of the final drink, as does the blend and ultimately the type and composition of the grains. The composition of the obtained drink, as presented below, is a compilation of available data from the Portfir database (Instituto Nacional de Saúde Doutor Ricardo Jorge, INSA), referring to pre-packed ground coffee prepared with tap water from three different brands available in Portugal, and the National Nutrient Database for Standard Reference (from USDA), which does not mention the number of brands, and includes fast food and home-brewed coffee under the same report number.

As expected, no major nutrients were found when analyzing the data from both databases. The presence of some starch (0.3 %) and proteins (0.2 %) is registered by INSA but are apparently not found in the US samples. On the other hand, the lipid fraction is negligible in Portuguese coffee samples; the only detected vitamin is niacin, and a short list of minerals is reported (INSA 2015). The USDA database reports a few more vitamins besides niacin, such as thiamine, riboflavin, pantothenic acid, α- and γ-tocopherols as well as traces of vitamin B_6 and phylloquinone. According to the same USDA database, the lipid fraction (0.02 %) is the sum of 0.002 g/100 g C16:0 with 0.015 g of C18:1c and 0.001 g of C18:2 (undifferentiated). Coffee (prepared by infusion) contains, on average, 40 mg/100 g of caffeine and 0.1 mg/100 g of proanthocyanidin monomers (USDA 2015).

Coffee is mainly consumed for its aroma and stimulant properties, which are highly variable and depend on the preparation method, which in turn determines the extent of extraction and hence the total caffeine content per serving. Caffeine concentration may vary from 25 to 85 mg per serving; decaffeinated coffee (where a small portion of caffeine remains) (Homan and Mobarhan 2006) was not herein considered.

Caffeine[1] is a two-ring purine classified as an alkaloid acting as a neurotransmitter agent (NCBI 2015a). Its chemical structure is represented in Fig. 8.1. Caffeine occurs naturally in tea and coffee, but can be prepared synthetically for commercial drug use. It is readily absorbed into the bloodstream; the mean time to reach peak concentration ranges from 30 min to 2 h. Caffeine has a plasma half-life of 3–5 h in adults, and an oral lethal dose (LD_{50}) in mice of 127 mg/kg of body weight (bw) (NCBI 2015a). The pharmacological dose of caffeine used to stimulate the central nervous system activity in humans is about 3 mg/kg bw and is observable at about 2 mg/kg bw, while the acute human fatal dose of caffeine appears to be greater than 170 mg/kg bw (FDA 1978). According to these figures, an average adult (70 kg) would need to ingest a total of 140–210 mg of caffeine to obtain the desired alertness; a quantity that will not be reached by a single espresso or Turkish coffee.

Caffeine induces a transient increase in blood pressure without long-term effects. In the past, coffee was generally viewed as a risk factor for cardiovascular disease (Corti et al. 2002; Vlachopoulos et al. 2005). Recent evidence suggests that moderate coffee consumption may reduce stroke risk (Larsson 2014; Lopez-Garcia et al. 2009).

Moderate coffee consumption is inversely associated with the risk of type II diabetes and moderately related to weight loss (van Dam and Feskens 2002; Goto et al. 2011; Greenberg et al. 2006), and noted as having protective effects against oxidative stress and related diseases, inflammatory diseases and Alzheimer's disease (Andersen et al. 2006; Arendash and Cao 2010).

Caffeic acid was found to mediate the inflammatory response in mice, and may exert a protective effect against cardiovascular and inflammatory diseases, as found in a cohort study of postmenopausal women (Andersen et al. 2006). Also, chlorogenic acid (the ester of caffeic acid), which is the most abundant polyphenol

[1] International Union of Pure and Applied Chemistry (IUPAC) name: 1,3,7-trimethylpurine-2,6-dione.

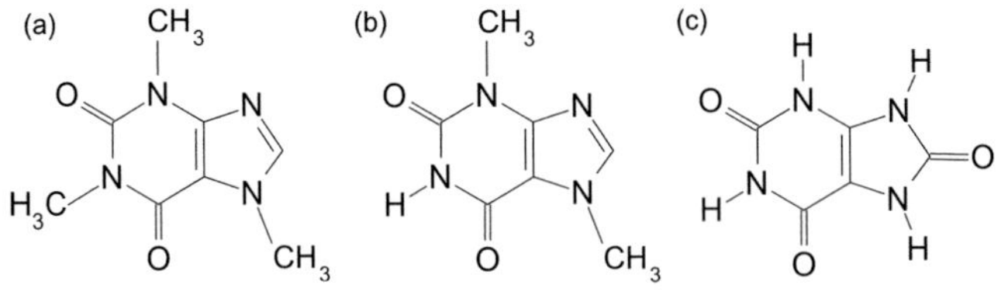

Fig. 8.1 Molecular structure of the purines caffeine (**a**), theobromine (**b**) and uric acid (**c**). Caffeine is a popular stimulant found in tea and coffee, which chemically is a methylxanthine, Theobromine is a xanthine derivative, acting as a vasodilator, diuretic and heart stimulant; Uric acid is an excretion metabolite with analogue structure. BKchem version 0.13.0, 2009 (http://bkchem.zirael.org/index.html) was used to draw these structures

in coffee and readily absorbed, is a potent anti-carcinogen (NCBI 2015b), through its significant reactive oxygen species (ROS)-scavenging activity. Chlorogenic acid has been shown to inhibit platelet secretion and aggregation in a dose-dependent manner, thus preventing arterial thrombosis formation (Fuentes et al. 2014).

Besides caffeine and related compounds, other minor constituents may contribute to the observed health beneficial effects of coffee; in addition, some of these compounds contribute to coffee's 'bouquet'. Some heterocyclic volatile compounds present in coffee have also been hypothesized as antioxidants, but their contribution to the overall antioxidant capacity of coffee is still unclear (Majer et al. 2005). Some authors have observed that coffee diterpenes, such as cafestol and kahweol, have a protective effect against *N*-nitrosodimethylamine and 2-amino-1-methyl-6-phenylimidazol(4,5-b)pyridine (PhIP), two carcinogenic amines often found in some foods (Homan and Mobarhan 2006; Majer et al. 2005). *N*-nitrosamines, such as *N*-nitrosodimethylamine, may accumulate in nitrite-preserved meat products, while PhIP can be formed in roasted chicken and other meats cooked at high temperatures. These observations suggest that coffee may, to some extent, have a protective action against these ingested biogenic amines, when taken after the meal.

Ludwig and colleagues (2014) tested the ROS activity of several model systems consisting of different combinations of compounds noted as effective in the literature (seven furans, three pyrroles and two thiophenes), as well as the scavenging activity of individual components. These same authors observed that models that combined all compounds showed increased activities. Among individual compounds, 2-methyl-tetrahydrofuran-3-one and 1-methylpyrrole were found to be the most active molecules, while aldehydes and ketones showed no relevant antioxidant activity under the same experimental conditions.

Lopez-Galilea and co-workers (2008) observed that coffees with high anti-oxidant capacity may show smaller amounts of volatile compounds. Thus, no evidence exists in support of a direct relationship between overall coffee aroma and its potential anti-cancer activity.

Coffee also exists in decaffeinated and instant forms. Instant coffee, also called soluble coffee, is a beverage produced from a concentrated solution of brewed coffee in a way that allows it to be later rehydrated. It is convenient for its easy and rapid preparation (simply add warm water) and has a long shelf life if kept dry. The first patent describing the production of instant coffee dates from 1890 (Strang 1890). It soon became popular in the USA, where the production process was further developed and instant coffee began to be marketed in 1910. In the 1930s, a well-known European multinational company launched a product that is still successful today, as a result of their proprietary innovative production method.

As with regular coffee, the green coffee bean itself is first roasted to bring out flavour and aroma. Rotating cylinders containing the green beans and hot combustion gases are used in most roasting plants. The roasting begins when the temperature of beans reaches 165 °C, accompanied by a popping sound similar to that produced by popcorn. These batch cylinders take about 8–15 min to complete roasting, with about 25–75 % efficiency. Coffee roasting using a fluidised bed only takes from 30 s to 4 min and operates at lower temperatures that allow greater retention of the aroma and flavour.

Subsequently, the ground coffee is dissolved in hot water and steam inside percolation columns where it undergoes an extraction to obtain a concentrated coffee solution. After filtration, water is removed via freeze-drying or spray-drying procedures. The final concentration of caffeine is generally lower than that of brewed coffee. In the analysis of composition data for brewed and instant coffee, retrieved from the food databases used in the present work, no major differences could be found, as the most important compounds are aroma compounds and other small molecules, not discriminated in food databases. Regarding the health benefits, available studies are not sufficient to allow for comparisons or to draw any type of conclusion.

To produce decaffeinated coffee, caffeine is isolated by solvent extraction from green coffee beans before the roasting process. An alternative high-pressure extraction procedure (Zosel method) involves supercritical carbon dioxide and allows a more selective and toxicologically safer decaffeination. Decaffeinated coffee, besides losing almost all caffeine, also loses some aroma; in addition the health-protective effects (mentioned above for regular coffee) may also decrease.

It is noteworthy that besides coffee and tea, many other drinks (iced tea, energy drinks, cola and other carbonated drinks) include caffeine as an ingredient. More than 20 different type of these beverages can be found in the US market (Homan and Mobarhan 2006); consequently, the daily ingestion of caffeine can be significant, and children are included among frequent consumers of some of these drinks.

8.3 Tea and Herbal Infusions

Tea is the drink obtained from the infusion of the leaves of *Camellia sinensis*, originally from China and Japan. It was brought to Europe by the Portuguese from Japan in the sixteenth century. The well-known British habit of drinking tea was introduced by a Portuguese princess, Catarina de Bragança, when she married Charles II of England. The drinking of tea, with its associated rituals, quickly became very popular among wealthy people in northern Europe. Currently in the UK, tea is perceived as a national beverage, while tea consumption is quite moderate in southern Europe, including Portugal. While black tea is predominant in Europe, green tea is preferred in the south of the Mediterranean sea. Green tea also originated from China, spread to the Middle East and became very popular in Arabic countries by the fifteenth century, where it still holds the status of a staple drink.

Besides black and green teas, several other types of teas are produced, depending on processing operations, such as 'white' and 'oolong' teas. Since they are not common in the Mediterranean basin, only green and black teas are discussed in the present work.

To produce green tea, after the leaves are harvested and only slightly wilted, they are steamed for colour fixation and then shaped and dried, or crushed if packed in tea bags.

On the other hand, to produce black tea, leaves are wilted indoors, after which they undergo an oxidation process (a polyphenol oxidase-mediated oxidation) also called 'fermentation', before leaves are rolled or crushed and finally dried. In both types of tea, the best-quality leaves are generally shaped and packed as rolled whole leaves, while damaged or poorer quality leaves are ground and subsequently packed in tea bags.

Black tea is marketed as single-variety products, named after the place of cultivation, such as Darjeeling, Assam or Ceylon, or blends and flavoured preparations, such as Earl Grey (consisting on the addition of bergamot—a citric fruit originating from Italy).

In Morocco and other Arabic countries, fresh mint leaves are added to hot green tea. Optionally, pine tree nuts (slightly fried in their own oil) are also placed in the bottom of the typical glasses, over which tea is poured.

Instant tea and industrial iced tea are alternative forms of marketing tea, though they have different compositions and properties.

8.3.1 Black Tea

An infusion of black tea, prepared with tap water, contains very low amounts of or no macronutrients, minerals or vitamins (USDA 2015). On the other hand, black tea contains, on average, 20 mg/100 g of caffeine and 2 mg/100 g of theobromine[2]

[2] IUPAC name: 3,7-dimethylpurine-2,6-dione.

(Fig. 8.1) as well the flavonoids (+)-catechin and epigallocatechin-3-gallate, in concentrations of 1.5 and 5.9 mg/100 g, respectively. In addition, black tea contains several derivatives of these compounds: 1.2 mg/100 g of (+)-gallocatechin, 8.0 mg/100 g of (−)-epigallocatechin, 2.1 mg/100 g of (−)-epicatechin and 5.9 mg/100 g of (−)-epicatechin-3-gallate (compounds present in many edible plants, as discussed in Sects. 5.5, 5.6 and 5.7). Non-polymeric catechins represent 20–30 % of the flavonoid fraction, because black tea primarily contains large polyphenols such as theaflavins and thearubigins, representing 60–70 % of the polyphenolic fraction, as well as many unidentified compounds (van Duynhoven et al. 2013). According to the same authors, these black tea-specific flavonoids result from the oxidation process and are generally present in the following concentrations: 1.6 mg/100 g of theaflavin (catechin dimer, formed from the condensation of flavan-3-ols in tea leaves during 'fermentation'), 1.7 mg/100 g of theaflavin-3,3′-digallate, 1.5 mg/100 g of theaflavin-3′-gallate as well as thearubigins that occur at a concentration of 81.3 mg/100 g (USDA 2015), and are thought to result from the condensation of two gallocatechins (epigallocatechin and epigallocatechin-3-gallate). These polymeric compounds are responsible for the characteristic colour and taste of black tea and have molecular weights that may reach 2 kDa (van Duynhoven et al. 2013).

Black tea contains the flavonols kaempferol (1.4 mg/100 g), myricetin (0.4 mg/100 g) and quercetin (2.2 mg/100 g) (Fig. 5.1), as well as proanthocyanidins in the form of monomers (9.3 mg/100 g), dimers (3.7 mg/100 g) and trimers (0.4 mg/100 g) (USDA 2015).

8.3.2 Green Tea

An infusion of green tea, prepared from bags, contains very low amounts of or no macronutrients, minerals or vitamins (USDA 2015). Green tea contains, on average, less caffeine (18 mg/100 g) than black tea and no theobromine. The available register from the USDA database is not as extensive as for black tea; no relevant data could be retrieved from the Portuguese database (INSA 2015). Thus, according to the available information, differences between green and black teas are mostly in the types of polycyclic compounds, namely flavonols, and in their concentration.

Green tea contains higher levels of catechins than black tea, accounting for 30–40 % of extractable solids and 80–90 % of total flavonoids, with (−)-epigallocatechin-3-gallate (EGCG) as the predominant catechin (Lambert and Elias 2010; van Duynhoven et al. 2013). The latter compound has been noted as having nutraceutical properties, as discussed below. Another compound with probable nutraceutical activity is the unusual amino acid, L-theanine, a natural analogue of glutamate, which has been reported in both green and black teas and associated with protective effects against neurodegenerative diseases (Butt et al. 2014; Di et al. 2010; Dimpfel et al. 2007).

Regular tea consumption (in the form of an infusion) has been associated with improved cardiovascular health (Hodgson et al. 2013), with a decrease in the prevalence of type II diabetes (Beresniak et al. 2012) and with controlling obesity (Beresniak et al. 2012; Chen et al. 2009). Moreover, tea is reported to have anti-inflammatory and anti-cancer activities (Di Paola et al. 2005; Geybels et al. 2013; Lambert and Elias 2010), as well as cardioprotective effects (Beresniak et al. 2012; Butt et al. 2014; Hodgson et al. 2013).

Most of the health-promoting effects have been attributed to tea polyphenols, particularly to EGCC (Beresniak et al. 2012; Chen et al. 2009; Dorchies et al. 2006; Lambert and Elias 2010), which is most abundant in green tea. EGCC has been noted as the tea compound with the highest antioxidant activity (Chen et al. 2009; Lambert and Elias 2010), although polymeric phenols found in black tea—theaflavins and thearubigins—are also free radical scavengers and, consequently, may show anti-carcinogenic activity (Geybels et al. 2013).

A major portion of tea polyphenols, mainly of higher molecular weight, predominantly persist in the colon, where they undergo extensive bioconversion by gut microbiota to metabolites that can be further absorbed by the human body. Recent studies point to the positive impact of this high-molecular-weight polyphenol on gut microbiota. As some bacterial groups are more resistant than others toward membrane denaturation by polyphenols, these resistant bacteria could take advantage of available niches left open by susceptible microbes (e.g. promoting *Bifidobacterium* over enteropathogens). This effect can beneficially affect the gut indigenous microbial composition and activity (van Duynhoven et al. 2013).

Tea polyphenols of smaller molecular weight are absorbed mainly at the small intestine and are then subject to phase II metabolism, which likely affects their bioavailability in the human body. On the other hand, microbial degradation may also play an important role in this process (van Duynhoven et al. 2013).

Biochemical and biological studies, prospective cohort studies and double-blind randomised clinical prevention trials tend to show convergent results for the beneficial preventive effects of tea components in various cancers (Beresniak et al. 2012). Black and green tea polyphenols were shown to be bioavailable in the human prostate (Geybels et al. 2013), and hence most likely reaching tumours. Geybels and colleagues (2013) found a significant correlation between regular black tea intake and positive outcomes in prostate cancer, including tumours in advanced stages. According to Lambert and Elias (2010), mechanisms involved may include direct antioxidative effects of catechins (with ROS scavenging and metal chelation activities), induction of endogenous antioxidants and pro-oxidant effects. According to the same authors, pro-oxidant effects of EGCG have been reported in large xenograft tumours, but whether such effects occur in normal or hyperplastic tissue at low doses has not been established.

On the other hand, Chen and co-workers (2009) reported effects on the decrease of body fat and differentiation in adipose tissue mediated by EGCC. The effect on body weight is obviously dependent on the manner of preparing and drinking tea, particularly the addition of sugar. The capacity of tea polyphenols to stimulate some endogenous pathways may explain the moderate weight control and most particularly the anti-cancer protective effects, as the endogenous antioxidant enzymes have a much higher capacity to deal with ROS than the (presumably low concentrated) tea polyphenols (Chen et al. 2009).

Protective effects of tea on cardiac function have been noted. Theaflavins and thearubigins have been observed to act as a safeguard against oxidative stress and help control blood pressure variations, thereby improving cardiac functioning (Butt et al. 2014; Hodgson et al. 2013). There is no consensus on these observations, as a meta-analysis did not find any support for a protective role of black tea on cardiovascular function (Wang et al. 2011). The limited data available on green tea indicate a possible association of green tea consumption with a reduced risk of cardiovascular diseases. According to Wang and colleagues (2011), more studies are needed to clarify this association.

Direct antioxidant effects of tea polyphenols may play an important role in inflammatory disorders. For example, in conditions of high oxidative stress (e.g. ulcerative colitis, hepatitis), tea polyphenols may be able to directly react with and scavenge free radicals and thus block tissue damage (Lambert and Elias 2010).

In addition to polyphenols, theanine has been noted as contributing to the beneficial effects of regular tea intake. It appears that L-theanine is absorbed from the small intestine of animals via a sodium-coupled active transport process crossing the blood–brain barrier (NCBI 2015c). Once in the brain, it acts as a neurotransmitter (Butt et al. 2014) and may provide effective prophylaxis and treatment for Alzheimer's disease (Di et al. 2010). Tea may also improve cognition. Dimpfel and colleagues (2007) measured in vivo electrical hippocampus activity of isolated compounds and mixtures of tea extracts. They observed consistent variations in a concentration-dependent manner, indicating the involvement of several active principles in the action of tea on electrical brain activity, concluding that tea may improve cognition and concomitant mental relaxation in humans.

These scientific studies and folk beliefs were the support for the development of such industrial products as tea extracts, which are consumed in concentrated forms rather than simply making an infusion. The number and variety of dietary supplements and food ingredients containing green tea extract has been increasing in the market, particularly in the USA.

It is noteworthy that tea (defined as an infusion of *C. sinensis* leaves) has been recognised by the US Food and Drug Administration (FDA) as a 'generally recognised as safe' (GRAS) food because of its long history of safe consumption. Green tea extracts (concentrated solid extracts of polyphenols—catechins, flavonoids, tannins—caffeine and other minor constituents) deriving from a GRAS foodstuff, legally also benefit from GRAS status in the USA, and no upper safe limit has been set. Tea extracts are further included as ingredients (at various concentrations) in dietary supplements, beverages, snacks, etc. It is noteworthy that cases of hepatic toxicity have been reported as caused by the continued ingestion of enormous quantities of these tea extracts (Patel et al. 2013).

In short, both antioxidant and pro-oxidative effects are reported for tea (in vitro and in vivo experiments) and attributed to tea polyphenols, the most promising effect involving the prevention and probable involvement in tumour remission, and attributed to EGCC. The underlying mechanisms and impact may depend on the stage of carcinogenesis. Increased endogenous antioxidant capacity could be more important prior to carcinogen exposure, whereas pro-oxidant cell-killing effects may be more important in clearing transformed cells from the body and limiting tumour growth. Despite many scientific studies, in 2015, the FDA and the European Food Safety Authority (EFSA) considered that there was (still) very little scientific evidence in support of the claim of anti-cancer activity for tea; more studies are required to provide solid evidence and explain the underlying mechanisms.

8.3.3 Herbal Infusions

Herbal infusions, sometimes called medicinal teas, have been used since immemorial times, mostly for medicinal purposes. Many herbs are currently dried and blended or concentrated as different extracts to be marketed by their alleged benefits in the alleviation of some ailments. Although dietary supplement manufacturers cannot legally make health claims without the approval of a new drug application, unsubstantiated medical uses for many botanical dietary supplements are well known and promoted by media and on the Internet. Ongoing studies on both beneficial health effects and risk assessment are expected to shed some light on these alternative medicines. Meanwhile, this expanding health food industry may trigger human toxicity cases from overdoses of herbal extracts. Databases by the EFSA (2012) ('Compendium of botanicals reported to contain naturally occurring substances of possible concern for human health when used in food and food supplements') and the FDA (2015) ('FDA poisonous plant database') are recommended as a source of updated information on the toxicity levels of compounds extracted from herbs, including those extracted and concentrated from common teas.

The most popular herbal infusions in the countries specified herein (Fig. 1.1) are chamomile (*Chamaemelum nobile*), lemon balm and balm mint (var. of *Melissa officinalis*), lemon verbena (*Aloysia triphylla*) and lemon peel (*Citrus limon*).

8.3.3.1 Chamomile (*Chamaemelum nobile*)

Chamomile is a low perennial plant with daisy-like white flowers. It is widely cultivated and mostly consumed as an herbal infusion and used by the cosmetic industry (for its essential oil) and food industry (as a flavouring agent). Chamomile flowers are noted as being rich in carbohydrates, protein and oils, as well as containing several microelements such as chromium, cobalt, manganese, selenium and iron, as well as inositol, and vitamins (ascorbic acid, β-carotene, thiamine and riboflavin) (Duke 1992). This information has more recently been confirmed and updated (Guimarães et al. 2013). Infusions of *Chamaemelum nobile* may provide tocopherols, carotenoids and essential fatty acids (C18:2 n6 and C18:3 n3), as well as phenolic compounds (flavonols and flavones, phenolic acids and derivatives).

Such infusions also contain organic acids (oxalic, quinic, malic, citric and fumaric acids), and 5-*O*-caffeoylquinic acid,[3] a neochlorogenic acid derivative also present in coffee. Some bioactivities have been attributed to 5-*O*-caffeoylquinic acid, such as anti-inflammatory, antioxidant, analgesic, anti-pyretic, anti-carcinogenic and anti-thrombotic (Fuentes et al. 2014; Jiang et al. 2000; Santos et al. 2006; Shibata et al. 1999).

Antioxidant properties have been noted for chamomile infusion (Guimarães et al. 2013; Kogiannou et al. 2013; Matić et al. 2013) and are generally attributed to the same phytochemicals as found in coffee, tea and other vegetable sources. It is noteworthy that chamomile infusion is noted to have anti-inflammatory activity via a mechanism similar to that attributed to non-steroidal anti-inflammatory drugs (Srivastava et al. 2009).

8.3.3.2 Balm (*Melissa officinalis*)

Balm exists in several varieties, from lemon balm to mint, with slightly different scents and flavours. It is a perennial herb in the mint family *Lamiaceae*, native to the Mediterranean region. Each different balm variety has its specific use, as is the case of mint in flavouring green tea, lemon balm in herbal infusions and other varieties in seasoning of certain dishes. The flavour of balm comes from the balance between citronellal,[4] geranial (or trans-citral),[5] linalyl acetate (or bergamol)[6] and isocaryophyllene[7] (the last of these chemicals is represented in Fig. 8.2).

Citronellal and geranial are food additives permitted for direct addition to food for human consumption, by the FDA (2014) and the EFSA (EC 2012).

Linalyl acetate or bergamol is an essential oil that is also present in lavender, being one of its main constituents (NCBI 2015d). Linalyl acetate is noted as having anaesthetic properties, particularly via inhalation. Nevertheless, when ingested, linalyl acetate is thought to be hydrolysed by gastric juice to linalool and acetic acid, with a half-life in the stomach of 5.5 min (and 52.6 min in the intestine). Therefore, it is expected that linalool will enter the systemic circulation upon oral uptake, after which it is probably converted to geraniol, further converted to the corresponding carboxylic acid presumably by an aldehyde dehydrogenase, to finally enter the Krebs cycle for complete degradation. The LD_{50} (oral, rat) of bergamol is 14.55 g/kg bw; beneficial properties are not substantiated.

On the other hand, isocaryophyllene is a natural bicyclic sesquiterpene that is a constituent of the essential oils of many plants, including *Cannabis sativa* (NCBI 2015e). Isocaryophyllene/caryophyllene is notable for having a cyclobutane ring, a rarity in nature (Fig. 8.2). Caryophyllene acts as a non-steroidal anti-inflammatory

---

[3] IUPAC name: (1R,3R,4S,5R)-3-[(E)-3-(3,4-dihydroxyphenyl)prop-2-enoyl]oxy-1,4,5-trihydroxy-cyclohexane-1-carboxylic acid.

[4] IUPAC name: 3,7-dimethyloct-6-enal.

[5] IUPAC name: (2*E*)-3,7-dimethylocta-2,6-dienal.

[6] IUPAC name: 3,7-dimethylocta-1,6-dien-3-yl acetate.

[7] IUPAC name: (1*R*,4Z,9*S*)-4,11,11-trimethyl-8-methylidenebicyclo[*7.2.0*]undec-4-ene.

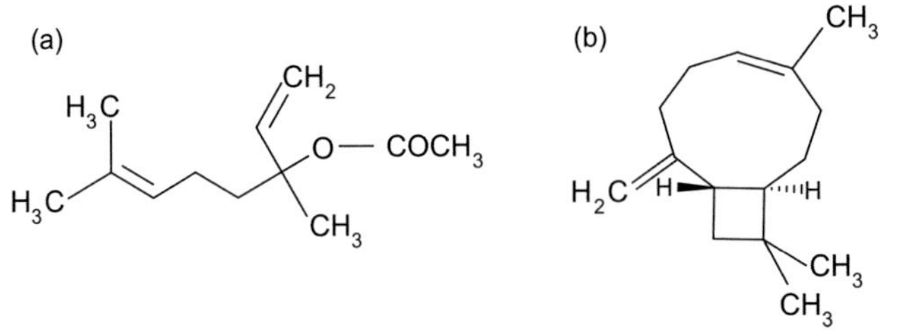

Fig. 8.2 Molecular structure of certain terpenes present in *Melissa officinalis* and mainly responsible for its aroma. Linalyl acetate or bergamol (**a**) is an essential oil, also present in lavender. Isocaryophyllene (**b**) is an authorized food additive, chemically a cannabinoid terpene, notable for having a cyclobutane ring. BKchem version 0.13.0, 2009 (http://bkchem.zirael.org/index.html) was used to draw these structures

agent; it also exhibits analgesic, anti-pyretic and platelet-inhibitory actions through the inhibition of prostaglandin synthesis (NCBI 2015f).

Some phenolic compounds have also been identified in balm infusions, such as caftaric acid, caffeic acid, *p*-coumaric acid, ferulic acid, luteolin, apigenin, quadranoside III, salvianic acid A and rosmarinic acid[8] (Hanganu et al. 2008; Mencherini et al. 2007). These polycyclic compounds share bioactive properties with similar chemical entities, herein described, such as potent ROS scavenging activity.

8.3.3.3 Verbena (*Aloysia triphylla*)

Lemon verbena is a perennial flowering plant in the family *Verbenaceae*, native to western South America. The Spanish and the Portuguese introduced it to southern Europe in the seventeenth century. It has been used since in herbal infusions and as a seasoning ingredient. It is also popular in folk medicine as a relaxing and anti-inflammatory agent.

Its aqueous extract contains considerable amounts of flavonoids and phenolic acids, mainly *trans*-citral, nerol or *trans*-geraniol, geraniol and verbascoside[9] (Abderrahim et al. 2011; Funes et al. 2009). Verbascoside is a polyphenolic compound derived from cinnamic acid by esterification, and noted as being an anti-neoplastic agent with antioxidant, anti-inflammatory and immunomodulatory activities (Abderrahim et al. 2011; Akbay et al. 2002; Funes et al. 2009; Speranza et al. 2010).

[8] IUPAC name: (2*R*)-3-(3,4-dihydroxyphenyl)-2-[(*E*)-3-(3,4-dihydroxyphenyl)prop-2-enoyl]oxypropanoic acid.

[9] IUPAC name: [(2*R*,3*R*,4*R*,5*R*,6*R*)-6-[2-(3,4-dihydroxyphenyl)ethoxy]-5-hydroxy-2-(hydroxymethyl)-4-[(2*S*,3*R*,4*R*,5*R*,6*S*)-3,4,5-trihydroxy-6-methyloxan-2-yl]oxyoxan-3-yl] (*E*)-3-(3,4-dihydroxyphenyl)prop-2-enoate.

8.3.3.4 Lemon Peel Infusion (*Citrus limon*)

The lemon tree is native to Asia and found worldwide. Lemon trees seem to have entered Europe near southern Italy during the times of the Roman Empire. The lemon was first recorded in the literature in a tenth century Arabic treatise on farming, and used as an ornamental plant in early Islamic gardens. It was distributed widely throughout the Arab world and the Mediterranean region between 1000 and 1150. The first substantial cultivation of lemons in Europe began in Genoa in the middle of the fifteenth century. Christopher Columbus later introduced the lemon to the Americas. It is noteworthy that nowadays Italy and Morocco are among the largest world exporters of lemons and limes, according to United Nations Food and Agriculture Organization (FAO) trade statistics (FAO 2015).

Lemon is highly appreciated for its juice and peel, with many culinary and industrial applications. Lemon peel is often used to make an infusion to which medicinal properties are attributed. The so-called lemon tea is a well-known home medicine, frequently made with honey and used to alleviate flu and cold symptoms, such as a sore throat.

Lemon peel (raw) contains some sugars (4.17 %), microelements such as zinc (Zn: 0.25 mg/100 g) and selenium (Se: 0.7 μg/100 g), vitamins such as ascorbic acid (129 mg/100 g), thiamine (0.06 mg/100 g), riboflavin (0.08 mg/100 g), niacin (0.4 mg/100 g), pantothenic acid (0.319 mg/100 g), vitamin B_6 (0.172 mg/100 g), folates (13 μg/100 g), as well as α- and β-carotenes (1 and 7 μg/100 g, respectively), lutein + zeaxanthin (18 μg/100 g) and α-tocopherol (0.25 mg/100 g). The lipid phase (0.3 %) is formed by saturated fatty acids (SFA) 14:0, 16:0 and 18:0, in a total amount of 0.039 g/100 g; monounsaturated fatty acids (MUFA) 16:1 and 18:1, in a total amount of 0.011 g/100 g; and polyunsaturated fatty acids (PUFA) 18:2 and 18:3 in a total quantity of 0.089 g/100 g (USDA 2015).

Lemon peel contains 35 mg/100 g of non-discriminated phytosterols, associated to the lipid fraction (USDA 2015). Terpenes constitute another important class of compounds that have been recovered from the lemon peel's essential oil, and include a mixture of optical isomers of limonene, with a clear predominance of D-isomer,[10] γ-terpinene,[11] β-pinene,[12] β-myrcene,[13] citral and geranial (Vekiari et al. 2002).

Guimarães and colleagues (2010) found that peel polar fraction showed higher contents of phenolics, flavonoids, ascorbic acid, carotenoids and reducing sugars than the volatile fraction. Therefore, these compounds are expected to be more abundant in the hot water infusion, although some volatiles are also likely to be extracted by the boiling water.

Thus, limonene is a non-polar cyclic terpene, extracted in the essential oil fraction (volatiles) of a variety of edible plants, most particularly citric fruits, as

[10] IUPAC name: (4*R*)-1-methyl-4-prop-1-en-2-ylcyclohexene.

[11] IUPAC name: 1-methyl-4-propan-2-ylcyclohexa-1,4-diene.

[12] IUPAC name: 6,6-dimethyl-4-methylidenebicyclo[*3.1.1*]heptanes.

[13] IUPAC name: 7-methyl-3-methylideneocta-1,6-diene.

is the case of lemon, from which the compound received its name (NCBI 2015g). D-isomer is responsible for the lemon scent, while L-isomer (present in limited amounts) has a pine/turpentine odour. Limonene has been considered as an anti-carcinogen. Humans easily metabolise it and their metabolites have been tested for their carcinogenic potency (NCBI 2015g).

Limonene is easily absorbed by humans and readily metabolised, perillic acid being the main metabolite (Chow et al. 2002). D-limonene and its derived metabolites have been shown to possess cancer chemotherapeutic and chemopreventive efficacy in various preclinical model systems; Chow and co-workers showed the bioavailability of perillic acid shortly after limonene intake. In a study with breast cancer patients, Miller and colleagues (2013) found that limonene accumulates in breast tissue, reaching high concentrations, whereas ferulic acid was mostly found in blood plasma. These authors observed a positive effect of limonene intake on some cancer biomarkers, and concluded that the direct effect on the tumour is most probably due to limonene, which reaches metastatic cells. On the other hand, perillic acid most likely has limited bioactivity in the target tissue. Other polar phenolic compounds may also contribute to cancer prevention.

In addition to the valuable contribution of the volatile fraction (where limonene is included), properties and health benefits of other lemon peel components such as ascorbic acid, carotenoids and polyphenolic compounds, herein discussed, may help explain most of the observed health-promoting effects of lemon tea.

In short, herbal infusions are closely related to folk medicine. Some health-beneficial effects have been consistently reported, while other beliefs have no scientific foundation. As mentioned previously, the composition of any vegetable food is rarely known in detail because it is a complex mixture of countless chemical entities. Moreover, plant composition varies with many factors such as agronomic practices, weather conditions or processing. Even for extensively studied plants, only a small fraction of the constituents have been isolated and identified; some active compounds probably remain unknown, while a few compounds have been extracted and used in pharmaceutical preparations. Despite marketing campaigns, herbal extracts of undefined composition (health supplements) cannot hold health claims and, in some cases, may pose toxicological issues, particularly when misused.

8.4 Wine

Wine has been consumed since ancient times, most probably because of its psychoactive effects, associated with health benefits. Hippocrates referred to the value of wine as a medicine, considering it to be vital to a healthy diet.

This alcoholic beverage is obtained from grapes of (mainly) *Vitis vinifera* fermented by the yeast *Saccharomyces cerevisiae* (sometimes including other related microorganisms, as well as lactic acid bacteria [LAB]), which consumes sugars and acids in the grape and converts them into alcohol, carbon dioxide and secondary metabolites. Further enzymatic and chemical reactions contribute to the aroma of wine.

The genus *Vitis* accounts for 60 species that can be found in temperate zones of the Northern Hemisphere (EOL 2015). *Vitis vinifera* is the only species native to Europe, and specifically to the Mediterranean region. Winemaking is a mixture of science, history and art, and is best known as oenology.

Vineyards are thought to be one of the oldest crops in the world, probably dating from the late Neolithic or Bronze age (EOL 2015). Many traditions and symbolisms are associated with wine, mostly of a religious nature.

In Ancient Egypt, red wine was used in religious ceremonies to represent blood. It played more important roles in Greek and Roman civilisations, associating respectively, the divinities Dionysius and Bacchus to the cult of wine. Wine has been at the centre of ceremonies, festivals and celebrations ever since. In Judaism, wine is used in sacred rites such as the Kiddush, a blessing recited over wine (or grape juice) to sanctify the Shabbat. On Pesach (Passover), it is an obligation of adults to drink four cups of wine. Wine (along with bread) plays a key role during Christian mass (in the sacred rite of Eucharist).

On the other hand, in Islam, alcoholic beverages, including wine, are forbidden under most interpretations of Islamic law, for the sake of avoiding alcohol intoxication.

Vineyards expanded in Europe by the action of the ancient Romans who planted vineyards near garrison towns; consequently, wine could be produced locally rather than shipped over long distances, thus expanding this crop to central Europe (France and southern Germany) and to the Middle East. Some of these areas are now world renowned for wine production. Later, in medieval Europe, these vineyards were preserved by the Catholic Church because wine was required for the Mass, and many vineyards became associated with monasteries and castles. Some of the designations (or brand names) still maintain. Portuguese and Spanish colonisers introduced the culture of wine (and *V. vinifera* cultivars) the American continent, from where it later reached Australia.

Nearly three-quarters of the world's commercial grape production is devoted to wine grapes (EOL 2015). Wine grapes are smaller, with harder skin and softer flesh than table grapes. They also contain more sugars and acids and can be red or white. On average, wine grapes contain a large amount of sugar (15–25 %), with roughly equal amounts of glucose and fructose and only a trace of sucrose. Vitamin C content is low (around 3 mg/100 g). Fruits also contain tartaric and malic acids (EOL 2015) in similar concentrations (around 0.5 g/100 g). The pigmentation is due to the presence of grape skin anthocyanins that are conveyed to red wines (Sect. 5.7.1).

Wine may be made from a single cultivar (varietal wine) or from a blend. There are about 10,000 known cultivars of *Vitis vinifera*, of which Alvarinho/Albariño, Cabernet Sauvignon, Garnacha, Periquita or Sangiovese are examples.

It should be noted that the basic difference between red and white wines is that the skins are included in the former during the wine-making process. Thus, only dark-skinned grapes can produce red wine, but dark-skinned grapes can also be used to produce white wines if the skin is removed.

The quality of grapes determines the quality of wines more than any other factor, and it is in its turn affected by weather during the growing season, soil minerals and acidity, time of harvest, etc.

To ensure a good wine, care is taken during harvesting, which is commonly done by hand. Manual harvesting uses knowledgeable labour to choose ripe clusters and to leave behind defective ones, thus acting as a primary quality-control step. Mechanical crushers are used and stems are separated before fermentation of the must to avoid excessive increases in tannin levels.

Alcoholic fermentation traditionally relies on yeasts that are normally present in the grapes and winery environments; to avoid unpredictable results and better control the fermentation, cultured yeast (isolated from previous successful fermentations) can be added to the must. The dominant yeast strains are vital for wine quality, because of their ability to inhibit undesirable microbial populations and their production of secondary metabolites that will contribute directly or indirectly to the wine ‘bouquet’. Different yeast strains are usually involved because of their different tolerance to ethanol, allowing the fermentation of sugars to be completed.

In certain wines, malolactic fermentation follows or partially overlaps primary alcoholic fermentation. This step is carried out by ethanol-tolerant LAB, also originating from grapes, such as *Oenococcus oeni*, and some strains of *Lactobacillus* spp. and *Pediococcus* spp. Malolactic fermentation consists on the conversion of malic acid (of tart/green apple taste) to lactic acid that communicates a softer buttery flavour and improves wine ‘body’. As with other fermentations, secondary metabolites are very important, even if present in only trace amounts.

Microbial fermentations traditionally take place in wine cellars during fall/autumn, at temperatures ranging from 15 to 20 °C, after which wine undergoes stabilisation and clarification to separate tartrate crystals that form easily in the cold. The wine is then evaluated and classified as to its destiny: immediate bottling or aging in oak wood. Oak can be added as chips to a stainless steel vessel or wine can be contained in a wooden barrel instead. The first process is mainly used in cheaper wines for its convenience. Preservatives, such as sulphur dioxide or potassium metabisulphite, are added during the early maturation stage and are later eliminated. Red wines usually undergo at least 1 year of ageing; white wines need shorter periods before bottling. Good-quality wines (judged on their organoleptic properties and some chemical parameters) can be preserved longer in oak, as ‘vintages’.

Good-quality wines are bottled in dark glass and sealed with cork to better protect their features. Alternative containers and seals (mainly made of polymeric materials) are only used for cheap wines as they cause short-term alterations in flavour and have not been considered suitable to preserve bottled wine for long periods. Winemaking is a good and profitable example of how to take advantage of crop variations, with selected crops and some vintages reaching high prices in the market.

In Europe, well-established mechanisms ensure the authenticity of prominent wines and regulate their manufacture and trade. This type of mechanism was first introduced in Portugal in the eighteenth century and first applied to 'Vinho do Porto'. Nowadays, national organisations are responsible for the control of the quantities of each type of wine and their specific manufacturing requisites, as well as for fraud prevention. European wines are mostly known after their region of origin and year of bottling, and are labelled accordingly.

The world most famous wines are French, Portuguese, Spanish and Italian; however, it is noteworthy that some wineries have been evolving in California (USA) and Australia to reach world-class status (EOL 2015). It is actually impossible to define a generic 'wine', since good wines are not industrial products and the composition is highly variable, for the reasons explained above. Nevertheless, for the sake of simplicity, only the most relevant wine compounds are described and their role in human health is discussed, as if an average wine existed.

8.4.1 White Wine

White wine (current, generic designation) provides 72–82 kcal/100 g, mainly from 1.2–2.6 % of reducing sugars (that remain after fermentation), ethanol (9.6–10.3 g/ 100 g) and some protein (<0.1 %). Minerals such as calcium (Ca), phosphorus (P), magnesium (Mg), iron (Fe), Zn, copper (Cu), manganese (Mn), Se and fluorine (F) are recorded in about the same range in both databases (INSA 2015; USDA 2015).

Water-soluble B vitamins are present in white wine: thiamine (0.005–0.02 mg/ 100 g), riboflavin (0.010–0.015 mg/100 g), niacin (0.060–0.108 mg/100 g), B_6 (0.020–0.050 mg/100 g) and pantothenic acid (0.045 mg/100 g); the USDA database also refers (in 2015) to the presence of phylloquinone (0.4 μg/100 g). No fibres or lipids are registered in any of the food composition databases. Indicative figures on flavonoids are provided for white wine (USDA 2015). Thus, malvidin (an anthocyanidin) is present at about 0.1 mg/100 g, and the flavonols catechin and epicatechin are present at concentrations of 0.8 and 0.6 mg/100 g, respectively. In addition, proanthocyanidins are found as monomers (0.6 mg/100 g) and dimers (0.2 mg/100 g).

***White wine**. Wine has been consumed since ancient times, most probably because of its psychoactive effects, associated with health benefits. Nowadays, producing and drinking wine involves science, culture and art. Wine contains flavonols and proanthocyanidins or tannins, which are at the centre of reactions occurring during ageing. In manufacturing white wine, the contact of extracted juice with the grape skins is minimized. Consequently, white wine contains less tannin than red wine and virtually no anthocyanins, which are coloured. Photo reprinted with kind permission from T. N. Wassermann*

The skin of grapes accumulates higher levels of polyphenols than fruits and teas, namely tannins and resveratrol, which protect the grapes from fungal diseases and sun damage. These compounds are then extracted to must during wine making, particularly in red wine because must is in contact with the grape skin for a long period of time, thus allowing a more complete extraction of these compounds, and also including many other polyphenols.

8.4.2 Red Wine

According to the above-mentioned databases, red wine (current and generic designation) provides 65–85 kcal/100 g from 0.20–0.62 % of reducing sugars, 0.07–0.1 % of proteins and 9.2–10.6 % of alcohol. As with white wine, no fibres or lipids are found, but the same water-soluble B-complex vitamins and minerals are registered. In addition, red wine contains some β-carotene (1 μg/100 g), lutein + zeaxanthin (6 μg/100 g) and phylloquinone (0.4 μg/100 g). The diversity and levels of flavonoids are much higher than in white wine, for the reasons described above. The USDA database (2015) registers the presence in red wines of the following anthocyanidins (Fig. 5.4): petunidin (2.0 mg/100 g), delphinidin (2.0 mg/100 g), malvidin (13.8 mg/100 g), peonidin (1.2 mg/100 g) and cyanidin (0.2 mg/100 g), as well as the flavonols catechin (7.1 mg/100 g), epicatechin (3.8 mg/100 g) and gallocatechin (0.1 mg/100 g). In addition, red wine contains the flavanones hesperetin (0.6 mg/100 g) and naringenin (1.8 mg/100 g), typical of citrus fruits and important for the 'bouquet' of wines. The flavonols kaempferol (0.1 mg/100 g), myricetin (0.4 mg/100 g) and quercetin (1.0 mg/100 g) are also present, as are the proanthocyanidins monomers (16.6 mg/100 g), dimers (20.5 mg/100 g), trimers (1.8 mg/100 g), 4–6mers (6.7 mg/100 g), polymers with 7–10 units (5.0 mg/100 g) and larger molecules >10 units (11.0 mg/100 g). Tannin fraction is at the centre of the chemical reactions that occur during wine ageing.

Red wine. *Compared with white wine, red wine contains more tannins, which undergo a series of chemical reactions during ageing that result in wine's 'bouquet'. The colour of red wine is acquired through contact with the skin of red-coloured grapes during wine-making operations. The level of anthocyanins and associated compounds is then higher than in white wines. The presence of resveratrol in red wines is also noteworthy because of its documented effects in the prevention of coronary heart disease. Photo reprinted with kind permission from T. N. Wassermann*

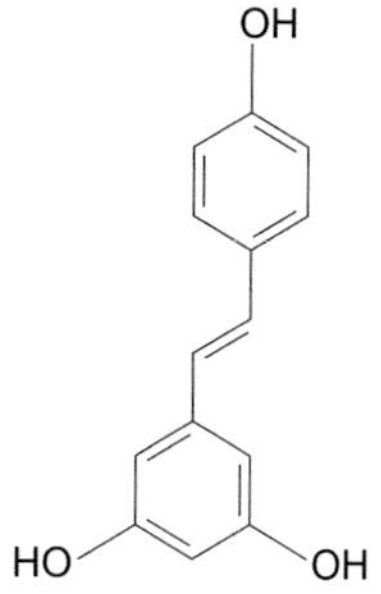

Fig. 8.3 Structure of resveratrol, a polyphenolic phytoalexin found in red wines, to which nutraceutical properties have been associated. BKchem version 0.13.0, 2009 (http://bkchem.zirael.org/index.html) was used to draw these structures

Resveratrol,[14] a polyphenolic phytoalexin, is naturally present in *cis*/*trans* isoforms, both of which may be glucosylated, with the *trans*-isomer being the biologically active one (Fig. 8.3). Resveratrol has been noted as primarily responsible for wine's health-promoting effects, and its concentration is higher in red than in white wines. Resveratrol benefits include the prevention of cardiovascular diseases, supported by in vitro and in vivo observations as well as by epidemiological studies (Goldfinger 2003; Opie et al. 2011; Wu and Hsieh 2011). Resveratrol has also been noted as activating the longevity gene and preventing several degenerative diseases (Lekli et al. 2010; Xie et al. 2015).

Red wine also contains melatonin (*N*-acetyl-5-methoxytryptamine), a biogenic amine with an (oral) $LD_{50} > 1230$ mg/kg bw in mice or >3200 mg/kg bw in rats. At low dosages, melatonin has been noted to have antioxidant and cardioprotective activities. Moreover, an oral administration of 0.1–0.3 mg/day at midday has been shown to lead to an increased duration of night sleep (European Commission 2010; Opie et al. 2011), acting via a different mechanism from those of medicines for the control of sleep disturbances (Arbon et al. 2015).

In addition, proanthocyanidins have been noted as having strong antioxidant activity, thus contributing to the health-protective effects attributed to wines. In vitro studies have reported that proanthocyanidins are potent scavengers of peroxyl and hydroxyl radicals, which are generated in the reperfused myocardium after ischemia in a dose-response manner, suggesting its involvement in the cardioprotective effects attributed to moderate wine intake (Pataki et al. 2002). It is noteworthy that proanthocyanidins are found in much higher concentrations in red wines than in white wines.

Wine (in its various kinds) is strongly linked to the MD, shaping the habits and the landscapes of southern European countries. On the other hand, the westernisation of the diet has led to an increase in beer consumption, partly replacing wine, although with different drinking patterns in terms of consumers' profiles and time of intake.

[14] IUPAC name: 5-[(*E*)-2-(4-hydroxyphenyl)ethenyl]benzene-1,3-diol.

Data from the FAO Food Balance Sheet (FBS) (FAO 2015) were analysed concerning the availability of wine and beer over time (from 1961 to 2011) in the countries that are the subject of the current review (Fig. 1.1). For reasons already mentioned, no time-based analysis was possible for Croatia.

It can be observed that in 1961, the per capita availability of wine was higher than in the 2000's first decades, and higher than beer: 65.5 kg/year in Portugal, 59.2 kg/year in Spain and 109.6 kg/year in Italy (world-class wine producers). In terms of total span, wine availability ranged from nearly zero in Morocco and 12.2 kg/year (Cyprus) to 109.6 kg/year (Italy). In the same time span (1960s), beer per capita availability was low in Portugal (4.6 kg/year), Italy (7 kg/year), Greece (5.3 kg/year) and Cyprus (8.5 kg/year) and slightly higher in Spain (13.3 kg/year).

When comparing these figures with 2011 data, marked changes can be seen in the availability of alcoholic drinks in these countries. In Portugal, wine availability decreased from 65.1 to 45.1 kg/capita/year, whereas beer availability increased from 4.6 to 50 kg/capita/year. In Spain, wine availability decreased from 59.2 to 17.2 kg/capita/year, whereas the availability of beer increased from 13.3 to 75.3 kg/capita/year. The increase in availability of beer is also noticeable in Greece (from 5.3 to 39 kg/capita/year) but less marked in Italy, although it is still significant (from 7 to 29.1 kg/capita/year).

Short-term epidemiological studies (Opie et al. 2011; Rosenkranz et al. 2002; Siedlinski et al. 2012; Vidavalur et al. 2006; Wu and Hsieh 2011) have suggested that moderate wine consumption (one to two glasses a day) reduces overall mortality and morbidity, mainly from cardiovascular disease (30–50 %); hypertension, diabetes and certain types of cancer, including colon and lung (reduced by 57 %), basal cell, ovarian and prostate carcinoma (reduced by 50 %). A J-shaped curve relates alcohol intake to mortality, favouring moderate alcohol drinkers compared with non-drinkers or heavy drinkers, independent of diet and other factors (Goldfinger 2003; Stranges et al. 2004). Moreover, some of these studies refer to the coexistence of high-risk dietary factors for coronary heart disease (CHD) with unanticipated low CHD incidence. This phenomenon, which was postulated as associated with low-to-moderate consumption of wine, is known as the 'French paradox'; it was first presented in 1992 by Renaud and De Lorgeril. These authors showed that the mortality rate for CHD in France was paradoxical and unexpectedly lower than in other industrialised countries, such as the USA and the UK, despite similarly high dietary intake of saturated and *trans*-fats. Their results strongly supported the view that in France, the untoward effects of saturated fats are counteracted by the intake of wine. Thus, mechanisms that explain underlying protection have been researched and have increased the public awareness of wine (Renaud and De Lorgeril 1992).

Most of the registered wine benefits are attributed to resveratrol. Since grape skins are removed early in white wine production, only red wines contain noticeable amounts of resveratrol. In vitro and in vivo effects of resveratrol in inhibiting low-density lipoprotein (LDL) oxidation, a key primary event in atherosclerosis, have been registered (Wu and Hsieh 2011).

Secondly, resveratrol was found to inhibit platelet aggregation: platelets are actively involved in the process of haemostasis, by which injury in the vascular endothelium is rapidly repaired, without compromising the fluidity of the blood. Improper regulation or over-reactivity of this repair system can lead to pathological thrombosis. Studies show that abnormal platelet aggregation is indirectly inhibited by resveratrol (Wu and Hsieh 2011), with quercetin also playing a role (Vidavalur et al. 2006).

Thirdly, resveratrol suppresses the proliferation of smooth muscle cells and pulmonary aortic endothelial cells, the migration and proliferation of which in the intima of susceptible vessels is a requisite for atherogenesis (Wu and Hsieh 2011). These authors hypothesised cardioprotection by resveratrol to occur via multi-active genomic mechanisms, as well as nongenomic activity that may involve direct and indirect interaction with target protein *N*-ribosyldihydronicotinamidequinone oxidoreductase. Molecular targets of resveratrol are thought to be diverse and include cytokines, transcription factors, growth factors and apoptosis/cell death-regulating proteins (Wong and Fiscus 2015).

Catalgol and colleagues (2012) refer to resveratrol as being involved in some other protective effects such as improving longevity, protecting against neurodegenerative diseases, helping avoid certain types of cancer and showing anti-diabetes and anti-inflammatory activities. The same authors review some mechanistic hypotheses and influencing parameters.

On the other hand, wine may not be less beneficial than red wine. Apparently, it plays a key role in delaying gastric emptying, possibly interacting with different signalling paths for infarct size reduction. White wine is also noted as promoting long-term survival (Opie et al. 2011).

Drinking patterns should not be neglected when analysing the benefits of consuming alcoholic beverages, including the French paradox: wine consumers tend to be educated and lean, and they view the wine as part of an enjoyable meal and usually they drink moderately all week long. Wine is generally more expensive than beer and some liquors; it also demands the domain of associated rituals and cultural aspects (e.g. the right temperature to serve a certain wine, or how to choose the right type of glass).

Conversely, beer drinkers tend to drink outside of meal times, consume large amounts of beer in convivial weekend meetings and provoke peaks of blood pressure. Moreover, beer drinkers tend to have poorer dietary habits than wine consumers (Ferrières 2004; Opie et al. 2011). Lean wine drinkers, such as the French, are known for eating smaller portions and to eat slower than, for example, Americans. These observations have been systematised and analysed (Rozin et al. 1999, 2003).

There is much evidence that excessive alcohol consumption is associated with increased morbidity and mortality, namely increases in hypertension (Stranges et al. 2004), as well as with work- and traffic-related accidents. Ethanol is essentially deleterious and, even in small doses, leads to the accumulation of homocysteine in plasma, a known cardiovascular risk factor (Badawy 2001). Some authors state that ethanol is deleterious to human health, even when moderately consumed, thus questioning the French paradox (Badawy 2001; Bleich et al. 2001).

Conversely, other authors note that ethanol itself has fundamental benefits on high-density lipoprotein cholesterol, platelets and haemostasis, making it difficult to differentiate benefits unique to wine from those that may also be associated to other alcoholic beverages (Goldfinger 2003). Other studies show that de-alcoholised red wine has cardiovascular-protective effects, hence pointing to a null or secondary role of ethanol in these processes, and making a distinction between different beverages (Opie et al. 2011). Thus, there is no consensus over these questions.

It is relevant to mention the extensive use of wine in Mediterranean cuisine as a seasoning ingredient, particularly in cooking meat and fish. Since ethanol evaporates during cooking, these dishes are also served to children, who soon come into contact with wine polyphenols and other distinctive compounds.

In short, many alcoholic beverages are produced from the fermentation of a wide variety of plants, but none has reached the status of wine, the moderate consumption of which with meals is viewed as elegnt and healthy. Wine consumers tend to drink wine during meals. Beer and spirits are mostly consumed outside meals and are more likely to be associated with addictive behaviours. The compound primarily associated with the recorded beneficial effects of wine is resveratrol, which is mainly present in red wines, to a lesser extent in white wines, and absent in other alcoholic drinks.

Thus, in contrast to other widely consumed alcoholic beverages, which have also been the object of scientific studies, extensive evidence of benefits has been collected for wine, although health claims are not yet recognised by the EFSA, which recommends more studies to substantiate the already collected evidence (EFSA 2010). Thus, whether the apparent protective effects of moderate alcohol intake are the same for all subsets of the population and for all types of alcoholic beverages is still unclear. Underlying mechanisms are still to be clarified.

Nevertheless, it is generally accepted that the MD confers health benefits and that wine with meals is part of that diet, taking into account cultural aspects. Olive oil and wine, coexisting in the MD, may have a synergistic postprandial haemodynamic response (Opie et al. 2011). It is noteworthy that, in general, French and southern Europeans appreciate food and are less concerned with energy intake than with celebrating the pleasure of cooking and serving a meal made with a wide variety of ingredients and flavours, in which wine generally plays a central role.

The Lyon Diet Heart Study showed that a Mediterranean-style diet may reduce the risk of recurrence after a first myocardial infarction by 50–70 %, thus emphasising the role of nutrition in the prevalence (and prevention) of CHD and of many cancers (De Lorgeril et al. 2002). According to the same authors, experts from the American Heart Association and the European Society of Cardiology are now in agreement in recommending the adoption of a Mediterranean-style diet for the prevention of CHD. It was realised that the Lyon trial illustrated the importance of a dietary pattern giving preference to fresh fruit and vegetables, bread and cereals, as well as to fish and plants rich in ALA, thus approximating concepts of the French paradox and the MD.

The present review points towards a complex association of dietary factors with CHD mortality. Differences in coronary mortality rates cannot be ascribed to one nutrient or to one food only. A comprehensive review of these complex interactions would have to be analysed in a broad geographic, climatic, agricultural, historical and socioeconomic context.

References

Abderrahim F, Estrella S, Susin C, Arribas SM, Gonzalez MC, Condezo-Hoyos L (2011) The antioxidant activity and thermal stability of lemon verbena (Aloysia triphylla) infusion. J Med Food 14(5):517–527. doi:10.1089/jmf.2010.0102

Akbay P, Calis I, Undeger U, Basaran N, Basaran AA (2002) In vitro immunomodulatory activity of verbascoside from Nepeta ucrainica L. Phytother Res 16(6):593–595. doi:10.1002/ptr.990

Andersen LF, Jacobs DR Jr, Carlsen MH, Blomhoff R (2006) Consumption of coffee is associated with reduced risk of death attributed to inflammatory and cardiovascular diseases in the Iowa Women's Health Study. Am J Clin Nutr 83(5):1039–1046

Arbon EL, Knurowska M, Dijk DJ (2015) Randomised clinical trial of the effects of prolonged-release melatonin, temazepam and zolpidem on slow-wave activity during sleep in healthy people. J Psychopharmacol 29(7):764–776. doi:10.1177/0269881115581963

Arendash GW, Cao C (2010) Caffeine and coffee as therapeutics against Alzheimer's disease. J Alzheimers Dis 20(Suppl 1):S117–S126. doi:10.3233/JAD-2010-091249

Badawy AA (2001) Moderate alcohol consumption as a cardiovascular risk factor: the role of homocysteine and the need to re-explain the 'French Paradox'. Alcohol Alcohol 36(3): 185–188. doi:10.1093/alcalc/36.3.185

Beresniak A, Duru G, Berger G, Bremond-Gignac D (2012) Relationships between black tea consumption and key health indicators in the world: an ecological study. BMJ Open 2. pii: e000648. doi:10.1136/bmjopen-2011-000648

Bleich S, Bleich K, Kropp S, Bittermann HJ, Degner D, Sperling W, Rüther E, Kornhuber J (2001) Moderate alcohol consumption in social drinkers raises plasma homocysteine levels: a contradiction to the 'French Paradox'? Alcohol Alcohol 36(3):189–192

Butt MS, Imran A, Sharif MK, Ahmad RS, Xiao H, Imran M, Rsool HA (2014) Black tea polyphenols: a mechanistic treatise. Crit Rev Food Sci Nutr 54(8):1002–1011. doi:10.1080/10408398.2011.623198

Catalgol B, Batirel S, Taga Y, Ozer NK (2012) Resveratrol: French paradox revisited. Front Pharmacol 3:141. doi:10.3389/fphar.2012.00141

Chen N, Bezzina R, Hinch E, Lewandowski PA, Cameron-Smith D, Mathai ML, Jois M, Sinclair AJ, Begg DP, Wark JD, Weisinger HS, Weisinger RS (2009) Green tea, black tea, and epigallocatechin modify body composition improve glucose tolerance and differentially alter metabolic gene expression in rats fed a high-fat diet. Nutr Res 29(11):784–793. doi:10.1016/j.nutres.2009.10.003

Chow HHS, Salazar D, Hakim IA (2002) Pharmacokinetics of perillic acid in humans after a single dose administration of a citrus preparation rich in d-limonene content. Cancer Epidemiol Biomarkers Prev 11(11):1472–1476

Corti R, Binggeli C, Sudano I, Spieker L, Hänseler E, Ruschitzka F, Chaplin WF, Lüscher TF, Noll G (2002) Coffee acutely increases sympathetic nerve activity and blood pressure independently of caffeine content role of habitual versus nonhabitual drinking. Circulation 106(23):2935–2940. doi:10.1161/01.CIR.0000046228.97025.3A

de Lorgeril M, Salen P, Paillard F, Laporte F, Boucher F, de Leiris J (2002) Mediterranean diet and the French paradox: two distinct biogeographic concepts for one consolidated scientific theory on the role of nutrition in coronary heart disease. Cardiovasc Res 54(3):503–515. doi:10.1016/S0008-6363(01)00545-4

Di Paola R, Mazzon E, Muià C, Genovese T, Menegazzi M, Zaffini R, Suzuki H, Cuzzocrea S (2005) Green tea polyphenol extract attenuates lung injury in experimental model of carrageenan-induced pleurisy in mice. Respir Res 6(1):66. doi:10.1186/1465-9921-6-66

Di X, Yan J, Zhao Y, Zhang J, Shi Z, Chang Y, Zhao B (2010) L-theanine protects the APP (Swedish mutation) transgenic SH-SY5Y cell against glutamate-induced excitotoxicity via inhibition of the NMDA receptor pathway. Neuroscience 168(3):778–786. doi:10.1016/j.neuroscience.2010.04.019

Dimpfel W, Kler A, Kriesl E, Lehnfeld R (2007) Theogallin and L-theanine as active ingredients in decaffeinated green tea extract: I. Electrophysiological characterization in the rat hippocampus in-vitro. J Pharm Pharmacol 59(8):1131–1136. doi:10.1211/jpp.59.8.0011

Dorchies OM, Wagner S, Vuadens O, Waldhauser K, Buetler TM, Kucera P, Ruegg UT (2006) Green tea extract and its major polyphenol (-)-epigallocatechin gallate improve muscle function in a mouse model for Duchenne muscular dystrophy. Am J Physiol Cell Physiol 290(2): C616–C625. doi:10.1152/ajpcell.00425.2005

Duke JA (1992) Handbook of phytochemical constituents of GRAS herbs and other economic plants. CRC, Boca Raton

EC (2012) Commission implementing Regulation (EU) No 872/2012, of 1 October 2012, adopting the list of flavouring substances provided for by Regulation (EC) No 2232/96 of the European Parliament and of the Council, introducing it in Annex I to Regulation (EC) No 1334/2008 of the European Parliament and of the Council and repealing Commission Regulation (EC) No 1565/2000 and Commission Decision 1999/217/EC. Off J Eur Union L 267(1):1–161

EFSA (2010) EFSA Panel on Dietetic Products, Nutrition and Allergies (NDA). Scientific Opinion on the substantiation of health claims related to various food(s)/food constituent(s) and protection of cells from premature aging, antioxidant activity, antioxidant content and antioxidant properties, and protection of DNA, proteins and lipids from oxidative damage pursuant to Article 13(1) of Regulation (EC) No 1924/20061. EFSA J 8(2):1489–1552. doi:10.2903/j.efsa.2010.1489

EFSA (2012) Compendium of botanicals reported to contain naturally occurring substances of possible concern for human health when used in food and food supplements. EFSA J 10(5): 2663–2723. doi:10.2903/j.efsa.2012.2663

EOL (2015) Vitis vinifera. Available at http://eol.org/pages/582304/details. Accessed 5 Nov 2015

European Commission (2010) Directorate-General for Health and Consumers. Scientific Committee on Consumer Safety. SCCS. Opinion on Melatonin. SCCS/1315/10, 24.3.10. http://ec.europa.eu/health/scientific_committees/consumer_safety/index_en.htm. Accessed 1 Apr 2015

FAO (2015) Trade. Crops and livestock products. FAO Statistics Division. http://faostat3.fao.org/download/T/TP/E. Accessed 20 Feb 2015

FDA (1978) Database of Select Committee on GRAS Substances (SCOGS) Reviews. Report n° 88. CFR section 182.1180

FDA (2014) CFR—Code of Federal Regulations Title 21: 21 CFR 172.515. http://www.accessdata.fda.gov/scripts/cdrh/cfdocs/cfcfr/cfrsearch.cfm?fr=101.22. Accessed 12 Jan 2015

FDA (2015) FDA Poisonous Plant Database. http://www.accessdata.fda.gov/scripts/plantox/. Accessed 1 Jun 2015

Ferrières J (2004) The French paradox: lessons for other countries. Heart 90(1):107–111

Fisk ID, Kettle A, Hofmeister S, Virdie A, Kenny JS (2012) Discrimination of roast and ground coffee aroma. Flavour 1:14. doi:10.1186/2044-7248-1-14

Fuentes E, Caballero J, Alarcón M, Rojas A, Palomo I (2014) Chlorogenic acid inhibits human platelet activation and thrombus formation. PLoS One 9(3):e90699. doi:10.1371/journal.pone.0090699

Funes L, Fernández-Arroyo S, Laporta O, Pons A, Roche E, Segura-Carretero A, Fernández-Gutiérrez A, Micol V (2009) Correlation between plasma antioxidant capacity and verbascoside levels in rats after oral administration of lemon verbena extract. Food Chem 117(4): 589–598. doi:10.1016/j.foodchem.2009.04.059

Geybels MS, Verhage Bas AJ, Arts ICW, van Schooten FJ, Goldbohm RA, van den Brandt PA (2013) Dietary flavonoid intake, black tea consumption, and risk of overall and advanced stage prostate cancer. Am J Epidemiol 177(12):1388–1398. doi:10.1093/aje/kws419

Goldfinger TM (2003) Beyond the French paradox: the impact of moderate beverage alcohol and wine consumption in the prevention of cardiovascular disease. Cardiol Clin 21(3):449–457

Goto A, Song Y, Chen BH, Manson JE, Buring JE, Simin Liu S (2011) Coffee and caffeine consumption in relation to sex hormone-binding globulin and risk of type 2 diabetes in post-menopausal women. Diabetes 60(1):269–275. doi:10.2337/db10-1193

Greenberg JA, Boozer CN, Geliebter A (2006) Coffee, diabetes, and weight control. Am J Clin Nutr 84:682–693

Guimarães R, Barros L, Barreira JC, Sousa MJ, Carvalho AM, Ferreira IC (2010) Targeting excessive free radicals with peels and juices of citrus fruits: grapefruit, lemon, lime and orange. Food Chem Toxicol 48:99–106

Guimarães R, Barros L, Duenas M, Calhelha RC, Carvalho AM, Santos-Buelga C, Queiroz MJ, Ferreira IC (2013) Nutrients, phytochemicals and bioactivity of wild Roman chamomile: A comparison between the herb and its preparations. Food Chem 136:718–725

Hanganu D, Vlase L, Filip L, Sand C, Mirel S, Indrei LL (2008) The study of some polyphenolic compounds from Melissa officinalis L. (Lamiaceae). Rev Med Chir Soc Med Nat Iasi 112(2): 525–529

Hodgson JM, Croft KD, Woodman RJ, Puddey IB, Fuchs D, Draijer R, Lukoshkova E, Head GA (2013) Black tea lowers the rate of blood pressure variation: a randomized controlled trial. Am J Clin Nutr 97(5):943–950. doi:10.3945/ajcn.112.051375

Homan DJ, Mobarhan S (2006) Coffee: good, bad, or just fun? a critical review of coffee's effects on liver enzymes. Nutr Rev 64(1):43–46. doi:10.1111/j.1753-4887.2006.tb00172.x

ICO (2015) About coffee. http://www.ico.org/. Accessed 12 Feb 2015

INSA (2015) Tabela da Composição de Alimentos (TCA). http://www.insa.pt/sites/INSA/Portugues/AreasCientificas/AlimentNutricao/AplicacoesOnline/TabelaAlimentos/PesquisaOnline/Paginas/PorPalavraChave.aspx. Accessed 5 Nov 2015

Jiang Y, Kusama K, Satoh K, Takayama E, Watanabe S, Sakagami H (2000) Induction of cytotoxicity by chlorogenic acid in human oral tumor cell lines. Phytomedicine 7(6):483–491. doi:10.1016/S0944-7113(00)80034-3

Kogiannou DA, Kalogeropoulos N, Kefalas P, Polissiou MG, Kaliora AC (2013) Herbal infusions; their phenolic profile, antioxidant and anti-inflammatory effects in HT29 and PC3 cells. Food Chem Toxicol 61:152–159. doi:10.1016/j.fct.2013.05.027

Lambert JD, Elias RJ (2010) The antioxidant and pro-oxidant activities of green tea polyphenols: a role in cancer prevention. Arch Biochem Biophys 501(1):65–72. doi:10.1016/j.abb.2010.06.013

Larsson SC (2014) Coffee, tea, and cocoa and risk of stroke. Stroke 45(1):309–314. doi:10.1161/STROKEAHA.113.003131

Lekli I, Ray D, Das DK (2010) Longevity nutrients resveratrol, wines and grapes. Genes Nutr 5(1): 55–60. doi:10.1007/s12263-009-0145-2

Lopez-Galilea I, de Peña MP, Cid C (2008) Application of multivariate analysis to investigate potential antioxidants in conventional and torrefacto roasted coffee. Eur Food Res Technol 227(1):141–149. doi:10.1007/s00217-007-0703-z

Lopez-Garcia E, Rodriguez-Artalejo F, Rexrode KM, Logroscino G, Hu FB, van Dam RB (2009) Coffee consumption and risk of stroke in women. Circulation 119(8):1116–1123. doi:10.1161/CIRCULATIONAHA.108.826164

Ludwig IA, Sánchez L, De Peña MP, Cid C (2014) Contribution of volatile compounds to the antioxidant capacity of coffee. Food Res Int 61:67–74. doi:10.1016/j.foodres.2014.03.045

Majer BJ, Hofer E, Cavin C, Lhoste E, Uhl M, Glatt HR, Meinl W, Knasmuller S (2005) Coffee diterpenes prevent the genotoxic effects of 2-amino-1-methyl-6-phenylimidazo[4,5-b]pyridine (PhIP) and N-nitrosodimethylamine in a human derived liver cell line (HepG2). Food Chem Toxicol 43(3):433–441. doi:10.1016/j.fct.2004.11.009

Matić IZ, Juranić Z, Savikin K, Zdunić G, Nađvinski N, Gođevac D (2013) Chamomile and marigold tea: chemical characterization and evaluation of anticancer activity. Phytother Res 27:852–858

Mencherini T, Picerno P, Scesa C, Aquino R (2007) Triterpene, antioxidant, and antimicrobial compounds from Melissa officinalis. J Nat Prod 70(12):1889–1894. doi:10.1021/np070351s

Miller JA, Lang JE, Ley M, Nagle R, Hsu C-H, Thompson PA, Cordova C, Waer A, Chow H-HS (2013) Human breast tissue disposition and bioactivity of limonene in women with early-stage breast cancer. Cancer Prev Res 6(6):577–584. doi:10.1158/1940-6207.CAPR-12-0452

NCBI (2015a) PubChem compound Database. CID 2519. http://pubchem.ncbi.nlm.nih.gov/compound/2519. Accessed 3 Nov 2015

NCBI (2015b) PubChem compound Database. CID 1794427. http://pubchem.ncbi.nlm.nih.gov/compound/1794427. Accessed 3 Nov 2015

NCBI (2015c) PubChem compound Database. CID 439378. http://pubchem.ncbi.nlm.nih.gov/compound/439378. Accessed 3 Nov 2015

NCBI (2015d) PubChem compound Database. CID 8294. http://pubchem.ncbi.nlm.nih.gov/compound/8294. Accessed 3 Nov 2015

NCBI (2015e) PubChem compound Database. CID 5281522. http://pubchem.ncbi.nlm.nih.gov/compound/5281522. Accessed 3 Nov 2015

NCBI (2015f) PubChem compound Database. CID 5354499. http://pubchem.ncbi.nlm.nih.gov/compound/5354499. Accessed 3 Nov 2015

NCBI (2015g) PubChem compound Database. CID 22311. http://pubchem.ncbi.nlm.nih.gov/compound/22311. Accessed 3 Nov 2015

Opie LH, Lamont K, Lecour S (2011) Wine and heart health: learning from the French paradox. SA Heart J 8(3):172–177

Pataki T, Bak I, Kovacs P, Bagchi D, Das DK, Tosaki A (2002) Grape seed proanthocyanidins improved cardiac recovery during reperfusion after ischemia in isolated rat hearts. Am J Clin Nutr 75(5):894–899

Patel SS, Beer S, Kearney DL, Phillips G, Carter BA (2013) Green tea extract: a potential cause of acute liver failure. World J Gastroenterol 19(31):5174–5177. doi:10.3748/wjg.v19.i31.5174

Renaud SD, de Lorgeril M (1992) Wine, alcohol, platelets, and the French paradox for coronary heart disease. Lancet 339:1523–1526. doi:10.1016/0140-6736(92)91277-F

Rosenkranz S, Knirel D, Dietrich H, Flesch M, Erdmann E, Böhm M (2002) Inhibition of the PDGF receptor by red wine flavonoids provides a molecular explanation for the 'French paradox'. FASEB J 16(14):1958–1960. doi:10.1096/fj.02-0207fje

Rozin P, Fischler C, Imada S, Sarubin A, Wrzesniewski A (1999) Attitudes to food and the role of food in life in the U.S.A., Japan, Flemish Belgium and France: possible implications for the diet-health debate. Appetite 33(2):163–180. doi:10.1006/appe.1999.0244

Rozin P, Kabnick K, Pete E, Fischler C, Shields C (2003) The ecology of eating: smaller portion sizes in France than in the United States help explain the French paradox. Psychol Sci 14(5): 450–454. doi:10.1111/1467-9280.02452

Santos MD, Almeida MC, Lopes NP, Souza GE (2006) Evaluation of the anti-inflammatory, analgesic and antipyretic activities of the natural polyphenol chlorogenic acid. Biol Pharm Bull 29(11):2236–2240. doi:10.1248/bpb.29.2236

Shibata H, Sakamoto Y, Oka M, Kono Y (1999) Natural antioxidant, chlorogenic acid, protects against DNA breakage caused by monochloramine. Biosci Biotechnol Biochem 63(7): 1295–1297. doi:10.1271/bbb.63.1295

Siedlinski M, Boer JM, Smit HA, Postma DS, Boezen HM (2012) Dietary factors and lung function in the general population: wine and resveratrol intake. Eur Respir J 39(2):385–391. doi:10.1183/09031936.00184110

Speranza L, Franceschelli S, Pesce M, Reale M, Menghini L, Vinciguerra I, De Lutiis MA, Felaco M, Grilli A (2010) Anti-inflammatory effects in THP-1 cells treated with verbascoside. Phytother Res 24(9):1398–1404. doi:10.1002/ptr.3173

Srivastava JK, Pandey M, Gupta S (2009) Chamomile, a novel and selective COX-2 inhibitor with anti-inflammatory activity. Life Sci 85(19–20):663–669. doi:10.1016/j.lfs.2009.09.007

Strang D (1890) Strang's patent soluble dry coffee-powder. 1890 First annual report, New Zealand, Patents, Designs and Trade-marks, patent number 3518, p 9, 25 Jan 1980. Available at http://atojs.natlib.govt.nz/cgi-bin/atojs?a=d&d=AJHR1890-I.2.3.2.1&e=-------10--1------0--. Accessed 3 Nov 2015

Stranges S, Wu T, Dorn JM, Freudenheim JL, Muti P, Farinaro E, Russell M, Nochajski TH, Trevisan M (2004) Relationship of alcohol drinking pattern to risk of hypertension a population-based study. Hypertension 44:813–819

UNESCO (2015) Culture. Intangible cultural heritage. Turkish coffee culture and tradition. Available at http://www.unesco.org/culture/ich/RL/00645. Accessed 2 Mar 2015

USDA (2015) Agricultural Research Service National Nutrient Database for Standard Reference Release 27, Software v.2.2.4, The National Agricultural Library. http://ndb.nal.usda.gov/ndb/foods. Accessed 5 May 2015

van Dam RM, Feskens EJ (2002) Coffee consumption and risk of type 2 diabetes mellitus. Lancet 360(9344):1477–1478. doi:10.1016/S0140-6736(02)11436-X

van Duynhoven J, Vaughan EE, van Dorsten F, Gomez-Roldan V, de Vos R, Vervoort J, van der Hooft JJJ, Roger L, Draijer R, Jacobs DM (2013) Interactions of black tea polyphenols with human gut microbiota: implications for gut and cardiovascular health. Am J Clin Nutr 98:1631S–1641S

Vekiari SA, Protopapadakis EE, Papadopoulou P, Papanicolaou D, Panou C, Vamvakias M (2002) Composition and seasonal variation of the essential oil from leaves and peel of a Cretan lemon variety. J Agric Food Chem 50(1):147–153. doi:10.1021/jf001369a

Vidavalur R, Otani H, Singal PK, Maulik N (2006) Significance of wine and resveratrol in cardiovascular disease: French paradox revisited. Exp Clin Cardiol 11(3):217–225

Vlachopoulos C, Panagiotakos D, Ioakeimidis N, Dima J, Stefanadis C (2005) Chronic coffee consumption has a detrimental effect on aortic stiffness and wave reflections. Am J Clin Nutr 81(6):1307–1312

Wang ZM, Zhou B, Wang YS, Gong QY, Wang QM, Jian-Jun Yan JJ, Gao W, Wang LS (2011) Black and green tea consumption and the risk of coronary artery disease: a meta-analysis. Am J Clin Nutr 93(3):506–515. doi:10.3945/ajcn.110.005363

Wong JC, Fiscus RR (2015) Resveratrol at anti-angiogenesis/anticancer concentrations suppresses protein kinase G signalling and decreases IAPs expression in HUVECs. Anticancer Res 35(1): 273–281

Wu JM, Hsieh T-C (2011) Resveratrol: a cardioprotective substance. Ann NY Acad Sci 1215: 16–21. doi:10.1111/j.1749-6632.2010.05854.x

Xie LX, Williams KJ, He CH, Weng E, Khong S, Rose TE, Kwon O, Bensinger SJ, Marbois BN, Clarke CF (2015) Resveratrol and para-coumarate serve as ring precursors for coenzyme Q biosynthesis. J Lipid Res 56(4):909–919. doi:10.1194/jlr.M057919

Part III

The Mediterranean Diet: Conclusions

9 Concluding Remarks

Abstract

The United Nations Education, Scientific and Cultural Organization's broad concept of the Mediterranean diet (MD) as an 'intangible cultural heritage' constitutes the starting point of this book, which is dedicated to the molecular biochemistry of this eating pattern. The concept of the MD is introduced and discussed, time trends in its evolution are documented, and research results on the associations between MD and health are summarised. The concept of a 'dietary pattern' is used as an integrated approach, enabling the identification and quantification of the associations between the overall diet and specific health/disease outcomes. The analysis of nutritional epidemiology studies, complemented by information provided by studies at a cellular and/or a molecular level, enabled the discussion of the multiple associations between the MD, health, well-being and longevity. This work includes analytical data on food composition, highlighting minor components, many of which are bioactive. Sections were organised by food groups, and examples of every representative food group within the MD were chosen: olive oil and table olives; greens and other vegetables—including grains, fruits, pulses, nuts and aromatic herbs; milk and dairy products; fish, meat and other animal protein sources; infusions and wines. Advances in food analysis are bringing to light a multitude of bioactive compounds. These chemical entities, their content in foods and disclosed mechanisms of action, as well as the effect of diet on microbiota are described here. Several global and governmental organizations acknowledge the MD as nutritionally adequate, health-promoting and sustainable because of its emphasis on biodiversity and the intake of small meat portions. In short, Mediterranean-style dietary patterns score high for health, as well as for estimated sustainability scores and can be followed in Mediterranean as well as in non-Mediterranean countries.

A.M. Delgado et al., *Chemistry of the Mediterranean Diet*,
DOI 10.1007/978-3-319-29370-7_9

9.1 The Mediterranean Diet: Concluding Remarks

The United Nations Education, Scientific and Cultural Organization's (UNESCO's) broad concept of the Mediterranean diet (MD) as an 'intangible cultural heritage' constitutes the starting point of this book, which is dedicated to the molecular biochemistry of this eating pattern.

'Good Mediterranean diet', 'prodigious Mediterranean diet', 'healthy Mediterranean diet', or simply 'Mediterranean diet': under these designations, there is a vast diversity of food patterns in the area surrounding the Mediterranean Sea. With reference to some claims stating 'there is no such thing as a Mediterranean diet', we would argue that, for the sake of accuracy and precision, the plural should be applied: Mediterranean diets. Actually, the MD of people living in Sicily certainly differs from that of people living in Marrakesh. So the broad 'Mediterranean diet' designation encompasses the numerous variants of food patterns spanning from countries facing the Atlantic, such as Portugal and Spain, to those neighbouring the Near East, such as Cyprus, or bordering the Sahara desert, such as Morocco. The strength of the concept behind the MD is its harmony with the people and the environment, as food habits are a product of ecological forces acting within the context of historical conditioning and belief systems (Fieldhouse 1995).

Food and eating are basic needs but also sources of pleasure and enjoyment and occasions for socialisation and bonding. Such links, represented by convivial meals, are more evident among peoples that actually share the food from the same plate. The *diaita*, as a Mediterranean way of living, is originally linked to rural landscapes, the seasons of the year and traditions and cultures, supporting sustainable agriculture and food habits. Many regional variations occur; however, there is 'unity in such diversity', with common features identifiable in the diets of the Mediterranean peoples that are central to its identity:

- Meals are opportunities for socialising
- Wine and infusions play a central role in these convivial meals
- Cooking methods are generally simple, and natural fresh ingredients are preferred over processed ones
- Olive oil is the main fat in the diet and is used for both cooking and seasoning
- Large daily intakes of seasonal vegetables, pulses and fruits are observed
- Tomato, onion and garlic are common ingredients through the Mediterranean area (many dishes starting from a base of fried onion, garlic and tomato in olive oil)
- Wheat is a staple food, providing starch and oligosaccharides, while simple sugars (consumed in desserts) are reserved for festivities and seldom consumed
- A low intake of dairy products, with a preference for cheese and yoghurt is observed
- A low intake of meat and fish (traditional preserves, as ham, chorizo, salami and anchovy paste are often used to flavour dishes where vegetables are predominant) is observed

Many factors determine eating patterns. Urbanisation has brought populations to cities, inducing its detachment from a natural pace. A large number of these individuals claim that they have no time or skills to cook. Agriculture evolved to energy- and water-intensive systems. Planted monoculture fields and large cattle exploitations replaced traditional farming systems, where plant diversity was paired with animal husbandries, or fishing, thus providing local diversified food resources. A global food industry became stronger and powerful, influencing the average consumer's demands (namely for sugar, saturated fats and salt) and marketing similar food items worldwide. Fast food, ready-to-eat and frozen pre-cooked meals are considered by many people an adequate response to save time and avoid cooking. In contrast with the MD, these globalised food patterns are based on meals that can be readily eaten at a desk while performing some job or on a couch while watching television. Major ingredients include high amounts of refined carbohydrates, sugars, salt, saturated and *trans*-fats and animal proteins. Artificial colourings and flavours are included to attract consumers. In general, vegetables and fruits are rarely included in this type of diet, which is also characterised by a lack of variety. Preferred drinks are of an industrial nature (such as soda, cola and beer) as opposed to wine and infusions.

A consequence of these food trends is that, in 2001, chronic diseases already contributed to approximately 60 % of the total reported deaths in the world and approximately 46 % of the global burden of disease (WHO 2003). Still according to the same source, chronic diseases are expected to increase to 57 % by 2020, overloading healthcare systems and increasing public and private expenses with health. Obesity and chronic non-communicable diseases, such as diabetes, certain types of cancer, cardiovascular and some neurodegenerative diseases are showing worrying trends, as they are being found earlier in life.

The concept of food or a dietary pattern includes the type, quantity, proportions, variety and combinations of foods and beverages consumed (DGAC 2015). It enables the identification and quantification of the associations between the 'overall diet' and specific health/disease outcomes. This integrated approach to diet looks at interactions amongst its several constituents and overcomes collinearity (between single foods, nutrients and other components). However, time distribution (be it circadian distribution, weekly and/or seasonal consumption), preparation and cooking methods as well combinations of foods and drinks consumed together should also be regarded as part of the dietary pattern, as they influence several outcomes.

Nutritional epidemiology has unveiled multiple associations between the MD, health, well-being and longevity. Biochemistry, molecular biology, genetics, physiology and other scientific disciplines widened and complemented such research by identifying and quantifying the compounds in foods and providing evidence on their physiological mechanisms and actions.

This book examines the MD at a molecular level, establishing connections to health and well-being. Findings from epidemiological studies in which associations between adherence to the MD and positive health outcomes have set the scene for the presentation of the main groups of foods that are central to the MD. Information about foods is conveyed by several sources, including labelling of pre-packaged, processed foods. However, such information only covers specific nutrients and is not easily available for fresh foods. Herein, we presented detailed compositions of foods resulting from the compilation of data from institutional databases and from bibliographies. Food processing strongly influences its composition, namely in minor constituents. Although fresh ingredients should be preferred, industrially processed foods constitute an important part of current diets, especially of urban populations. Food processing may contribute to the preservation of nutrients (e.g. vitamins) and increase the bioavailability of active components (e.g. lycopene). On the other hand, the incorporation of sugars and fat in many foods through industrial processing has led to a wide availability of high-energy-dense foods at affordable prices. This contributes to the imbalance between energy requirements and energy intake, a major determinant of overweight/obesity.

In the first sections of this book, the concept of the MD was introduced and discussed, time trends in its evolution were documented, and research results on the associations between the MD and health were summarised. Nutritional epidemiology research has been crucial to establish associations between the 'prodigious Mediterranean diet' (Peres 1997), well-being and health, and longevity of the populations following this food pattern, as first studied by Ancel Keys, a well-known American physician.

The further sections of this book are organised by food groups and bring a complementary view of the MD by focusing on its chemical aspects. In each section, examples of every representative food group within the MD have been chosen (olive oil and table olives; greens and other vegetables—including grains, fruits, pulses, nuts and aromatic herbs; dairy products; animal protein sources; infusions and wine) to be presented and discussed. The composition of the selected food items are mainly based on data from two public food databases: the US Department of Agriculture (USDA) food database, which covers food items available in the USA, and the Portuguese Food Information Resource (PortFIR) from the Portuguese National Institute of Health (Instituto Nacional de Saúde Doutor Ricardo Jorge [INSA]), which covers food items available in the Portuguese market.

Advances in food analysis are bringing to light a multitude of components with biological activity, and an increasing number of minor constituents of food matrices have been identified and quantified.

It is worth noting that the reported food composition in nutrients, as well as in bioactive components, depends on many factors—from the location and season to the analytical methodology. Moreover, food composition is not static and, therefore, figures provided via chemical analysis should be regarded only as indicative values. Furthermore, analytical methods are not standardised between Europe and the USA, giving leeway for different findings. However, we found data to be most often convergent, which helps in understanding the figures.

Many minor components of foods, although present in small amounts, have important biological functions. Synergistic effects amongst them have subsequently been hypothesised and proven by research at molecular, cellular and tissue levels.

This book describes these chemical entities and their level in foods as well as their mechanisms of action.

Besides minor constituents of foods, such as phytosterols, essential fatty acids or proanthocyanidins (discussed in the corresponding sections), diet/food patterns influence the gut microbiome, which in turn strongly influences health in several ways. Gut microbiota composition is influenced by many factors, of which breastfeeding is of utmost importance, strongly affected by the diet and body mass index of the mother (Cabrera-Rubio et al. 2012). Diet during childhood helps to build the microbiota, which reaches a steady state in adulthood and is thought to determine immunity, tendency to obesity, and even mood (Kelly and Mulder 2012; Logan 2015; Susuki and Worobei 2014; Voreades et al. 2014). Some food components are known to benefit gut health, as is the case for fibres, oligosaccharides and polyphenols. On the other hand, excessive intake of simple sugars, sulphur-containing amino acids (from animal proteins), aspartame and some food preservatives, have been proven deleterious to these microbial communities. This field of study is developing rapidly, aiming to understand how environment, diet and host genetics influence the microbiota/microbiome and identify its relationships with health outcomes.

In addition, the book succinctly touches on the negative impacts of harmful ingredients and the risks of excess salt and trans-fat intake.

This evidence contributes to an explanation of why the MD constitutes a paradigm, a valuable resource for the formulation of nutritional theoretical models and applied healthy eating patterns. In fact, this dietary or food pattern is one of the recommended patterns to the general US population (DGAC 2015; USDA and USDHHS 2010).

Recently, the Scientific Report of the 2015 Dietary Guidelines Advisory Committee (DGAC) identified the MD (in the report, *Healthy Mediterranean-style Pattern*) as one of the three patterns with available research data on nutritional adequacy and associated health benefits. Features of a healthy dietary pattern include a high level of vegetables, fruits, whole grains, low- or non-fat dairy, seafood, legumes, and nuts; moderate consumption of alcohol (among adults); low consumption of red and processed meat; as well as low consumption of sugar (sucrose), sweetened foods and drinks and refined grains.

Along with the healthiness of diets, scientists and policy makers are increasingly acknowledging the issue of their sustainability (Burlingame and Dernini 2011). The UN Food and Agriculture Organization (FAO) has identified the MD as an example of a sustainable diet due to its emphasis on biodiversity and smaller meat portions (FAO 2008, 2010). The UNESCO recognises that the MD is rooted in respect for the territory and biodiversity. The most recent report from the US DGAC analyses for the first time the impact of foods and drinks on environmental outcomes, recognising that dietary recommendations should promote both health and sustainability (DGAC 2015). The identification of dietary patterns that are nutritionally adequate, promote health and also protect natural resources showed that the MD or a Mediterranean-style diet (in Mediterranean as well as in non-Mediterranean countries) had favourable environmental outcomes such as reduced greenhouse gas emissions, agricultural land use and energy and water consumption. Of great significance is the fact that Mediterranean-style dietary patterns scored high for health as well as for estimated sustainability scores (DGAC 2015). These findings reinforce the concept of the MD as a cultural pattern that is both deeply rooted in the past and soundly facing the future.

References

Burlingame B, Dernini S (2011) Sustainable diets: the Mediterranean diet as an example. Public Health Nutr 14(12A):2285–2287. doi:10.1017/S1368980011002527

Cabrera-Rubio R, Collado MC, Laitinen K, Salminen S, Isolauri E, Mira A (2012) The human milk microbiome changes over lactation and is shaped by maternal weight and mode of delivery. Am J Clin Nutr 96(3):544–551. doi:10.3945/ajcn.112.037382, Epub 2012 Jul 25

DGAC (2015) Scientific report of the 2015 dietary guidelines advisory committee. Advisory report to the Secretary of Health and Human Services and the Secretary of Agriculture. United States Department of Agriculture and United States Department of Health and Human Services. Available http://health.gov/dietaryguidelines/2015-scientific-report/pdfs/scientific-report-of-the-2015-dietary-guidelines-advisory-committee.pdf. Accessed 30 Oct 2015

FAO (2008) Report of the FAO Regional Conference for Europe, Twenty-Sixth Session, Innsbruck, 26–27 June 2008. ERC/08/REP. Food and Agriculture Organization, Rome. Available at ftp://ftp.fao.org/docrep/fao/meeting/014/k3400e.pdf. Accessed 30 Oct 2015

FAO (2010) Report of the international scientific symposium 'Biodiversity and sustainable diets—united against hunger', Rome, 3–5 Nov 2010. Available at http://www.fao.org/ag/humannutrition/28506-0efe4aed57af34e2dbb8dc578d465df8b.pdf. Accessed 5 July 2015

Fieldhouse P (1995) Food and nutrition: Customs and culture, 2nd edn. Chapman and Hall, London

Kelly D, Mulder IE (2012) Microbiome and immunological interactions. Nutr Rev 70(Suppl 1): S18–S30. doi:10.1111/j.1753-4887.2012.00498.x

Logan AC (2015) Dysbiotic drift: mental health, environmental grey space, and microbiota. J Physiol Anthropol 34:23. doi:10.1186/s40101-015-0061-7

Peres E (1997) Bem comidos e Bem Bebidos. Editorial Caminho, Lisboa

Suzuki TA, Worobey M (2014) Geographical variation of human gut microbial composition. Biol Lett 10:20131037. doi:10.1098/rsbl.2013.1037

USDA and USDHHS (2010) Dietary guidelines for Americans. 7th edn. U.S. Government Printing Office, Washington, DC. Available at www.dietaryguidelines.gov. Accessed 30 Oct 2015

Voreades N, Kozil A, Weir TL (2014) Diet and the development of the human intestinal microbiome. Front Microbiol 5:494. doi:10.3389/fmicb.2014.00494

WHO (2003) Food based dietary guidelines in the WHO European Region. Nutrition and Food Security Programme. EUR/03/5045414E79832, World Health Organization Regional Office for Europe, Copenhagen. Available at http://www.euro.who.int/__data/assets/pdf_file/0017/150083/E79832.pdf. Accessed 30 Oct 2015

Index

A.M. Delgado et al., *Chemistry of the Mediterranean Diet*,
DOI 10.1007/978-3-319-29370-7

D

E

K

L

M

N